Lecture Notes in Physics

New Series m: Monographs

The Editorial Policy for Monographs

The series Lecture Notes in Physics reports new developments in physical research and teaching - quickly, informally, and at a high level. The type of material considered for publication in the New Series m includes monographs presenting original research or new angles in a classical field. The timeliness of a manuscript is more important than its form, which may be preliminary or tentative. Manuscripts should be reasonably self-contained. They will often present not only results of the author(s) but also related work by other people and will provide sufficient motivation, examples, and applications.
The manuscripts or a detailed description thereof should be submitted either to one of the series editors or to the managing editor. The proposal is then carefully refereed. A final decision concerning publication can often only be made on the basis of the complete manuscript, but otherwise the editors will try to make a preliminary decision as definite as they can on the basis of the available information.
Manuscripts should be no less than 100 and preferably no more than 400 pages in length. Final manuscripts should preferably be in English, or possibly in French or German. They should include a table of contents and an informative introduction accessible also to readers not particularly familiar with the topic treated. Authors are free to use the material in other publications. However, if extensive use is made elsewhere, the publisher should be informed.
Authors receive jointly 50 complimentary copies of their book. They are entitled to purchase further copies of their book at a reduced rate. As a rule no reprints of individual contributions can be supplied. No royalty is paid on Lecture Notes in Physics volumes. Commitment to publish is made by letter of interest rather than by signing a formal contract. Springer-Verlag secures the copyright for each volume.

The Production Process

The books are hardbound, and quality paper appropriate to the needs of the author(s) is used. Publication time is about ten weeks. More than twenty years of experience guarantee authors the best possible service. To reach the goal of rapid publication at a low price the technique of photographic reproduction from a camera-ready manuscript was chosen. This process shifts the main responsibility for the technical quality considerably from the publisher to the author. We therefore urge all authors to observe very carefully our guidelines for the preparation of camera-ready manuscripts, which we will supply on request. This applies especially to the quality of figures and halftones submitted for publication. Figures should be submitted as originals or glossy prints, as very often Xerox copies are not suitable for reproduction. In addition, it might be useful to look at some of the volumes already published or, especially if some atypical text is planned, to write to the Physics Editorial Department of Springer-Verlag direct. This avoids mistakes and time-consuming correspondence during the production period.
As a special service, we offer free of charge LaTeX and TeX macro packages to format the text according to Springer-Verlag's quality requirements. We strongly recommend authors to make use of this offer, as the result will be a book of considerably improved technical quality. The typescript will be reduced in size (75% of the original). Therefore, for example, any writing within figures should not be smaller than 2.5 mm.
Manuscripts not meeting the technical standard of the series will have to be returned for improvement.
For further information please contact Springer-Verlag, Physics Editorial Department II, Tiergartenstrasse 17, W-6900 Heidelberg, FRG.

Daniele Funaro

Polynomial Approximation of Differential Equations

Springer-Verlag Berlin Heidelberg GmbH

Author

Daniele Funaro
Dipartimento di Matematica, Università degli Studi
Strada Nuova, 65, I-27100 Pavia, Italy

ISBN 978-3-662-13878-6 ISBN 978-3-540-46783-0 (eBook)
DOI 10.1007/978-3-540-46783-0

Originally published by Springer-Verlag Berlin Heidelberg New York in 1992
Softcover reprint of the hardcover 1st edition 1992

Typesetting: Camera ready by author

58/3140-543210 - Printed on acid-free paper

PREFACE

This book is devoted to the analysis of approximate solution techniques for differential equations, based on classical orthogonal polynomials. These techniques are popularly known as *spectral methods*. In the last few decades, there has been a growing interest in this subject. As a matter of fact, spectral methods provide a competitive alternative to other standard approximation techniques, for a large variety of problems. Initial applications were concerned with the investigation of periodic solutions of boundary value problems using trigonometric polynomials. Subsequently, the analysis was extended to algebraic polynomials. Expansions in orthogonal basis functions were preferred, due to their high accuracy and flexibility in computations.

The aim of this book is to present a preliminary mathematical background for beginners who wish to study and perform numerical experiments, or who wish to improve their skill in order to tackle more specific applications. In addition, it furnishes a comprehensive collection of basic formulas and theorems that are useful for implementations at any level of complexity. We tried to maintain an elementary exposition so that no experience in functional analysis is required.

The book is divided into thirteen chapters. In the first chapter, the reader can find definitions and properties relative to different families of polynomials. Orthogonality, zeroes, Gaussian quadrature formulas, basis transformations, and many other algebraic relations are studied in chapters 2, 3, 4. With the help of the short introduction to functional analysis given in chapter 5, a survey of results in approximation theory is provided in chapter 6. Chapter 7 is devoted to the study of the discretization of derivative operators. The eigenvalues of the corresponding matrices are analyzed in chapter 8. The study of the differential equations begins in chapter 9. Ordinary differential equations and time-dependent partial differential equations in one space variable are examined in chapters 9 and 10 respectively. Domain-decomposition methods and related solution techniques are considered in chapter 11. Numerical experiments on four upgraded model problems in one dimension are presented in chapter 12. Finally, suggestions for the treatment of problems in two variables are presented in the last chapter.

Acknowledgments

The author is grateful for the helpful scientific support of Claudio Canuto and his critical revision of the entire manuscript. Useful comments and suggestions were provided by David Gottlieb for chapter 10 and Alfio Quarteroni for chapter 11.

Special thanks are due to the editors, Yousuff Hussaini and Gordon Erlebacher, for the uncommon effort of making the exposition more fluent.

Final thanks go to the Istituto di Analisi Numerica del C.N.R., Pavia, for funding this research in part.

Daniele Funaro
Pavia, December 1991

CONTENTS

1
SPECIAL FAMILIES OF POLYNOMIALS

Approximating functions in spectral methods are related to polynomial solutions of eigenvalue problems in ordinary differential equations, known as Sturm-Liouville problems. These originate from applying the method of separation of variables in the analysis of boundary-value problems. We outline both basic and remarkable properties of the most commonly used families of polynomials of this kind.

1.1 Sturm-Liouville problems

Let I denote an open interval in **R**. We consider the continuous functions $a : \bar{I} \to \mathbf{R}$, $b : I \to \mathbf{R}$, $w : I \to \mathbf{R}$, satisfying $a \geq 0$ in $\bar{I}$ and $w > 0$ in I. We are concerned with the solutions (λ, u) to the following eigenvalue problem:

$$(1.1.1) \qquad -(au')' + bu = \lambda w u \qquad \text{in } I, \quad \lambda \in \mathbf{R}.$$

A large variety of boundary conditions can be assumed in order to determine uniquely, up to a constant factor, the eigenfunction u. Problem (1.1.1) is said to be *regular* if $a > 0$ in $\bar{I}$. When a vanishes at least at one point in $\bar{I}$, then we have a *singular* problem. The literature about this subject gives plenty of results. For a general theory, we can refer for instance to COURANT and HILBERT(1953), YOSIDA(1960), BRAUER and NOHEL

(1967), DETTMAN(1969). We only remark that, in most of the cases, one can show that the set of eigenvalues form a divergent sequence of real positive numbers. Many families of eigenfunctions have been widely studied. Among these, Bessel functions are the most representative (see section 12.4). However, we are mainly concerned with *polynomial* solutions of (1.1.1). In particular, we are interested in introducing and characterizing sequences $\{(\lambda_n, u_n)\}_{n\in\mathbf{N}}$ of solutions to (1.1.1) with $\lambda_n > 0$, $\lim_{n\to\infty} \lambda_n = +\infty$, where the u_n's are polynomials of degree n, suitably normalized. Many examples will be given in the following. Here we state a first general result, which will be proven in section 2.2.

Theorem 1.1.1 - *If* $\{u_n\}_{n\in\mathbf{N}}$ *is a sequence of solutions to* (1.1.1), *where* u_n *is a polynomial of degree* n, *then it is possible to find three real sequences* $\{\rho_n\}$, $\{\sigma_n\}$, $\{\tau_n\}$, $n \geq 2$ *such that*

$$u_n(x) = (\rho_n x + \sigma_n)\, u_{n-1}(x) + \tau_n\, u_{n-2}(x), \qquad \forall n \geq 2,\ \forall x \in I. \tag{1.1.2}$$

■

According to (1.1.2), we define $\rho_1, \sigma_1 \in \mathbf{R}$, such that $u_1(x) = (\rho_1 x + \sigma_1)u_0(x)$. Theorem 1.1.1 allows us to recursively compute the nth polynomial at a given point $x \in I$, starting from the values $u_0(x)$ and $u_1(x)$. By differentiating the expression (1.1.2), we get a similar relation for the derivative, i.e.

$$u'_n(x) = (\rho_n x + \sigma_n)\, u'_{n-1}(x) + \rho_n\, u_{n-1}(x) + \tau_n\, u'_{n-2}(x), \qquad \forall n \geq 2,\ \forall x \in I. \tag{1.1.3}$$

The simultaneous evaluation of (1.1.2) and (1.1.3) gives u'_n at the point x. Evaluation of higher order derivatives can be clearly carried out in the same way.

1.2 The Gamma function

Before continuing our analysis of polynomial eigenfunctions, it is convenient to review some properties of the well-known *Euler Gamma* function which, for any real positive number x, is defined as

$$\Gamma(x) := \int_0^{+\infty} t^{x-1}\, e^{-t}\, dt.$$

Proving the following functional equation is straightforward after integration by parts:

$$\Gamma(x+1) = x\,\Gamma(x)\ , \quad x > 0. \tag{1.2.1}$$

In particular, when n is an integer, we have a fundamental relation obtained from (1.2.1) by induction, i.e. $\Gamma(n+1) = n!$.
Another important relation is

$$\Gamma(x)\Gamma(1-x) = \frac{\pi}{\sin \pi\, x}\ , \qquad 0 < x < 1. \tag{1.2.2}$$

From (1.2.2) one can easily obtain $\Gamma\left(\frac{1}{2}\right) = \sqrt{\pi}$. A useful equality, which can be deduced from the definition of the *Beta function* (see for instance LUKE (1969), Vol.1), is

$$\int_{-1}^{1} (1+t)^{x-1}\, (1-t)^{y-1}\, dt = 2^{x+y-1}\, \frac{\Gamma(x)\Gamma(y)}{\Gamma(x+y)}\ , \qquad x > 0,\ y > 0. \tag{1.2.3}$$

The binomial coefficients are generalized according to

$$\binom{x}{k} := \frac{\Gamma(x+1)}{k!\ \Gamma(x-k+1)}\ , \qquad k \in \mathbf{N},\ \ x > k-1. \tag{1.2.4}$$

The formulas considered here, as well as many other properties of the Gamma function are widely discussed in LUKE(1969), Vol.1; SRIVASTAVA and MANOCHA(1984), HOCHSTADT(1971).

The numerical evaluation of the Gamma function is generally performed by reducing the computation shifting backwards to the interval $]0,1]$ using (1.2.1). The interval can be further reduced to $]\frac{1}{2},\ 1]$ by taking into account (1.2.2). Finally, suitable series expansions centered about 1 can be implemented. For other theoretical and computational aspects we can refer for instance to LUKE(1969), Vol.2. We only mention the following expansion:

(1.2.5) $$\Gamma(x) = \left(x+\frac{9}{2}\right)^{x-\frac{1}{2}} e^{1-x} \left(\tau_0 + \sum_{k=1}^{\infty} \tau_k H_k(x)\right),$$

where

$$H_k(x) := \frac{(x-1)(x-2)\cdots(x-k)}{x(x+1)\cdots(x+k-1)}, \qquad k \geq 1,$$

$$\tau_0 := \sqrt{\frac{2}{11}}, \quad \tau_k := 2(-1)^k k \sum_{m=0}^{k} \frac{(k+m-1)!}{m!\,(k-m)!} \frac{(-1)^m e^m}{(m+\frac{11}{2})^{m+1/2}}, \quad k \geq 1.$$

This is basically the expression given in LUKE(1969), Vol.1, page 30. More theoretical details and tabulated values are contained in that book.

1.3 Jacobi polynomials

We introduce the first family of polynomial solutions to (1.1.1) called Jacobi polynomials. They depend on two parameters α, $\beta \in \mathbf{R}$ with $\alpha > -1$, $\beta > -1$. As we will see in the following, an appropriate choice of these parameters leads to other well-known families, such as Legendre or Chebyshev polynomials.

Let us set $I =]-1,1[$ and let us take a, b, w in (1.1.1) such that

$$a(x) = (1-x)^{\alpha+1}(1+x)^{\beta+1}, \qquad \forall x \in \bar{I},$$
$$b(x) = 0, \quad w(x) = (1-x)^{\alpha}(1+x)^{\beta}, \qquad \forall x \in I,$$
$$\alpha > -1, \ \beta > -1.$$

After simplification, this choice yields the *singular* eigenvalue equation

(1.3.1) $$-(1-x^2)\,u'' + ((\alpha+\beta+2)x + \alpha - \beta)u' = \lambda u.$$

By applying the Frobenius method (see for instance DETTMAN(1969)), we get the following result.

Theorem 1.3.1 - *The solution to* (1.3.1) *is a polynomial of degree* n *only when* $\lambda = n(n+\alpha+\beta+1)$, $n \in \mathbf{N}$.

■

We shall define the n-degree Jacobi polynomial $P_n^{(\alpha,\beta)}$ to be the unique solution of (1.3.1) normalized by

$$P_n^{(\alpha,\beta)}(1) := \binom{n+\alpha}{n} = \frac{\Gamma(n+\alpha+1)}{n!\,\Gamma(\alpha+1)}, \qquad n \in \mathbf{N}, \tag{1.3.2}$$

or equivalently by

$$P_n^{(\alpha,\beta)}(-1) := (-1)^n \binom{n+\beta}{n}, \qquad n \in \mathbf{N}. \tag{1.3.3}$$

We observe that no boundary conditions are imposed in (1.3.1). These are replaced by requiring the solutions to be polynomials. Condition (1.3.2) (or (1.3.3)) is only imposed to select a unique eigenfunction, otherwise defined to within a multiplicative constant.

Many theorems and properties for this family of polynomials are well-known. Here we just collect some basic results and we refer to SZEGÖ(1939), LUKE(1969), ASKEY(1975), SRIVASTAVA and MANOCHA(1984) for a detailed essay. First we give the Rodrigues' formula:

$$(1-x)^\alpha\,(1+x)^\beta\,P_n^{(\alpha,\beta)}(x) = \frac{(-1)^n}{2^n\,n!}\,\frac{d^n}{dx^n}\left[(1-x)^{n+\alpha}\,(1+x)^{n+\beta}\right], \; n \in \mathbf{N}. \tag{1.3.4}$$

A more explicit form is

$$P_n^{(\alpha,\beta)}(x) = 2^{-n}\sum_{k=0}^{n}\binom{n+\alpha}{k}\binom{n+\beta}{n-k}(x-1)^{n-k}(x+1)^k \tag{1.3.5}$$

$$= \frac{\Gamma(2n+\alpha+\beta+1)}{2^n\,n!\,\Gamma(n+\alpha+\beta+1)}\left[x^n + \frac{(\alpha-\beta)n}{2n+\alpha+\beta}\,x^{n-1} + \cdots\right], \qquad n \in \mathbf{N}.$$

Both (1.3.4) and (1.3.5) can be checked by direct substitution in (1.3.1). A lot of formulas relate Jacobi polynomials corresponding to different choices of the pair of parameters $(\alpha,\ \beta)$. We show one of the most representative, i.e.

$$\text{(1.3.6)} \qquad \frac{d}{dx}\left[P_n^{(\alpha,\beta)}\right] = \frac{1}{2}\,(n+\alpha+\beta+1)\,P_{n-1}^{(\alpha+1,\beta+1)}, \qquad n \geq 1.$$

By (1.3.4) or (1.3.5) it is clear that

$$\text{(1.3.7)} \qquad P_0^{(\alpha,\beta)}(x) = 1\,, \qquad P_1^{(\alpha,\beta)}(x) = \frac{1}{2}(\alpha+\beta+2)\,x + \frac{1}{2}\,(\alpha-\beta).$$

Since (1.3.4) and (1.3.5) are quite unpracticable, higher degree polynomials can be determined using theorem 1.1.1. More precisely, we have

$$\text{(1.3.8)} \qquad \rho_n = \frac{(2n+\alpha+\beta)\,(2n+\alpha+\beta-1)}{2n\,(n+\alpha+\beta)},$$

$$\sigma_n = \frac{(\alpha^2-\beta^2)\,(2n+\alpha+\beta-1)}{2n\,(n+\alpha+\beta)\,(2n+\alpha+\beta-2)},$$

$$\tau_n = -\frac{(n+\alpha-1)\,(n+\beta-1)\,(2n+\alpha+\beta)}{n\,(n+\alpha+\beta)\,(2n+\alpha+\beta-2)}, \qquad n \geq 2.$$

Moreover, one has: $\rho_1 = \frac{1}{2}(\alpha+\beta+2)$, $\sigma_1 = \frac{1}{2}(\alpha-\beta)$. Derivatives can be recovered using (1.1.3) or (1.3.6).

Finally, we present the following estimate in the interval $\bar{I}$ (see SZEGÖ (1939)):

$$\text{(1.3.9)} \quad \max_{-1\leq x\leq 1} |P_n^{(\alpha,\beta)}(x)| = \max\left\{|P_n^{(\alpha,\beta)}(\pm 1)|\right\} = \max\left\{\binom{n+\alpha}{n}, \binom{n+\beta}{n}\right\}, \qquad n \in \mathbf{N}.$$

Further properties can be proven for the so called *ultraspherical* (or *Gegenbauer*) polynomials. These are Jacobi polynomials where $\alpha = \beta$. We soon see two celebrated examples.

1.4 Legendre polynomials

Legendre polynomials are Jacobi ultraspherical polynomials with $\alpha = \beta = 0$. To simplify the notation it is standard to set $P_n := P_n^{(0,0)}$. We now review the basic properties. According to theorem 1.3.1, we have the differential equation

$$(1.4.1) \qquad (1 - x^2)\,P_n'' - 2x\,P_n' + n(n+1)\,P_n = 0\,, \qquad n \in \mathbf{N}.$$

Conditions (1.3.2), (1.3.3) give respectively $P_n(1) = 1$, $P_n(-1) = (-1)^n$, $n \in \mathbf{N}$. The recursion formula is

$$(1.4.2) \qquad P_n(x) = \frac{2n-1}{n}\,xP_{n-1}(x) - \frac{n-1}{n}\,P_{n-2}(x), \qquad x \in \bar{I}, \quad n \geq 2.$$

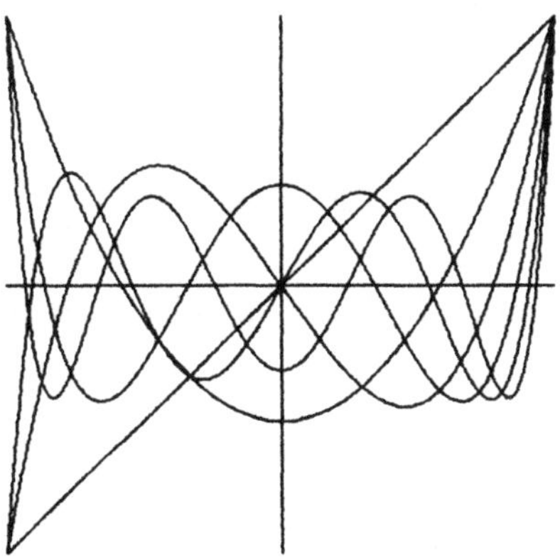

Figure 1.4.1 - Legendre polynomials for $1 \leq n \leq 6$.

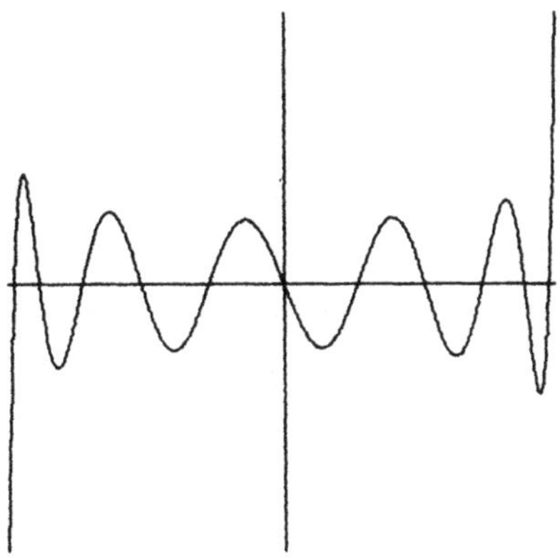

Figure 1.4.2 - The eleventh Legendre polynomial.

It is easy to check that P_n is an even (odd) function if and only if n is even (odd). Moreover, from (1.4.1) one gets

$$(1.4.3) \qquad P_n'(\pm 1) = -\,\frac{n(n+1)}{2}\,(\pm 1)^n, \qquad n \in \mathbf{N}.$$

Furthermore, (1.3.9) leads to

(1.4.4) $$|P_n(x)| \leq 1\,, \qquad |x| \leq 1, \quad n \in \mathbf{N}.$$

Combining (1.3.6) and (1.3.9), we can also estimate the derivative according to

(1.4.5) $$|P_n'(x)| \leq \tfrac{1}{2}n\,(n+1)\,, \qquad |x| \leq 1\,, \quad n \in \mathbf{N}.$$

Another useful relation can be easily established by taking $x = 0$ in (1.4.2), i.e.

(1.4.6) $$P_n(0) = \begin{cases} 0 & \text{if } n \text{ is odd,} \\ n!\; 2^{-n} \left[\left(\frac{n}{2}\right)!\right]^{-2} & \text{if } n \text{ is even.} \end{cases}$$

Therefore, one deduces that $\lim_{n \to +\infty} P_n(0) = 0$.

In Figure 1.4.1 the polynomials P_n, $1 \leq n \leq 6$, are plotted, while Figure 1.4.2 shows the behavior of P_{11}.

In order to give some general ideas of the behavior of Legendre polynomials, we state two other results (see SZEGÖ(1939) and JACKSON(1930) respectively).

Theorem 1.4.1. - *For any $n \geq 5$, when x increases from 0 to 1, the successive relative maximum values of $|P_n(x)|$ also increase.*

■

Theorem 1.4.2. - *For any $n \in \mathbf{N}$ and any $x \in \bar{I}$, one has*

(1.4.7) $$P_n(x) = \frac{1}{\pi} \int_0^\pi \left(x + \mathbf{i}\sqrt{1-x^2}\,\cos\theta\right)^n d\theta.$$

■

Though the right hand side in (1.4.7) contains the imaginary unity (i.e., $\mathbf{i} = \sqrt{-1}$), the global integral is real. Taking the absolute value on both sides of (1.4.7), we can prove the estimate:

(1.4.8)
$$|P_n(x)| \le \frac{1}{\pi}\int_0^\pi \left[x^2+(1-x^2)\cos^2\theta\right]^{n/2} d\theta$$

$$\le \frac{1}{\pi}\int_0^\pi \left[x^2+(1-x^2)\cos^2\theta\right] d\theta = \frac{1+x^2}{2}, \quad x\in\bar{I},\ n\ge 2.$$

This shows that the Legendre polynomials in $\bar{I}$ are uniformly bounded between the parabolas $y=-\frac{1}{2}(1+x^2)$ and $y=\frac{1}{2}(1+x^2)$.

1.5 Chebyshev polynomials

The Chebyshev polynomials (of the *first kind*) are related to the ultraspherical polynomials with $\alpha=\beta=-\frac{1}{2}$. In fact, they are defined by

(1.5.1)
$$T_n := \delta_n\, P_n^{(-\frac{1}{2},-\frac{1}{2})}, \qquad n\in\mathbf{N},$$

where
$$\delta_n := \frac{(n!\,2^n)^2}{(2n)!} = \frac{n!\,\sqrt{\pi}}{\Gamma(n+\frac{1}{2})} = \left[\binom{n-\frac{1}{2}}{n}\right]^{-1}.$$

Consequently, they are solutions of the Sturm-Liouville problem

(1.5.2)
$$(1-x^2)\,T_n'' - x\,T_n' + n^2\,T_n = 0, \qquad n\in\mathbf{N}.$$

By taking $\alpha=\beta=-\frac{1}{2}$ in (1.3.8), after appropriate scaling, we arrive at the recursion formula

(1.5.3)
$$T_n(x) = 2x\,T_{n-1}(x) - T_{n-2}(x), \quad x\in\bar{I},\ n\ge 2,$$

where $T_0(x)=1$ and $T_1(x)=x$.

In general, we obtain $T_n(\pm 1) = (\pm 1)^n$, $n \in \mathbf{N}$. Evaluating at the points $x = \pm 1$, (1.5.2) yields

$$T_n'(\pm 1) = -(\pm 1)^n \, n^2 , \quad n \in \mathbf{N}. \tag{1.5.4}$$

Moreover, T_n is even (odd) if and only if n is even (odd). An explicit expression for the Chebyshev polynomial of degree n is

$$T_n(x) = \sum_{k=0}^{[n/2]} \left((-1)^k \sum_{m=k}^{[n/2]} \binom{n}{2m} \binom{m}{k} \right) x^{n-2k} \tag{1.5.5}$$

$$= 2^{n-1} x^n - n 2^{n-3} x^{n-2} + \tfrac{1}{2} n(n-3) \, 2^{n-5} x^{n-4} + \cdots, \qquad x \in \bar{I}, \ n \in \mathbf{N}.$$

In (1.5.5), $[\bullet]$ denotes the integer part of $\bullet$.

The most remarkable characterization is given by the simple relation

$$T_n(\cos\theta) = \cos n\theta, \qquad \theta \in [0, \pi], \quad n \in \mathbf{N}. \tag{1.5.6}$$

The above expression relates algebraic and trigonometric polynomials. Such an important peculiarity can be proven by noting that

$$\frac{dT_n}{dx}(\cos\theta) = \frac{n \sin n\theta}{\sin\theta}, \tag{1.5.7}$$

$$\frac{d^2 T_n}{dx^2}(\cos\theta) = \frac{n \sin n\theta \cos\theta}{\sin^3\theta} - \frac{n^2 \cos n\theta}{\sin^3\theta}, \tag{1.5.8}$$

where $x = \cos\theta$, $\theta \in [0, \pi]$, and $n \in \mathbf{N}$. Thus, (1.5.2) is satisfied by replacing x by $\cos\theta$.

A lot of properties are direct consequence of (1.5.6). In particular we have $|T_n(x)| \le 1$, $\forall n \in \mathbf{N}$, $\forall x \in \bar{I}$. Moreover, T_n vanishes n times in $\bar{I}$, therefore the derivative T_n' vanishes $n-1$ times in $\bar{I}$. The function $|T_n|$ attains the maximum value of 1, $n+1$ times in $\bar{I}$. We give the plot in Figure 1.5.1 of the T_n's for $1 \le n \le 6$. In addition, T_{11} is plotted in Figure 1.5.2.

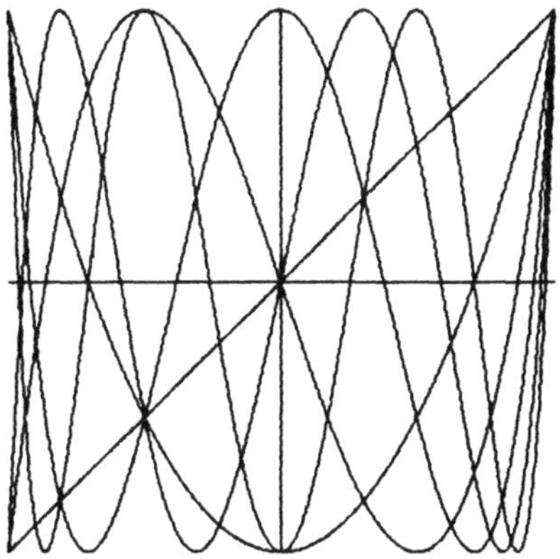

Figure 1.5.1 - Chebyshev polynomials for $1 \leq n \leq 6$.

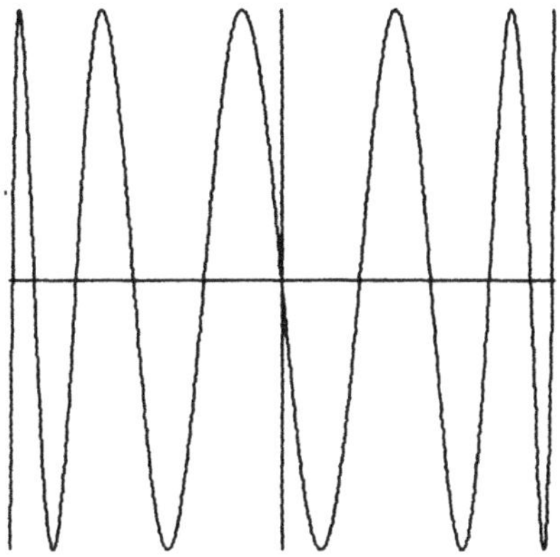

Figure 1.5.2 - The eleventh Chebyshev polynomial.

Other simple relations can be easily established:

$$T_n(x) = \frac{1}{2}\left(e^{in\theta} + e^{-in\theta}\right) = \frac{1}{2}\left[\left(x + \sqrt{x^2-1}\right)^n + \left(x - \sqrt{x^2-1}\right)^n\right], \quad x = \cos\theta \in \bar{I}, \quad n \in \mathbf{N}, \tag{1.5.9}$$

$$T_n = \frac{1}{2}\left[\frac{T'_{n+1}}{n+1} - \frac{T'_{n-1}}{n-1}\right], \quad n \geq 2, \tag{1.5.10}$$

$$T_m(T_n(x)) = T_{mn}(x), \quad \forall x \in \bar{I}, \quad \forall n, m \in \mathbf{N}. \tag{1.5.11}$$

We refer to LUKE(1969) and RIVLIN(1974) for more details.

Another family of polynomials, known as Chebyshev polynomials of the *second kind* is defined as

$$U_n := \frac{1}{n+1}\, T'_{n+1}, \qquad n \in \mathbf{N}. \tag{1.5.12}$$

These are also ultraspherical polynomials. By virtue of (1.3.6) and (1.5.1), we have

$$U_n = \delta_{n+1}\, P_n^{(\frac{1}{2},\frac{1}{2})}, \qquad n \in \mathbf{N}. \tag{1.5.13}$$

The U_n's satisfy similar properties to those of Chebyshev polynomials of the first kind.

1.6 Laguerre polynomials

In this section, we introduce another family of polynomial solutions of the Sturm-Liouville problem. Let $\alpha > -1$ and $I =]0, +\infty[$, then define the coefficients in (1.1.1) to be

$$a(x) = x^{\alpha+1}\, e^{-x}, \qquad \forall x \in \bar{I},$$

$$b(x) = 0, \qquad w(x) = x^{\alpha} e^{-x}, \qquad \forall x \in I.$$

This choice leads to the eigenvalue problem

$$xu'' + (\alpha + 1 - x)\, u' + \lambda\, u \;=\; 0. \tag{1.6.1}$$

This admits polynomial solutions only if $\lambda = n$, $n \in \mathbf{N}$. Therefore, we define the n-degree Laguerre polynonial $L_n^{(\alpha)}$ to be the unique solution of the *singular* problem (1.6.1), satisfying the condition

$$L_n^{(\alpha)}(0) := \binom{n+\alpha}{n}, \qquad n \in \mathbf{N},\ \alpha > -1. \tag{1.6.2}$$

We have the Rodrigues' formula

$$e^{-x}\, x^{\alpha}\, L_n^{(\alpha)}(x) \;=\; \frac{1}{n!}\,\frac{d^n}{dx^n}\left(e^{-x}\, x^{n+\alpha}\right), \qquad n \in \mathbf{N},\ x \in I. \tag{1.6.3}$$

The following expression also holds:

$$L_n^{(\alpha)}(x) \;=\; \sum_{k=0}^{n} \binom{n+\alpha}{n-k}\, \frac{(-1)^k}{k!}\, x^k, \qquad n \in \mathbf{N},\ x \in \bar{I}. \tag{1.6.4}$$

By virtue of theorem 1.1.1, we have the three-term recursion formula

$$(1.6.5) \qquad L_n^{(\alpha)}(x) = \frac{2n+\alpha-1-x}{n} L_{n-1}^{(\alpha)}(x) - \frac{n+\alpha-1}{n} L_{n-2}^{(\alpha)}(x), \qquad \forall n \geq 2,$$

where $L_0^{(\alpha)}(x) = 1$ and $L_1^{(\alpha)}(x) = 1 + \alpha - x$.

Various equations relate Laguerre polynomials corresponding to different values of the parameter α. Some examples are (see also SZEGÖ (1939))

$$(1.6.6) \qquad \frac{d}{dx} L_{n+1}^{(\alpha)} = -L_n^{(\alpha+1)}, \qquad n \in \mathbf{N},\ \alpha > -1,$$

$$(1.6.7) \qquad L_{n+1}^{(\alpha)} = L_{n+1}^{(\alpha+1)} - L_n^{(\alpha+1)}, \qquad n \in \mathbf{N},\ \alpha > -1.$$

From (1.6.7) one obtains

$$(1.6.8) \qquad L_n^{(\alpha+1)} = \sum_{k=0}^{n} L_k^{(\alpha)}, \qquad n \in \mathbf{N},\ \alpha > -1.$$

An asymptotic formula relates Laguerre and Jacobi polynomials, namely:

$$(1.6.9) \qquad L_n^{(\alpha)}(x) = \lim_{\beta \to +\infty} \left[P_n^{(\alpha,\beta)} \left(1 - \frac{2x}{\beta} \right) \right], \qquad \forall x \in [0, +\infty[.$$

This can be checked with the help of the differential equations (1.3.1) and (1.6.1).

A statement similar to that of theorem 1.4.1 also holds (see SZEGÖ(1939), p.171).

Theorem 1.6.1 - *For any $\alpha > -1$ and $n \geq 4$, the successive values of the relative maxima of $|L_n^{(\alpha)}(x)|$ are decreasing when $x < \alpha + \frac{1}{2}$ and increasing when $x > \alpha + \frac{1}{2}$.*

■

The plots of figures 1.6.1 and 1.6.2 show respectively $L_n^{(0)}$, $1 \leq n \leq 6$, and $L_7^{(0)}$. The size of the window is $[0, 20] \times [-600, 600]$ in figure 1.6.1 and $[0, 20] \times [-900, 900]$ in figure 1.6.2. When n grows, graphical representations of Laguerre polynomials are difficult to obtain. An initial gentle behavior explodes to severe oscillations, for increasing values of x.

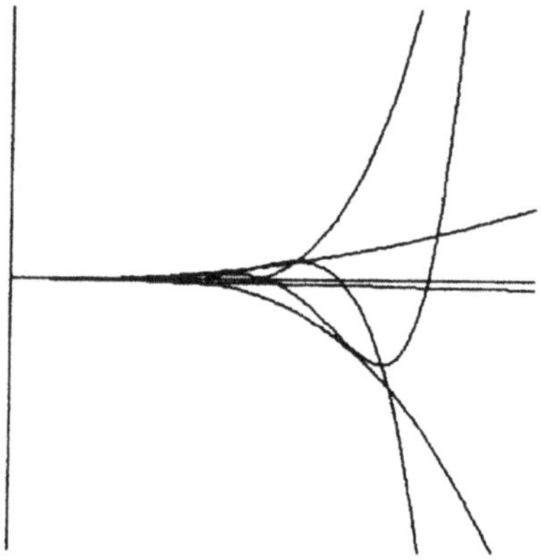

Figure 1.6.1 - Laguerre polynomials for $\alpha = 0$ and $1 \leq n \leq 6$.

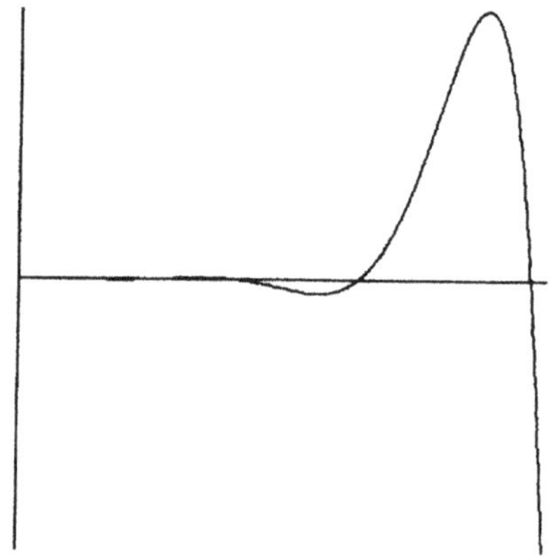

Figure 1.6.2 - The Laguerre polynomial $L_7^{(0)}$.

Further insight into the asymptotic behavior can be given. We recall two results (see SZEGÖ(1939), p.171 and p.234). The first one shows the behavior when $x \to +\infty$. In the second one, the case $n \to +\infty$ is considered. It is convenient to define $Q_n^{(\alpha)}(x) := e^{-x/2}\, x^{(\alpha+1)/2}\, L_n^{(\alpha)}(x)$.

Theorem 1.6.2 - *For any* $\alpha > -1$ *and* $n \geq 2$, *the values of the relative maxima of* $|Q_n^{(\alpha)}(x)|$ *form an increasing sequence when* $x \in]x_0, +\infty[$, *where*

$$x_0 := 0 \quad \textit{if } \alpha \leq 1, \qquad x_0 := \frac{\alpha^2 - 1}{2n + \alpha + 1} \quad \textit{if } \alpha > 1.$$

■

Theorem 1.6.3 - *Let* $\alpha > -1$, *then for any* $\gamma > 0$ *and* $0 < \eta < 4$ *we can find two positive constants* C_1, C_2 *such that*

$$\max_{x \in [\gamma, (4-\eta)n]} |Q_n^{(\alpha)}(x)| \approx C_1\, n^{\alpha/2}, \tag{1.6.10}$$

$$\max_{x \in [\gamma, +\infty[} |Q_n^{(\alpha)}(x)| \approx C_2\, \sqrt[6]{n}\; n^{\alpha/2}. \tag{1.6.11}$$

■

In order to allow more flexible numerical computations, it is useful to introduce some suitable scaling functions $S_n^{(\alpha)} : \bar{I} \to \mathbf{R}$ and to define

$$\hat{L}_n^{(\alpha)} := S_n^{(\alpha)} L_n^{(\alpha)}, \qquad n \in \mathbf{N}, \ \alpha > -1. \tag{1.6.12}$$

The aim is to avoid ill-conditioned operations when evaluating point values of Laguerre polynomials. According to FUNARO(1990a) an effective choice of $S_n^{(\alpha)}$ is

$$S_0^{(\alpha)}(x) := 1, \qquad S_n^{(\alpha)}(x) := \left[\binom{n+\alpha}{n} \prod_{k=1}^{n} \left(1 + \frac{x}{4k}\right) \right]^{-1}, \qquad n \geq 1. \tag{1.6.13}$$

We will refer to $\hat{L}_n^{(\alpha)}$, $n \in \mathbf{N}$, as the family of *scaled Laguerre functions.* It is clear that these are not polynomials. Plots of $\hat{L}_n^{(0)}$, $1 \leq n \leq 12$, are given in figure 1.6.3. Now the window size is $[0, 50] \times [-300, 300]$.

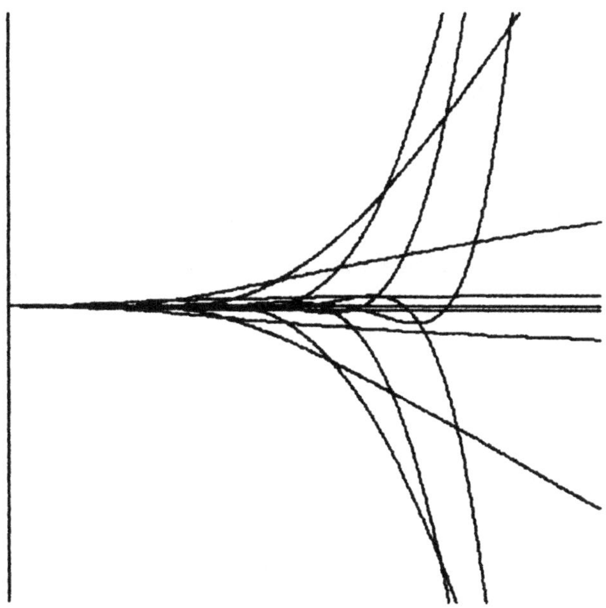

Figure 1.6.3 - Scaled Laguerre functions for $\alpha = 0$ and $1 \leq n \leq 12$.

It is evident that the scaled Laguerre functions have a milder behavior. We note that $\hat{L}_n^{(\alpha)}(0) = 1$, $\forall n \in \mathbf{N}$. More details are given in FUNARO(1990a). By substitution in (1.6.5), after simplification, one obtains $\forall n \geq 2$ the recursion formula

(1.6.14)
$$\hat{L}_n^{(\alpha)}(x) = \frac{4n}{(n+\alpha)(4n+x)}\left[(2n+\alpha-1-x)\hat{L}_{n-1}^{(\alpha)}(x) - \frac{4(n-1)^2}{4n+x-4}\hat{L}_{n-2}^{(\alpha)}(x)\right],$$

with $\hat{L}_0^{(\alpha)}(x) = 1$ and $\hat{L}_1^{(\alpha)}(x) = \frac{4(\alpha+1-x)}{(\alpha+1)(x+4)}$.

Moreover, for the derivatives one has $\forall n \geq 2$

(1.6.15)
$$\frac{d}{dx}\hat{L}_n^{(\alpha)}(x) = \frac{4n}{(n+\alpha)(4n+x)}\left[(2n+\alpha-1-x)\frac{d}{dx}\hat{L}_{n-1}^{(\alpha)}(x)\right.$$
$$\left. -\frac{6n+\alpha-1}{4n+x}\hat{L}_{n-1}^{(\alpha)}(x) + \frac{4(n-1)^2}{4n+x-4}\left(\frac{2(4n+x-2)}{(4n+x)(4n+x-4)}\hat{L}_{n-2}^{(\alpha)}(x) - \frac{d}{dx}\hat{L}_{n-2}^{(\alpha)}(x)\right)\right],$$

with $\frac{d}{dx}\hat{L}_0^{(\alpha)}(x) = 0$ and $\frac{d}{dx}\hat{L}_1^{(\alpha)}(x) = -\frac{\alpha+5}{\alpha+1}\frac{4}{(x+4)^2}$.

Applications will be examined in the following chapters (see sections 3.10 and 7.5).

1.7 Hermite polynomials

The set of Hermite polynomials H_n, $n \in \mathbf{N}$, is the last family we consider. Following the notations adopted in SZEGÖ(1939) or COURANT and HILBERT(1953), the H_n's are solutions of the *non singular* Sturm-Liouville problem obtained by setting in (1.1.1):

$$a(x) = w(x) = e^{-x^2}, \qquad b(x) = 0, \qquad \forall x \in I \equiv \mathbf{R}.$$

Therefore, one easily gets the differential equation

(1.7.1) $$H_n''(x) - 2xH_n'(x) + 2nH_n(x) = 0, \quad n \in \mathbf{N}, \ x \in \mathbf{R}.$$

The normalizing condition is

(1.7.2) $$H_n(0) := (-1)^{n/2} \frac{n!}{(n/2)!} \qquad \text{if } n \text{ is even},$$

(1.7.3) $$H_n'(0) := (-1)^{(n-1)/2} \frac{(n+1)!}{((n+1)/2)!} \qquad \text{if } n \text{ is odd}.$$

The nth degree polynomial H_n is an even or odd function according to the parity of n. As usual, we have the Rodrigues' formula

(1.7.4) $$H_n(x) = (-1)^n \, e^{x^2} \frac{d^n}{dx^n} e^{-x^2}, \quad n \in \mathbf{N}, \ x \in \mathbf{R}.$$

More explicitly, we can write

(1.7.5) $$H_n(x) = n! \sum_{m=0}^{[n/2]} \frac{(-1)^m}{m!} \frac{(2x)^{n-2m}}{(n-2m)!}$$

$$= 2^n x^n - n(n-1)2^{n-2}x^{n-2} + n(n-1)(n-2)(n-3)2^{n-5}x^{n-4} - \cdots,$$

$$n \in \mathbf{N}, \ x \in \mathbf{R}.$$

The corresponding three-term recursion formula is

(1.7.6) $$H_n(x) = 2xH_{n-1}(x) - 2(n-1)H_{n-2}(x), \quad \forall n \geq 2,$$

where $H_0(x) = 1$ and $H_1(x) = 2x$.
By differentiating (1.7.4) we also get

(1.7.7) $$H_n'(x) = 2xH_n(x) - H_{n+1}(x), \quad n \in \mathbf{N}, \ x \in \mathbf{R}.$$

Together with (1.7.6), we obtain a very simple relation to evaluate the derivatives, i.e.

(1.7.8) $$H_n'(x) = 2nH_{n-1}(x), \qquad n \geq 1, \quad x \in \mathbf{R}.$$

We give the plots of Hermite polynomials in figures 1.7.1 and 1.7.2. The sizes of the windows are respectively $[-5,5] \times [-900,900]$ and $[-5,5] \times [-450000,450000]$.

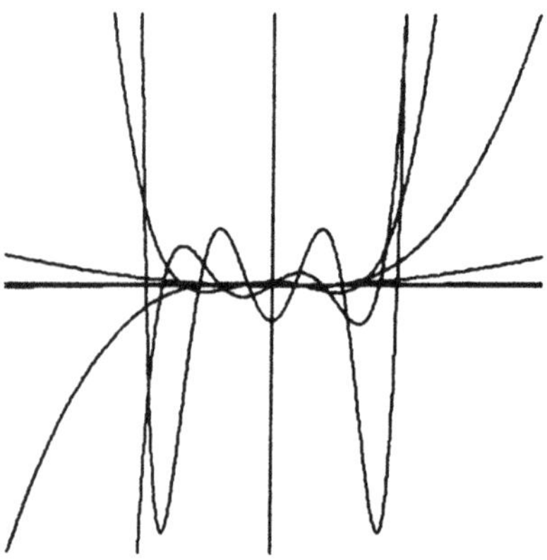

Figure 1.7.1 - Hermite polynomials for $1 \leq n \leq 6$.

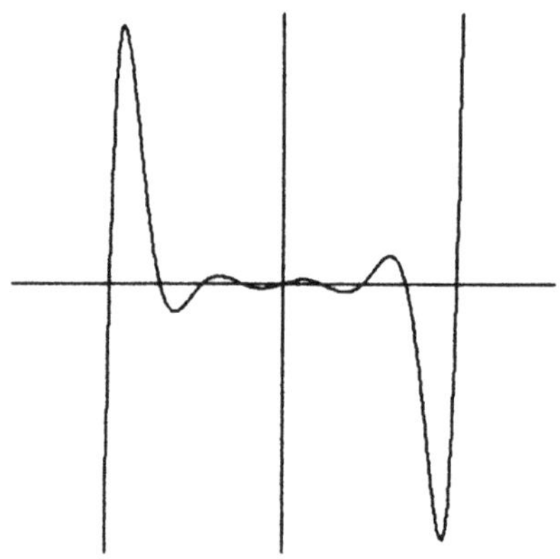

Figure 1.7.2 - The nineth Hermite polynomial.

It is worthwhile to mention the following result.

Theorem 1.7.1 - *For any* $n \geq 4$, *the successive values of the relative maxima of* $|H_n(x)|$ *are increasing for* $x \geq 0$.

■

Other asymptotic properties can be found in SZEGÖ(1939).

The Hermite polynomials can be expressed in term of the Laguerre polynomials according to

(1.7.9) $$H_n(x) = (-1)^{n/2}\, 2^n\, (n/2)!\, L_{n/2}^{(-1/2)}(x^2), \quad \text{if } n \text{ is even},$$

(1.7.10) $$H_n(x) = (-1)^{(n-1)/2}\, 2^n\, ((n-1)/2)!\, x\, L_{(n-1)/2}^{(1/2)}(x^2), \quad \text{if } n \text{ is odd}.$$

The formulas above are easily checked with the help of (1.6.1)-(1.6.2) and (1.7.1)-(1.7.2)-(1.7.3).

Inspired by (1.7.9) and (1.7.10), recalling the definition (1.6.12)-(1.6.13), it is natural to define the *scaled Hermite functions* by

(1.7.11) $$\hat{H}_n(x) := \hat{L}_{n/2}^{(-1/2)}(x^2), \quad \text{if } n \text{ is even},$$

(1.7.12) $$\hat{H}_n(x) := x\, \hat{L}_{(n-1)/2}^{(1/2)}(x^2), \quad \text{if } n \text{ is odd}.$$

Consequently

(1.7.13) $$\hat{H}_n(0) = 1, \quad \text{if } n \text{ is even},$$

(1.7.14) $$\hat{H}'_n(0) = 1, \quad \text{if } n \text{ is odd}.$$

Derivatives can be computed via (1.6.14) and (1.6.15).

Finally we show the plots of $\hat{H}_n$, $1 \leq n \leq 9$, in figure 1.7.3. The window measures $[-5,5] \times [-3,3]$.

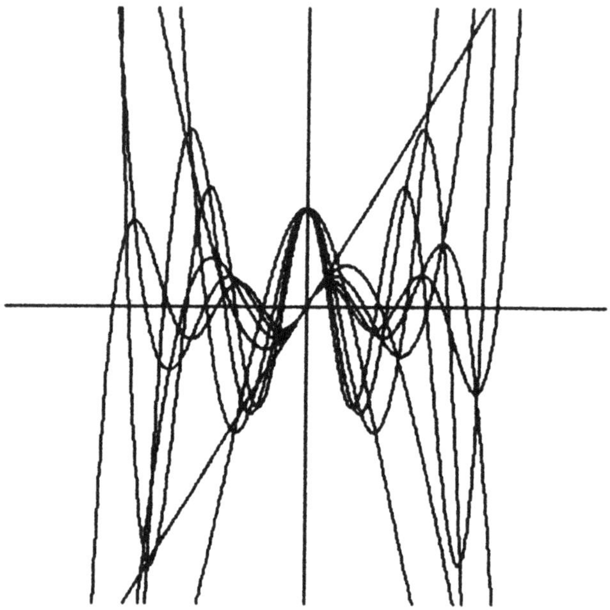

Figure 1.7.3 - Scaled Hermite functions for $1 \leq n \leq 9$.

* * * * * * * * * * * * *

Our survey is now concluded. Of course, other families of polynolmials could be considered. However, the ones presented here are those actually used in spectral methods.

2
ORTHOGONALITY

Inner products and orthogonal functions are fundamental concepts in approximation theory. All the families introduced in the first chapter form an orthogonal set of functions with respect to a suitable inner product. This property will reveal many other interesting features.

2.1 Inner products and norms

In the following, $\mathbf{X}$ will denote a *real vector space*, i.e. a set in which addition and scalar multiplication are defined with the usual basic properties (for the unfamiliar reader a wide bibliography is available; we suggest, for instance, HOFFMAN and KUNZE(1971), LOWENTHAL(1975)).

An *inner product* $(\cdot\,,\cdot):\mathbf{X}\times\mathbf{X}\to\mathbf{R}$ is a *bilinear application*. This means that, for any couple of vectors in $\mathbf{X}$, their inner product is a real number and the application is linear for each one of the two variables. Moreover, we require *symmetry*, i.e.

(2.1.1) $$(u,v) = (v,u), \qquad \forall u,v\in\mathbf{X},$$

and the application has to be *positive-definite*, i.e. (here $\mathbf{0}$ is the zero of $\mathbf{X}$):

(2.1.2) $$(u,u)\geq 0, \quad \forall u\in\mathbf{X}, \qquad (u,u)=0 \iff u\equiv\mathbf{0}.$$

A *norm* $\|\cdot\| : \mathbf{X} \to \mathbf{R}^+$ is a real positive function in $\mathbf{X}$ with the properties:

(2.1.3) $$\|u\| \geq 0, \quad \forall u \in \mathbf{X}, \qquad \|u\| = 0 \iff u \equiv \mathbf{0},$$

(2.1.4) $$\|\lambda u\| = |\lambda| \, \|u\|, \qquad \forall u \in \mathbf{X}, \ \ \forall \lambda \in \mathbf{R},$$

(2.1.5) $$\|u + v\| \leq \|u\| + \|v\|, \quad \forall u, v \in \mathbf{X} \ \ (\textit{triangle inequality}).$$

Inner products and norms are, in general, independent concepts. Nevertheless, whenever an inner product is available in $\mathbf{X}$, then a norm is automatically defined by setting

(2.1.6) $$\|u\| := \sqrt{(u,u)}, \qquad \forall u \in \mathbf{X}.$$

Checking that (2.1.6) gives actually a norm is an easy exercise. In particular (2.1.5) is a byproduct of the well-known *Schwarz inequality*

(2.1.7) $$|(u,v)| \leq \|u\| \, \|v\|, \qquad \forall u, v \in \mathbf{X}.$$

Detailed proofs of (2.1.7) and of many other properties of inner products and norms are widely available in all the basic texts of linear algebra.

We give an example. Let $\mathbf{X} = C^0(\bar{I})$ be the linear space of continuous functions in the interval $\bar{I}$. Let $w : I \to \mathbf{R}$ be a continuous integrable function satisfying $w > 0$. Then, when I is bounded, an inner product $(\cdot\,,\cdot)_w$ and its corresponding norm $\|\cdot\|_w$ are defined by

(2.1.8) $$(u,v)_w := \int_I uv \; w \; dx, \qquad \forall u, v \in C^0(\bar{I}),$$

(2.1.9) $$\|u\|_w := \left(\int_I u^2 \; w \; dx \right)^{\frac{1}{2}}, \qquad \forall u \in C^0(\bar{I}).$$

The function w is called *weight function.* If I is not bounded, we just have to be careful that the integrals in (2.1.8) and (2.1.9) are finite. We will be more precise in chapter five.

With the help of this short introduction, we are ready to analyze more closely the solutions of Sturm-Liouville problems.

2.2 Orthogonal functions

Two functions $u, v \in C^0(\bar{I})$ are said to be *orthogonal* when $(u,v)_w = 0$ for some weight function w.

Let us assume that the function a in (1.1.1) vanishes at the endpoints of the interval I (if I is not bounded, $\lim_{x\to\pm\infty} a(x) = 0$). Then, we have the following fundamental result.

Theorem 2.2.1 - *Let $\{u_n\}_{n\in\mathbf{N}}$ be a sequence of solutions of* (1.1.1) *corresponding to the eigenvalues $\{\lambda_n\}_{n\in\mathbf{N}}$. Let us require that $\lambda_n \neq \lambda_m$ if $n \neq m$. Then one has*

$$\int_I u_n u_m \, w \, dx = 0, \qquad \forall n, m \in \mathbf{N} \quad \text{with } n \neq m. \tag{2.2.1}$$

Proof - Let us multiply the equation (1.1.1) by a differentiable function v and integrate in I. Recalling the assumptions on a, after integration by parts, one gets

$$\int_I a u' v' \, dx + \int_I b u v \, dx = \lambda \int_I u v w \, dx. \tag{2.2.2}$$

Let $n \neq m$, thus relation (2.2.2) is satisfied either when $u = u_n$, $v = u_m$, $\lambda = \lambda_n$, or when $u = u_m$, $v = u_n$, $\lambda = \lambda_m$. In both cases the left-hand sides coincide. Having $\lambda_n \neq \lambda_m$, this implies (2.2.1).

■

In particular, theorem 2.2.1 shows that all the families of polynomials introduced in chapter one are orthogonal with respect to the inner product (2.1.8), where w is their corresponding weight function. More precisely, for any $n, m \in \mathbf{N}$, $n \neq m$, one has

$$(2.2.3)\ (\textit{Jacobi}) \qquad \int_{-1}^{1} P_n^{(\alpha,\beta)}(x) P_m^{(\alpha,\beta)}(x)\ (1-x)^{\alpha}(1+x)^{\beta}\ dx \ = \ 0,$$

$$(2.2.4)\ (\textit{Legendre}) \qquad \int_{-1}^{1} P_n(x) P_m(x)\ dx \ = \ 0,$$

$$(2.2.5)\ (\textit{Chebyshev}) \qquad \int_{-1}^{1} T_n(x) T_m(x)\ \frac{dx}{\sqrt{1-x^2}} \ = \ 0,$$

$$(2.2.6)\ (\textit{Laguerre}) \qquad \int_{0}^{+\infty} L_n^{(\alpha)}(x) L_m^{(\alpha)}(x)\ x^{\alpha}\ e^{-x}\ dx \ = \ 0,$$

$$(2.2.7)\ (\textit{Hermite}) \qquad \int_{-\infty}^{+\infty} H_n(x) H_m(x)\ e^{-x^2}\ dx \ = \ 0.$$

Note that scaled Laguerre (or Hermite) functions are not orthogonal.
Moreover, considering that $b \equiv 0$, (2.2.1) and (2.2.2) imply that the orthogonal polynomials also satisfy

$$(2.2.8) \qquad \int_I a u_n' u_m'\ dx \ = \ 0, \qquad \forall n, m \in \mathbf{N} \quad \text{with } n \neq m.$$

This shows that the derivatives are orthogonal polynomials with respect to the weight function a.
For any $n \geq 1$, if p is a polynomial of degree at most $n-1$ we have $\int_I p u_n w\ dx \ = \ 0$, since p is a linear combination of the u_k's for $k \leq n-1$.
By taking $x = \cos\theta$ in (2.2.5) and recalling (1.5.6), one obtains the well-known orthogonality relation for trigonometric functions (see for instance ZYGMUND(1988)):

$$(2.2.9) \qquad \int_{0}^{\pi} \cos n\theta\ \cos m\theta\ d\theta \ = \ 0, \qquad \forall n, m \in \mathbf{N}, \quad n \neq m.$$

This will allows us further characterize Chebyshev polynomials.

As we promised, we give the proof of theorem 1.1.1.

Proof - For $n = 2$ we have to determine three degrees of freedom ρ_2, σ_2, τ_2, in order to evaluate a second degree polynomial. Therefore, (1.1.2) is trivially satisfied. If $n > 2$, we write $p := u_n - \rho_n x u_{n-1}$, where ρ_n is such that p is a polynomial whose degree is at most $n-1$. Thus, due to the orthogonality condition

$$\int_I p u_m w \, dx \;=\; \int_I u_n u_m w \, dx \;-\; \rho_n \int_I u_{n-1}(x u_m) w \, dx \;=\; 0, \quad \forall m < n-2$$

(note that the degree of $x u_m$ is less than $n-1$). This implies that p is only a linear combination of the terms u_{n-1} and u_{n-2}, i.e., $p = \sigma_n u_{n-1} + \tau_n u_{n-2}$. The proof is concluded.

■

Finally, norms can be also evaluated.

Theorem 2.2.2 - *For any $n \in \mathbf{N}$, we have*

(2.2.10) (*Jacobi*)

$$\int_{-1}^{1} \left[P_n^{(\alpha,\beta)}(x)\right]^2 (1-x)^\alpha (1+x)^\beta \, dx$$

$$= \begin{cases} 2^{\alpha+\beta+1} \dfrac{\Gamma(\alpha+1)\,\Gamma(\beta+1)}{\Gamma(\alpha+\beta+2)} & \text{if } n = 0, \\[2ex] \dfrac{2^{\alpha+\beta+1}}{(2n+\alpha+\beta+1)\; n!} \; \dfrac{\Gamma(n+\alpha+1)\,\Gamma(n+\beta+1)}{\Gamma(n+\alpha+\beta+1)} & \text{if } n > 0, \end{cases}$$

(2.2.11) (*Legendre*)

$$\int_{-1}^{1} P_n^2(x)\, dx \;=\; \frac{2}{2n+1},$$

(2.2.12) (*Chebyshev*)

$$\int_{-1}^{1} T_n^2(x)\, \frac{dx}{\sqrt{1-x^2}} \;=\; \begin{cases} \pi & \text{if } n = 0, \\ \frac{\pi}{2} & \text{if } n > 0, \end{cases}$$

(2.2.13) (*Laguerre*) $$\int_0^{+\infty} \left[L_n^{(\alpha)}(x)\right]^2 x^\alpha \, e^{-x} \, dx \;=\; \frac{\Gamma(n+\alpha+1)}{n!},$$

(2.2.14) (*Hermite*) $$\int_{-\infty}^{+\infty} H_n^2(x) \, e^{-x^2} \, dx \;=\; 2^n \, n! \, \sqrt{\pi}.$$

Proof - Taking $u = v = u_n$ in (2.2.2), if $b \equiv 0$ we have

(2.2.15) $$\|u_n\|_w^2 \;=\; \lambda_n^{-1} \, \|u_n'\|_a^2.$$

Let us consider first the Jacobi case. We get $\lambda_n = n(n+\alpha+\beta+1)$ by theorem 1.3.1. Moreover, recalling (1.3.6) and substituting in (2.2.15), by iterating n times this procedure one obtains

(2.2.16) $$\int_{-1}^{1} \left[P_n^{(\alpha,\beta)}(x)\right]^2 (1-x)^\alpha (1+x)^\beta \, dx$$

$$= \frac{n+\alpha+\beta+1}{4n} \int_{-1}^{1} \left[P_{n-1}^{(\alpha+1,\beta+1)}(x)\right]^2 (1-x)^{\alpha+1}(1+x)^{\beta+1} \, dx \;=\; \cdots \;=$$

$$= \frac{(n+\alpha+\beta+1)\cdots(n+\alpha+\beta+n)}{4^n \, n!} \int_{-1}^{1} \left[P_0^{(\alpha+n,\beta+n)}(x)\right]^2 (1-x)^{\alpha+n}(1+x)^{\beta+n} \, dx$$

$$= \frac{\Gamma(2n+\alpha+\beta+1)}{2^{2n} \, n! \, \Gamma(n+\alpha+\beta+1)} \int_{-1}^{1} (1-x)^{\alpha+n}(1+x)^{\beta+n} \, dx.$$

We arrive at (2.2.10) by using (1.2.3) with $y = \alpha + n + 1$ and $x = \beta + n + 1$. Then, setting $\alpha = \beta = 0$ we have (2.2.11). By choosing $\alpha = \beta = -1/2$ and considering (1.5.1) we have (2.2.12) (otherwise recall (1.5.6) and take $x = \cos\theta$). To prove (2.2.13) and (2.2.14) we argue exactly as we did for the Jacobi case. This time we use the relations (1.6.6) and (1.7.8) respectively.

■

Let us now define

$$(2.2.17) \qquad \tilde{u}_n := \gamma_n^{-1} u_n, \qquad \text{where} \qquad \gamma_n := \lim_{x \to +\infty} \frac{u_n(x)}{x^n}.$$

Of course, we get $\tilde{u}_n(x) = x^n + \{lower\ degree\ terms\}$. The values of γ_n can be deduced in the different cases from (1.3.5), (1.5.5), (1.6.4) and (1.7.5). Then, we have the following minimizing property.

Theorem 2.2.3 - *For any $n \in \mathbf{N}$ and for any polynomial p of degree n such that $p(x) = x^n + \{lower\ degree\ terms\}$, we have*

$$(2.2.18) \qquad \|\tilde{u}_n\|_w \le \|p\|_w.$$

Proof - Let us note that $\tilde{u}_n - p$ has degree at most $n-1$. Consequently, $\tilde{u}_n - p$ is orthogonal to $\tilde{u}_n$. Hence

$$(2.2.19) \qquad \|\tilde{u}_n\|_w^2 = (\tilde{u}_n, p)_w.$$

Applying the Schwarz inequality (2.1.7) to the right-hand side of (2.2.19) we get (2.2.18). ■

2.3 Fourier coefficients

Hereafter, $\mathbf{P}_n$ denotes the linear space of polynomials whose degree is at most n. The dimension of $\mathbf{P}_n$ is clearly $n+1$. The fact that the polynomials u_k, $0 \le k \le n$, are orthogonal with respect to some inner product implies that they form a basis for $\mathbf{P}_n$. Therefore, for any $p \in \mathbf{P}_n$, one can determine in a unique way $n+1$ coefficients c_k, $0 \le k \le n$, such that

$$(2.3.1) \qquad p = \sum_{k=0}^{n} c_k u_k.$$

The c_k's are called the *Fourier coefficients* of p with respect to the designated basis. By knowing the coefficients of p, the value of $p(x)$ at a given x can be efficiently evaluated by (1.1.2) and (2.3.1) with a computational cost proportional to n.

If $p = \sum_{k=0}^{n} c_k u_k$ and $q = \sum_{k=0}^{n} b_k u_k$, then the following relations are trivially proven:

(2.3.2) $$p + q = \sum_{k=0}^{n} (c_k + b_k) u_k,$$

(2.3.3) $$\lambda p = \sum_{k=0}^{n} (\lambda c_k) u_k, \qquad \forall \lambda \in \mathbf{R},$$

(2.3.4) $$pq = \sum_{k=0}^{n} \sum_{j=0}^{n} (c_k b_j) u_k u_j.$$

By integrating (2.3.4) in I, orthogonality implies

(2.3.5) $$(p, q)_w = \sum_{m=0}^{n} c_m b_m \; \|u_m\|_w^2.$$

Hence

(2.3.6) $$\|p\|_w = \left(\sum_{m=0}^{n} c_m^2 \; \|u_m\|_w^2 \right)^{\frac{1}{2}}.$$

The quantities $\|u_m\|_w^2$, $m \in \mathbf{N}$, have been computed in theorem 2.2.2 for the different families of polynomials. Formula (2.3.6) allows us to evaluate the weighted norm of p when its coefficients are known. In particular, if we take $q \equiv u_k$, $0 \le k \le n$, in (2.3.5) (then $b_m = \delta_{km}$), the following explicit expression for the coefficients is available:

(2.3.7) $$c_k = \frac{(p, u_k)_w}{\|u_k\|_w^2}, \qquad p \in \mathbf{P}_n, \quad 0 \le k \le n.$$

Given $g : \mathbf{R^2} \to \mathbf{R}$, one can try to characterize the Fourier coefficients of $g(x, p(x))$, where p is a polynomial in $\mathbf{P}_n$, in terms of the coefficients of p. This turns out to be useful for the approximation of nonlinear terms in differential equations. For instance, we can easily handle the case $g(x, p(x)) = xp(x)$. Namely, from the coefficients of the expression $p = \sum_{k=0}^{n} c_k u_k$ we want to determine those of the expression $xp = \sum_{k=0}^{n+1} b_k u_k$. First of all we deduce from (1.1.2)

(2.3.8) $$x u_k = \frac{1}{\rho_{k+1}} (u_{k+1} - \sigma_{k+1} u_k - \tau_{k+1} u_{k-1}), \qquad 1 \le k \le n.$$

Therefore, one obtains

$$b_k = \begin{cases} \dfrac{c_n}{\rho_{n+1}} & k = n+1, \\[2ex] \dfrac{c_{n-1}}{\rho_n} - \dfrac{c_n \sigma_{n+1}}{\rho_{n+1}} & k = n, \\[2ex] \dfrac{c_{k-1}}{\rho_k} - \dfrac{c_k \sigma_{k+1}}{\rho_{k+1}} - \dfrac{c_{k+1} \tau_{k+2}}{\rho_{k+2}} & 1 \le k \le n-1, \\[2ex] -\dfrac{c_0 \sigma_1}{\rho_1} - \dfrac{c_1 \tau_2}{\rho_2} & k = 0. \end{cases} \tag{2.3.9}$$

For the various families of polynomials, the sequences $\{\rho_n\}$, $\{\sigma_n\}$, $\{\tau_n\}$ are respectively given by (1.3.8), (1.4.2), (1.5.2), (1.6.5) and (1.7.6) (ρ_1 and σ_1 are defined according to u_0 and u_1). For ultraspherical and Hermite polynomials we have $\sigma_n = 0$, $\forall n \ge 1$.

For more general functions g the answer can be very complicated and more explicit expressions are often unavailable. An interesting case is $g(x, p(x)) = p^2(x)$. For ultraspherical polynomials the following relation holds

$$P_k^{(\alpha,\alpha)} P_j^{(\alpha,\alpha)} = \left\{ \frac{\nu\, \Gamma(k+\nu+\frac{1}{2})\, \Gamma(j+\nu+\frac{1}{2})\, \Gamma(\nu+1)}{2\, \Gamma(k+2\nu)\, \Gamma(j+2\nu)\, \Gamma(\nu+\frac{1}{2})} \right. \tag{2.3.10}$$

$$\times \sum_{m=0}^{\min(k,j)} \frac{k+j+\nu-2m}{(m+\nu)(k-m+\nu)(j-m+\nu} \binom{m+\nu}{m} \binom{k-m+\nu}{k-m} \binom{j-m+\nu}{j-m}$$

$$\times \left[\binom{k+j-2m+2\nu}{k+j-2m} \right]^{-1} \frac{\Gamma(k+j-m+2\nu)}{\Gamma(k+j-m+\nu+1)} \frac{\Gamma(k+j-2m+2\nu+1)}{\Gamma(k+j-2m+\nu+\frac{1}{2})} \left. P_{k+j-2m}^{(\alpha,\alpha)} \right\},$$

$$\alpha > -1,\ \alpha \ne -\tfrac{1}{2},\quad k, j \in \mathbf{N},$$

where $\nu = \alpha + \frac{1}{2}$.

If $\alpha = -\frac{1}{2}$, by (1.5.6) we get

(2.3.11) $$T_kT_j = \frac{1}{2}\left(T_{k+j} + T_{|k-j|}\right), \qquad k,j \in \mathbf{N}.$$

Besides, for Hermite polynomials we have

(2.3.12) $$H_kH_j = \sum_{m=0}^{\min(k,j)} 2^m\, m! \binom{k}{m}\binom{j}{m} H_{k+j-2m}, \qquad k,j \in \mathbf{N}.$$

Taking $p = q$ in (2.3.4), we can finally evaluate the expansion coefficients of p^2 in terms of those of p. The reader can find more results, suggestions and references in ASKEY(1975), lecture 5.

2.4 The projection operator

More generally, in analogy with (2.3.7), if f is a given function such that fw is integrable in I (when I is not bounded fw will be also required to decay to zero at infinity), then we define its Fourier coefficients by setting

(2.4.1) $$c_k := \frac{1}{\|u_k\|_w^2} \int_I f u_k w\, dx, \qquad k \in \mathbf{N}.$$

Let us introduce now the operator $\Pi_{w,n} : C^0(\bar{I}) \to \mathbf{P}_n$, $n \in \mathbf{N}$. This operator acts as follows. For any continuous function f, we compute its coefficients according to (2.4.1). Then, $\Pi_{w,n} f \in \mathbf{P}_n$ is defined to be the polynomial $p_n := \sum_{k=0}^{n} c_k u_k$. It turns out that $\Pi_{w,n}$ is a linear operator. We call p_n the *orthogonal projection* of f onto $\mathbf{P}_n$, through the inner product $(\cdot,\cdot)_w$. Of course, one has

(2.4.2) $$\Pi_{w,n} p = p, \qquad \forall p \in \mathbf{P}_n,$$

(2.4.3) $$\Pi_{w,n} u_m = \mathbf{0}, \qquad \forall m > n.$$

From (2.3.8), we get for instance

$$(2.4.4) \qquad \Pi_{w,n}(xu_n) = -\frac{\sigma_{n+1}}{\rho_{n+1}}u_n - \frac{\tau_{n+1}}{\rho_{n+1}}u_{n-1}, \qquad n \geq 1.$$

Taking a truncation of the sum in (2.3.10) or (2.3.12) in order that $k + j - 2m \leq n$, we obtain an expression for $\Pi_{w,n}(u_k u_j)$, $k, j \in \mathbf{N}$. Therefore, by the linearity of $\Pi_{w,n}$, one obtains the coefficients of $\Pi_{w,n}p^2$ in terms of the coefficients of p.

The main properties of the projection operator are investigated in section 6.2. There, under suitable hypotheses, we shall see that $\lim_{n\to+\infty} \Pi_{w,n}f = f$, where the limiting process must be correctly interpreted.

2.5 The maximum norm

Let us assume that I is a bounded interval. In the linear space of bounded functions in $\bar{I}$, we define the norm

$$(2.5.1) \qquad \|u\|_\infty := \max_{x\in\bar{I}} |u(x)|, \qquad \forall u \text{ bounded.}$$

One easily verifies that $\|\cdot\|_\infty$ actually satisfies all the properties (2.1.3)-(2.1.5). We call $\|u\|_\infty$ the *maximum norm* of u in $\bar{I}$. It can be proven that it is not possible to define an inner product that generates the maximum norm through a relation like (2.1.6).

Two celebrated theorems give a bound to the maximum norm of the derivative of polynomials. The first one is due to A.Markoff and is successively generalized to higher order derivatives by his brother (see MARKOFF(1889) and MARKOFF(1892)). The second one is due to S.Bernstein (BERNSTEIN(1912a)). Other references can be found in JACKSON(1930), CHENEY(1966) as well in ZYGMUND(1988), Vol.2. Further refinements are analyzed in section 3.9. Applications are illustrated in chapter six.

Theorem 2.5.1 *(Markoff) - Let $\bar{I} = [-1, 1]$, then for any $n \in \mathbf{N}$ we have*

$$(2.5.2) \qquad \|p'\|_\infty \leq n^2 \|p\|_\infty, \qquad \forall p \in \mathbf{P}_n.$$

■

Theorem 2.5.2 *(Bernstein) - Let $\bar{I} = [-1,1]$, then for any $n \in \mathbf{N}$ we have*

$$|p'(x)| \leq \frac{n}{\sqrt{1-x^2}} \|p\|_\infty, \quad \forall x \in I, \quad \forall p \in \mathbf{P}_n. \tag{2.5.3}$$

■

Recalling definition (2.2.17), we finally state one of the main results characterizing Chebyshev polynomials (compare with theorem 2.2.3).

Theorem 2.5.3 - *For any $n \in \mathbf{N}$ and for any polynomial p in $\bar{I} = [-1,1]$ of degree n such that $p(x) = x^n + \{lower\ degree\ terms\}$, we have*

$$\|\tilde{T}_n\|_\infty \leq \|p\|_\infty. \tag{2.5.4}$$

■

The proof is given for instance in RIVLIN(1974). Note that

$$\|\tilde{T}_n\|_\infty = \begin{cases} 1 & \text{if } n = 0, \\ 2^{1-n} & \text{if } n \geq 1. \end{cases} \tag{2.5.5}$$

2.6 Basis transformations

An interesting question is how to transform the Fourier coefficients of a given polynomial corresponding to an assigned orthogonal basis, into the coefficients of another basis orthogonal with respect to a different weight function. The goal is to determine the so-called *connection coefficients* of the expansion of any element of the first basis in terms of the elements of the second basis. For a survey of the major results, we suggest that the interested reader consult lecture 7 from the book of ASKEY(1975). Here we only mention the relation existing between some Jacobi polynomials and Chebyshev polynomials, namely

$$P_n^{(\alpha,\alpha)} = \left\{ \frac{\Gamma(2\alpha+1)\,\Gamma(n+\alpha+1)}{\Gamma(\alpha+1)\,\Gamma(n+2\alpha+1)} \right. \tag{2.6.1}$$

$$\left. \times \sum_{k=0}^{n} \frac{\Gamma(k+\alpha+\frac{1}{2})\,\Gamma(n-k+\alpha+\frac{1}{2})}{\left[\Gamma(\alpha+\frac{1}{2})\right]^2\, k!\,(n-k)!}\, T_{|n-2k|} \right\}, \qquad n \in \mathbf{N},\ \alpha > -\tfrac{1}{2}.$$

A problem related to the previous one, is the determination of the weighted norm of a polynomial, when the expansion coefficients, corresponding to a basis orthogonal with respect to another weight function, are known. We give an example. Sometimes, in practical applications it is useful to evaluate the integral $\int_{-1}^{1} p^2 dx$, $p \in \mathbf{P}_n$. If p is known in terms of its Legendre coefficients, (2.3.6) is the formula we are looking for. The situation becomes more complicated when p is given in terms of the Fourier coefficients of another basis, say the Chebyshev basis. Let us assume that

$$p = \sum_{k=0}^{n} c_k T_k. \tag{2.6.2}$$

Then

$$\int_{-1}^{1} p^2\, dx = \sum_{k=0}^{n}\sum_{j=0}^{n} c_k c_j I_{kj}, \quad \text{where } I_{kj} = \int_{-1}^{1} T_k T_j\, dx. \tag{2.6.3}$$

The help of some trigonometry leads to the formula

$$I_{kj} = \int_0^{\pi} \cos k\theta\ \cos j\theta\ \sin\theta\ d\theta \tag{2.6.4}$$

$$= \begin{cases} 0 & \text{if } k+j \text{ is odd,} \\ \dfrac{1}{1-(k+j)^2} + \dfrac{1}{1-(k-j)^2} & \text{if } k+j \text{ is even.} \end{cases}$$

We note that (2.6.3) has a computational cost proportional to n^2, while the cost of (2.3.6) is only proportional to n.

Generalizations to other bases can be also considered, but they are outside the aims of this book.

3
NUMERICAL INTEGRATION

As we shall see in this chapter, orthogonal polynomials allow us to generate in a very elegant way high order integration formulas, known as Gauss formulas. Since the theory is based on the knowledge of the zeroes of the polynomials, the first part is devoted to the characterization of their main properties.

3.1 Zeroes of orthogonal polynomials

It is well-known that a polynomial of degree n has at most n distinct complex zeroes. In general, very little can be said about real zeroes. An amazing theorem gives a precise and complete answer in the case of orthogonal polynomials.

Theorem 3.1.1 - *Let* $\{u_n\}_{n\in\mathrm{N}}$ *be a sequence of solutions of* (1.1.1), *where* u_n *is a polynomial of degree* n, *satisfying the orthogonal relation* (2.2.1). *Then, for any* $n \geq 1$, u_n *has exactly* n *real distinct zeroes in* I.

Proof - We first note that $\int_I u_n(x)w(x)dx = 0$. Therefore u_n changes sign in I, hence it has at least one real zero $\xi_1 \in I$. If $n = 1$ the proof is ended. If $n > 1$, then we can find another zero $\xi_2 \in I$ of u_n, with $\xi_2 \neq \xi_1$, since if u_n vanished only at ξ_1, then the polynomial $(x - \xi_1)u_n$ would not change sign in I, which is in contradiction with the relation $\int_I(x - \xi_1)u_n(x)w(x)dx = 0$, obtained by orthogonality. In a similar fashion, one considers the polynomial $(x-\xi_1)(x-\xi_2)u_n$ and, if $n > 2$, deduces the existence of a third zero, and so on. The procedure ends when n zeroes are finally obtained.

■

For any $n \geq 1$ we denote the n zeroes of u_n by $\xi_k^{(n)}$, $1 \leq k \leq n$. We assume that these are in increasing order (although many authors prefer the reverse order). It is obvious that the polynomial u'_n has $n-1$ real distinct zeroes in I. We denote these zeroes by $\eta_k^{(n)}$, $1 \leq k \leq n-1$. One can prove that u_n and u_{n-1} do not have common zeroes. Moreover, between any two consecutive zeroes of u_{n-1}, there exists one and only one zero of u_n.

Many general theorems characterize the zeroes of orthogonal polynomials. It suffices to recall the following statement.

Theorem 3.1.2 - *Let $\{u_n\}_{n\in\mathbf{N}}$ be a sequence of orthogonal polynomials in I. Then, for any interval $[a,b] \subset I$, $a < b$, it is possible to find $m \in \mathbf{N}$ such that u_m has at least one zero in $[a,b]$.*

■

In other words, this theorem states that the set $J = \bigcup_{n\geq 1} \bigcup_{k=1}^{n} \{\xi_k^{(n)}\}$ is dense in $\bar{I}$.

We are going to review some properties of the zeroes of classical orthogonal polynomials. We refer for instance to SZEGÖ(1939) for proofs and further results.

Zeroes in the Jacobi case - We restrict ourselves to the case where $-\frac{1}{2} \leq \alpha \leq \frac{1}{2}$ and $-\frac{1}{2} \leq \beta \leq \frac{1}{2}$. Under these conditions, for $n \geq 1$ we have the estimates

$$(3.1.1) \qquad -1 \leq -\cos\frac{k+(\alpha+\beta-1)/2}{n+(\alpha+\beta+1)/2}\pi \leq \xi_k^{(n)} \leq -\cos\frac{k}{n+(\alpha+\beta+1)/2}\pi \leq 1,$$

$$1 \leq k \leq n.$$

This shows that asymptotically the distance between two consecutive zeroes is proportional to $1/n$ for points located in the central part of the interval $I =]-1,1[$, and proportional to $1/n^2$ for points located near the endpoints of the same interval.

In the case of ultraspherical polynomials ($-\frac{1}{2} \le \alpha = \beta \le \frac{1}{2}$), the inequalities can be refined as follows:

$$(3.1.2) \qquad -1 < -\cos \frac{k + \frac{\alpha}{2} - \frac{1}{4}}{n + \alpha + \frac{1}{2}}\pi \le \xi_k^{(n)} \le -\cos \frac{k}{n + \alpha + \frac{1}{2}}\pi \le 0, \quad 1 \le k \le \left[\tfrac{n}{2}\right],$$

$$(3.1.3) \qquad 0 \le \cos \frac{n - k + 1}{n + \alpha + \frac{1}{2}}\pi \le \xi_k^{(n)} \le \cos \frac{n - k + \frac{\alpha}{2} + \frac{3}{4}}{n + \alpha + \frac{1}{2}}\pi < 1,$$

$$n + 1 - \left[\tfrac{n}{2}\right] \le k \le n.$$

Of course, if n is odd we have $\xi_{(n+1)/2}^{(n)} = 0$.

Particularly interesting are the following cases, in which an exact expression of the zeroes is known:

$$(3.1.4) \qquad \xi_k^{(n)} = -\cos \frac{2k-1}{2n}\pi \quad \text{if } \alpha = \beta = -\frac{1}{2},$$

$$(3.1.5) \qquad \xi_k^{(n)} = -\cos \frac{k}{n+1}\pi \quad \text{if } \alpha = \beta = \frac{1}{2},$$

$$(3.1.6) \qquad \xi_k^{(n)} = -\cos \frac{2k}{2n+1}\pi \quad \text{if } \alpha = \frac{1}{2},\ \beta = -\frac{1}{2},$$

$$(3.1.7) \qquad \xi_k^{(n)} = -\cos \frac{2k-1}{2n+1}\pi \quad \text{if } \alpha = -\frac{1}{2},\ \beta = \frac{1}{2},$$

where $1 \le k \le n$.

Zeroes in the Legendre case - Estimates indicating the location of the zeroes are obtained by setting $\alpha = 0$ in (3.1.2) and (3.1.3). Figure 3.1.1 shows their distribution in the interval $I =]-1, 1[$ for $n = 7$ and $n = 10$.

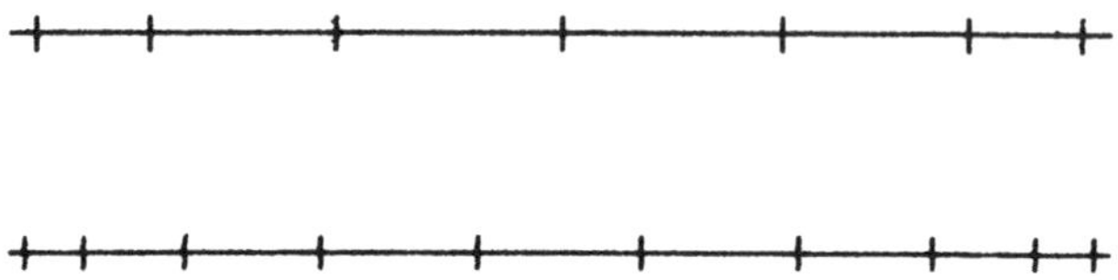

Figure 3.1.1 - Legendre zeroes for $n = 7$ and $n = 10$.

Zeroes in the Chebyshev case - Due to (1.5.1) the zeroes are those of ultraspherical polynomials with $\alpha = \beta = -\frac{1}{2}$. Hence, they are exactly given by (3.1.4). On the other hand, thanks to equality (1.5.6), they can be also obtained from the roots of the equation: $\cos n\theta = 0, \quad n \geq 1, \quad \theta \in]0, \pi[$. In figure 3.1.2 we give the distribution of the Chebyshev zeroes for $n = 7$ and $n = 10$.

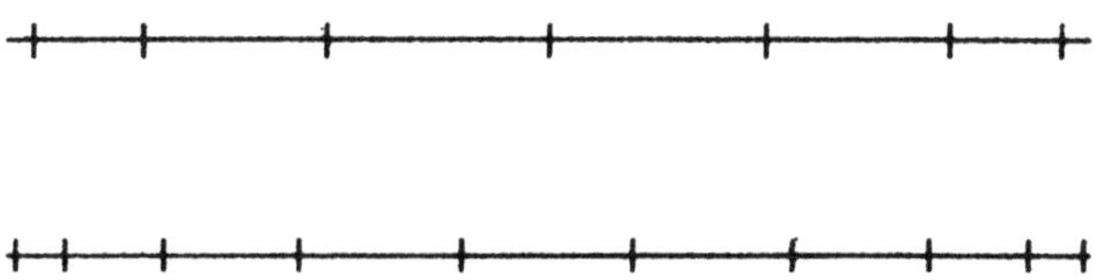

Figure 3.1.2 - Chebyshev zeroes for $n = 7$ and $n = 10$.

Zeroes in the Laguerre case - According to TRICOMI(1954), p.153, and GATTESCHI(1964), a good approximation of the zeroes of Laguerre polynomials is obtained by the following procedure. For any $n \geq 1$, we first find the roots of the equation

(3.1.8) $$y_k^{(n)} - \sin y_k^{(n)} = 2\pi \, \frac{n - k + \frac{3}{4}}{2n + \alpha + 1}, \qquad 1 \leq k \leq n.$$

Then we set $\hat{y}_k^{(n)} := \left[\cos \frac{1}{2} y_k^{(n)}\right]^2$, and define for $1 \leq k \leq n$,

(3.1.9) $$z_k^{(n)} := 2(2n+\alpha+1)\hat{y}_k^{(n)} - \frac{1}{6(2n+\alpha+1)}\left[\frac{5}{4(1-\hat{y}_k^{(n)})^2} - \frac{1}{1-\hat{y}_k^{(n)}} - 1 + 3\alpha^2\right].$$

Let us permute the values of $z_k^{(n)}$, $1 \le k \le n$, in such a way they are in increasing order. After this rearrangement, we denote the new sequence by $\hat{z}_k^{(n)}$, $1 \le k \le n$. Finally, we have

(3.1.10) $$\xi_k^{(n)} \approx \hat{z}_k^{(n)}, \quad 1 \le k \le n.$$

By virtue of theorem 3.1.2, one also gets $\lim_{n\to+\infty} \xi_n^{(n)} = +\infty$, $\lim_{n\to+\infty} \xi_1^{(n)} = 0$. Moreover, the largest zero tends to infinity like $4n$, while the smallest zero behaves like $\frac{\pi^2}{4n}$. We plot in figure 3.1.3 the Laguerre zeroes for $\alpha = 0$ and $n = 10$ in the interval $]0, 40[$.

Figure 3.1.3 - Laguerre zeroes for $\alpha = 0$ and $n = 10$.

Zeroes in the Hermite case - It is sufficient to recall the relations (1.7.9) and (1.7.10). Therefore, when n is even, positive zeroes are approximated by $\sqrt{\hat{z}_k^{(n/2)}}$, $1 \le k \le \frac{n}{2}$, where these quantities are obtained with the procedure described above with $\alpha = -\frac{1}{2}$. When n is odd, positive zeroes are approximated by $\sqrt{\hat{z}_k^{((n-1)/2)}}$, $1 \le k \le \frac{n-1}{2}$, where these terms are evaluated for $\alpha = \frac{1}{2}$. Moreover $\xi_{(n+1)/2}^{(n)} = 0$. For all n, the zeroes are symmetrically distributed around $x = 0$. A plot in the interval $]-6, 6[$ is given in figure 3.1.4 for $n = 15$.

Figure 3.1.4 - Hermite zeroes for $n = 15$.

In the numerical computations, the following algorithm is generally adopted. For each zero, an initial guess is obtained as described above (for the Jacobi case the average between the upper and lower bound in the estimates (3.1.1) or (3.1.2)-(3.1.3) can be taken into account). A few iterations of a Newton method is then sufficient to determine accurately the zero. The point values of the polynomial and its derivative can be computed from (1.1.2) and (1.1.3).

Concerning the case of Laguerre (or Hermite), when n grows, rounding errors occur in the procedure, due to the sharp oscillations exhibited by the polynomials. Considerable improvements are obtained operating with scaled Laguerre (or Hermite) functions. Of course, the values of the zeroes are the same, but the derivatives in the neighbourhood of the zeroes are now moderate.

Similar arguments hold for the determination of the zeroes of the derivatives of orthogonal polynomials. An initial approximation can obtained by considering that each zero $\eta_k^{(n)}$, $1 \leq k \leq n-1$, of u_n' lies between two consecutive zeroes of u_n. In alternative, one can use relations (1.3.6), (1.6.6), (1.7.8).

As it will be clear in the following sections, for $n \geq 1$, in the Jacobi case it is convenient to set $\eta_0^{(n)} := -1$ and $\eta_n^{(n)} := 1$ (though these are not in general zeroes). Particularly interesting are the situations where an explicit expression is known, i.e.

(3.1.11) $$\eta_k^{(n)} = -\cos\frac{k\pi}{n} \qquad 0 \leq k \leq n, \quad \text{if } \alpha = \beta = -\frac{1}{2},$$

(3.1.12) $$\eta_k^{(n)} = -\cos\frac{2k+1}{2n+2}\pi \qquad 1 \leq k \leq n-1, \quad \text{if } \alpha = \beta = \frac{1}{2},$$

(3.1.13) $$\eta_k^{(n)} = -\cos\frac{2k+1}{2n+1}\pi \qquad 1 \leq k \leq n, \quad \text{if } \alpha = \frac{1}{2}, \ \beta = -\frac{1}{2},$$

(3.1.14) $$\eta_k^{(n)} = -\cos\frac{2k}{2n+1}\pi \qquad 0 \leq k \leq n-1, \quad \text{if } \alpha = -\frac{1}{2}, \ \beta = \frac{1}{2}.$$

We give the plots of the zeroes of the derivative of Legendre and Chebyshev polynomials for $n = 10$, in figure 3.1.5.

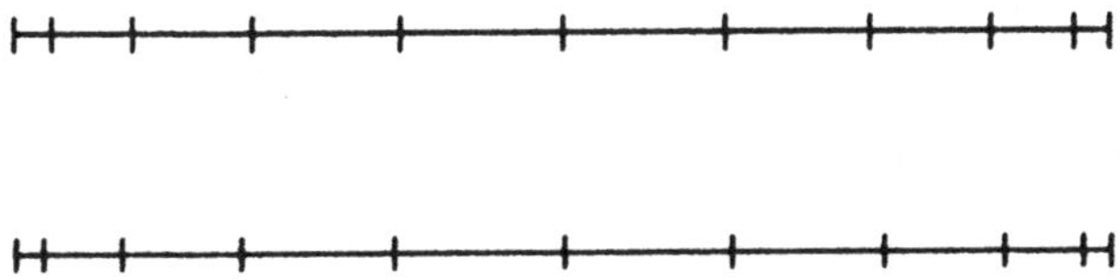

Figure 3.1.5 - The points $\eta_k^{(10)}$, $0 \le k \le 10$, in the cases of Legendre and Chebyshev.

By virtue of (1.5.6) and (1.5.7), Chebyshev polynomials satisfy for $n \ge 1$

(3.1.15) $$T_n(\eta_j^{(n)}) = (-1)^{n+j}, \qquad 0 \le j \le n,$$

(3.1.16) $$T_n'(\xi_j^{(n)}) = \frac{n\,(-1)^{n+j}}{\sqrt{1-[\xi_j^{(n)}]^2}}, \qquad 1 \le j \le n.$$

Other useful relations are ($n \ge 1$, $\alpha \ge -1$, $\beta \ge -1$)

(3.1.17) $$\frac{d}{dx}P_n^{(\alpha,\beta)}(1) = \frac{(n+\alpha+\beta+1)\,\Gamma(n+\alpha+1)}{2\,(n-1)!\,\Gamma(\alpha+2)},$$

(3.1.18) $$\frac{d}{dx}P_n^{(\alpha,\beta)}(-1) = -\frac{(-1)^n\,(n+\alpha+\beta+1)\,\Gamma(n+\beta+1)}{2\,(n-1)!\,\Gamma(\beta+2)},$$

(3.1.19) $$\frac{d}{dx}L_n^{(\alpha)}(0) = -\frac{\Gamma(n+\alpha+1)}{(n-1)!\,\Gamma(\alpha+2)}.$$

These equalities are easily obtained from (1.3.1) and (1.6.1).

3.2 Lagrange polynomials

As remarked in section 2.3, the set $\{u_k\}_{0\le k\le n}$ is a basis in the space $\mathbf{P}_n$ of polynomials of degree at most n. When $n+1$ distinct points are given in $\bar{I}$, another basis in $\mathbf{P}_n$ is generated in a natural way. This is the basis of *Lagrange polynomials* with respect to the prescribed points. An element of the basis attains the value 1 at a certain point and vanishes in the remaining n points.

Let us analyse first the Jacobi case. We examine two interesting examples. On one hand we have the set of the Lagrange polynomials in $\mathbf{P}_{n-1}$ relative to the n points $\xi_k^{(n)}$, $1\le k\le n$, i.e., the zeroes of $P_n^{(\alpha,\beta)}$. On the other we have the set of Lagrange polynomials in $\mathbf{P}_n$ relative to the $n+1$ points $\eta_k^{(n)}$, $0\le k\le n$, i.e., the zeroes of $\frac{d}{dx}P_n^{(\alpha,\beta)}$ plus the two points -1 and 1.
In the first case, the elements of the basis are denoted by $l_j^{(n)}$, $1\le j\le n$. These polynomials in $\mathbf{P}_{n-1}$ are uniquely defined by the conditions

$$(3.2.1)\qquad l_j^{(n)}(\xi_i^{(n)}) = \begin{cases} 1 & \text{if } i=j \\ 0 & \text{if } i\neq j \end{cases}, \qquad 1\le j\le n.$$

They actually form a basis because any polynomial $p\in\mathbf{P}_{n-1}$ can be written as follows:

$$(3.2.2)\qquad p = \sum_{j=1}^{n} p(\xi_j^{(n)})\, l_j^{(n)}.$$

Therefore, p is a linear combination of the Lagrange polynomials. Such a combination is uniquely determined by the coefficients $p(\xi_j^{(n)})$, $1\le j\le n$.
The following expression is easily proven:

$$(3.2.3)\qquad l_j^{(n)}(x) = \prod_{\substack{k=1\\k\neq j}}^{n} \frac{x-\xi_k^{(n)}}{\xi_j^{(n)}-\xi_k^{(n)}}, \qquad 1\le j\le n, \quad x\in[-1,1].$$

For future applications, it is more convenient to consider the alternate expression

$$(3.2.4)\qquad l_j^{(n)}(x) = \begin{cases} \dfrac{u_n(x)}{u_n'(\xi_j^{(n)})\,(x-\xi_j^{(n)})} & \text{if } x\neq\xi_j^{(n)}, \\ \\ 1 & \text{if } x=\xi_j^{(n)}, \end{cases}$$

where $1 \le j \le n$ and $u_n = P_n^{(\alpha,\beta)}$. Of course, we have

$$\lim_{x \to \xi_j^{(n)}} l_j^{(n)}(x) = \lim_{x \to \xi_j^{(n)}} \frac{u_n'(x)}{u_n'(\xi_j^{(n)})} = 1, \qquad 1 \le j \le n. \tag{3.2.5}$$

Plots of the Lagrange basis with respect to the Legendre and Chebyshev zeroes are given in figures 3.2.1 and 3.2.2 for $n = 7$.

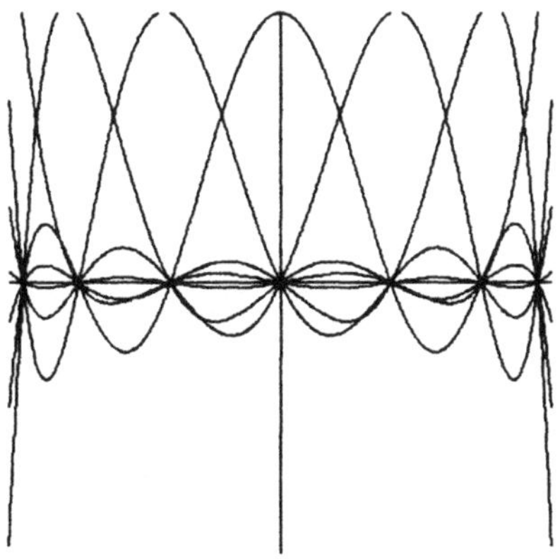

Figure 3.2.1 - Legendre Lagrange basis with respect to $\xi_i^{(7)}$, $1 \le i \le 7$.

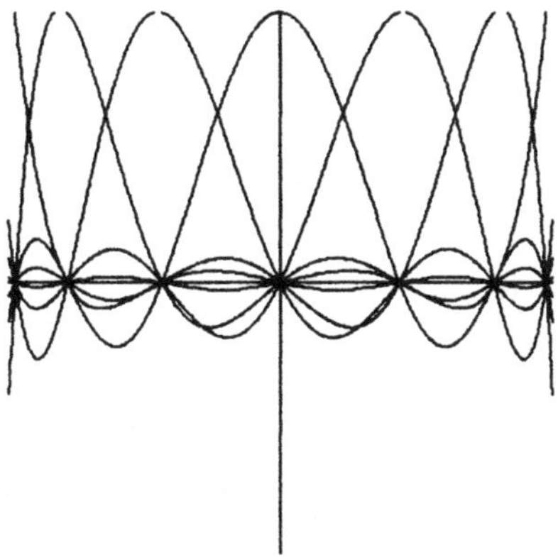

Figure 3.2.2 - Chebyshev Lagrange basis with respect to $\xi_i^{(7)}$, $1 \le i \le 7$.

The other basis we mentioned above is denoted by $\{\tilde{l}_j^{(n)}\}_{0 \le j \le n} \subset \mathbf{P}_n$, and it is defined by

$$\tilde{l}_j^{(n)}(\eta_i^{(n)}) = \delta_{ij}, \qquad 0 \le i \le n,\ 0 \le j \le n. \tag{3.2.6}$$

Indeed, for any polynomial $p \in \mathbf{P}_n$, one has

$$p = \sum_{j=0}^{n} p(\eta_j^{(n)})\, \tilde{l}_j^{(n)}. \tag{3.2.7}$$

In analogy with (3.2.4), we prove the following proposition.

Theorem 3.2.1 - *For any $n \geq 1$, we have*

$$(3.2.8) \qquad \tilde{l}_j^{(n)}(x) = \begin{cases} \dfrac{(-1)^n\ (n-1)!\ \Gamma(\beta+2)}{(n+\alpha+\beta+1)\ \Gamma(n+\beta+1)}(x-1)u_n'(x) & \text{if } j=0, \\[2ex] \dfrac{(x^2-1)u_n'(x)}{n(n+\alpha+\beta+1)\ u_n(\eta_j^{(n)})\ (x-\eta_j^{(n)})} & \text{if } 1 \leq j \leq n-1, \\[2ex] \dfrac{(n-1)!\ \Gamma(\alpha+2)}{(n+\alpha+\beta+1)\ \Gamma(n+\alpha+1)}(x+1)u_n'(x) & \text{if } j=n, \end{cases}$$

with $x \neq \eta_j^{(n)}$ and $u_n = P_n^{(\alpha,\beta)}$. Besides, we have $\tilde{l}_j^{(n)}(\eta_j^{(n)}) = 1, \quad 0 \leq j \leq n$.

Proof - If $j = 0$, then $\gamma(x-1)u_n'$, $\gamma \in \mathbf{R}$, is a polynomial in $\mathbf{P}_n$ vanishing at $\eta_k^{(n)}$, $1 \leq k \leq n$. After evaluating in $x=-1$, the constant γ has to be determined so that $-2\gamma u_n'(-1) = 1$. Finally, the value $u_n'(-1)$ is found from (3.1.18). A similar procedure applies when $j = n$ (this time recalling (3.1.17)). If $1 \leq j \leq n-1$, for some $\gamma \in \mathbf{R}$ we have

$$(3.2.9) \qquad \lim_{x\to\eta_j^{(n)}} \gamma\ \frac{(x^2-1)u_n'(x)}{x-\eta_j^{(n)}} = \lim_{x\to\eta_j^{(n)}} \gamma[(x^2-1)u_n''(x) + 2xu_n'(x)]$$

$$= \gamma([\eta_j^{(n)}]^2 - 1)u_n''(\eta_j^{(n)}) = \gamma n(n+\alpha+\beta+1)u_n(\eta_j^{(n)}).$$

The last equality in (3.2.9) is a consequence of (1.3.1). The required γ is then obtained by equating the last expression in (3.2.9) to 1.

■

Recalling (1.5.1), (1.5.4) and (3.1.15), relation (3.2.8) becomes in the Chebyshev case (see also GOTTLIEB, HUSSAINI and ORSZAG(1984)):

$$\tilde{l}_j^{(n)}(x) = \begin{cases} \frac{(-1)^n}{2n^2}\,(x-1)T_n'(x) & \text{if } j=0, \\ \frac{(-1)^{j+n}}{n^2}\,\frac{(x^2-1)T_n'(x)}{x-\eta_j^{(n)}} & \text{if } 1 \le j \le n-1, \\ \frac{1}{2n^2}\,(x+1)T_n'(x) & \text{if } j=n. \end{cases} \tag{3.2.10}$$

The values $\eta_j^{(n)}$, $0 \le j \le n$, are given in (3.1.11).
Lagrange bases with respect to zeroes of derivatives of Legendre and Chebyshev polynomials are shown for $n = 7$, respectively in figures 3.2.3 and 3.2.4.

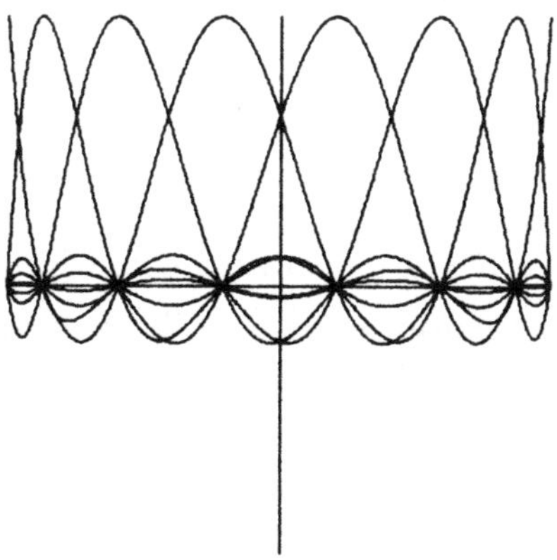

Figure 3.2.3 - Legendre Lagrange basis with respect to $\eta_i^{(7)}$, $0 \le i \le 7$.

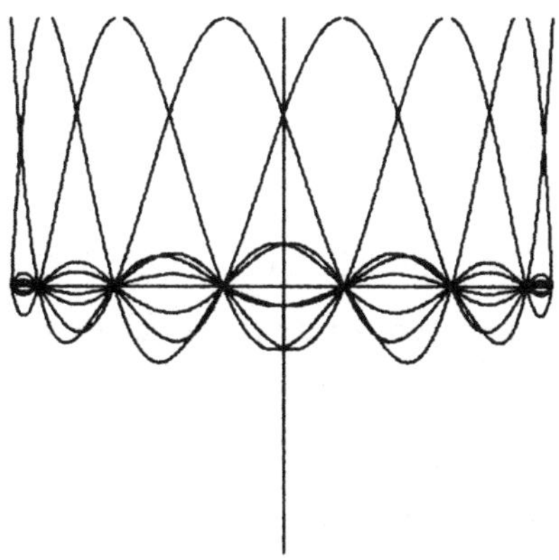

Figure 3.2.4 - Chebyshev Lagrange basis with respect to $\eta_i^{(7)}$, $0 \le i \le 7$.

Application of a similar procedure to the case of Laguerre (or Hermite) polynomials is straightforward. Again, we denote by $\{l_j^{(n)}\}_{1\le j\le n}$ the Lagrange basis with respect to the Laguerre (respectively Hermite) zeroes. Relations (3.2.2) and (3.2.3) are still valid. Formula (3.2.4) also holds provided $u_n = L_n^{(\alpha)}$ (respectively $u_n = H_n$).

In particular, for the Laguerre case, another set of Lagrange polynomials in $\mathbf{P}_{n-1}$ will be useful in what follows. We first define $\eta_0^{(n)} := 0$, and then we construct the Lagrange basis $\{\tilde{l}_j^{(n)}\}_{0\leq j\leq n-1}$ with respect to $\eta_i^{(n)}$, $0 \leq i \leq n-1$. With a proof similar to that of theorem 3.2.1 (this time using (1.6.1) and (3.1.19)), one gets, for $n \geq 1$,

(3.2.11)
$$\tilde{l}_j^{(n)}(x) = \begin{cases} -\dfrac{(n-1)!\,\Gamma(\alpha+2)}{\Gamma(n+\alpha+1)}\, u_n'(x) & \text{if } j=0, \\ -\dfrac{x\, u_n'(x)}{n\, u_n(\eta_j^{(n)})\,(x-\eta_j^{(n)})} & \text{if } 1\leq j\leq n-1, \end{cases}$$

where $u_n = L_n^{(\alpha)}$ and $x \in [0, +\infty[$.

3.3 The interpolation operators

Following the guideline of section 2.4, we introduce some new operators defined in the space of continuous functions. They are named *interpolation operators*. We first begin by considering the set of polynomials $\{u_k\}_{k\in\mathbf{N}}$, orthogonal with respect to the weight w. Next, we choose $n \geq 1$ and evaluate the zeroes $\xi_k^{(n)}$, $1 \leq k \leq n$, of u_n. At this point, we define $I_{w,n} : C^0(I) \to \mathbf{P}_{n-1}$ to be the operator mapping a continuous function f to the unique polynomial $p_n \in \mathbf{P}_{n-1}$ satisfying $p_n(\xi_j^{(n)}) = f(\xi_j^{(n)})$, $1 \leq j \leq n$. This polynomial is the *interpolant* of f with respect to the zeroes of u_n. The operator $I_{w,n}$ is linear. Moreover, one can check that

(3.3.1)
$$I_{w,n}p = p, \qquad \forall p \in \mathbf{P}_{n-1}.$$

A question arose in sections 2.3 and 2.4 on how to characterize a certain expression $g(x,p(x))$ in terms of a given polynomial p. A very simple answer is obtained when p is determined by the values attained in a set of given points. For example, we have

(3.3.2)
$$[I_{w,n}g(x,p(x))]_{x=\xi_j^{(n)}} = g(\xi_j^{(n)}, p(\xi_j^{(n)})), \quad 1\leq j\leq n,$$

for any polynomial $p \in \mathbf{P}_{n-1}$.

This means that, by knowing the values $p(\xi_j^{(n)})$, $1 \leq j \leq n$, we can for instance recover the point values of $I_{w,n}p^2$. These are $p^2(\xi_j^{(n)})$, $1 \leq j \leq n$. Hence, the value in x of the interpolant of p^2 is $\sum_{j=1}^n p^2(\xi_j^{(n)}) l_j^{(n)}(x)$.
This trivial remark is a key point in the treatment of nonlinear terms in the approximation of differential equations (see section 9.8).

Another interpolant operator is defined in relation to the zeroes of $\frac{d}{dx}P_n^{(\alpha,\beta)}$. Thus for $n \geq 1$, let us consider $\tilde{I}_{w,n} : C^0([-1,1]) \to \mathbf{P}_n$ to be the operator associating to the function f, the unique polynomial $p_n \in \mathbf{P}_n$ satisfying $p_n(\eta_j^{(n)}) = f(\eta_j^{(n)})$, $0 \leq j \leq n$. The operator is linear and we get

$$\tilde{I}_{w,n}p \;=\; p, \qquad \forall p \in \mathbf{P}_n. \tag{3.3.3}$$

For Laguerre polynomials $\tilde{I}_{w,n} : C^0([0,+\infty[) \to \mathbf{P}_{n-1}$ is characterized by $(\tilde{I}_{w,n}f)(\eta_j^{(n)})$ $= f(\eta_j^{(n)})$, $0 \leq j \leq n-1$, where $\eta_0^{(n)} = 0$.

As already mentioned in section 2.4, the analysis when n tends to $+\infty$, of the operators introduced above, is one of the primary issues of chapter six.

3.4 Gauss integration formulas

We have reached the crucial subject of this chapter, that is the formulas for numerical integration. Given a polynomial $p \in \mathbf{P}_n$, one is concerned with finding the value of $\int_I pw\,dx$. If the Fourier expansion of $p = \sum_{k=0}^{n-1} c_k u_k$ is known, the answer has been given in chapter two, i.e., $\int_I pw\,dx = c_0 \int_I u_0 w\,dx$. On the other hand, when p is known through its values at the zeroes of u_n, it is sufficient to integrate (3.2.2) obtaining

$$\int_I pw\;dx \;=\; \sum_{j=1}^n p(\xi_j^{(n)})\; w_j^{(n)}, \tag{3.4.1}$$

where $w_j^{(n)} := \int_I l_j^{(n)} w\,dx$, $1 \leq j \leq n$, are the *weights* of the integration formula.

Expression (3.4.1) is known as *Gauss quadrature formula.* We will also refer to the $\xi_j^{(n)}$'s as the *Gauss integration nodes.* Of course, once the weights are evaluated, (3.4.1) holds for any polynomial $p \in \mathbf{P}_{n-1}$.

Until now, we have not made use of the property that the nodes are related to zeroes of orthogonal polynomials. This assumption is fundamental in proving the next remarkable result.

Theorem 3.4.1 - *Formula* (3.4.1) *is true for any polynomial p of degree less or equal* $2n-1$.

Proof - We take $q := I_{w,n}p$. Hence, $q \in \mathbf{P}_{n-1}$ and $p-q$ vanishes at the points $\xi_j^{(n)}$, $1 \le j \le n$. Then $p - q = u_n r$, where r is a polynomial of degree at most $n-1$. Finally, by orthogonality and by using the exactness of (3.4.1) for polynomials of degree $n-1$, we get

$$\int_I pw\ dx = \int_I qw\ dx + \int_I u_n rw\ dx = \int_I qw\ dx$$

$$= \sum_{j=1}^{n} q(\xi_j^{(n)})\ w_j^{(n)} = \sum_{j=1}^{n} p(\xi_j^{(n)})\ w_j^{(n)}.$$

■

We leave as exercise the proof of the theorem's converse, i.e., if formula (3.4.1) holds for any $p \in \mathbf{P}_{2n-1}$ then the nodes are the zeroes of u_n. The degree of the integration formula cannot be improved further. In fact, if one takes $p = u_n^2 \in \mathbf{P}_{2n}$ the right-hand side of (3.4.1) vanishes. On the contrary, the left-hand side is different from zero.

Explicit expressions are known for the quantities $w_j^{(n)}$, $1 \le j \le n$, in the various cases. We summarize the formulas.

Jacobi case - For $\alpha > -1$, $\beta > -1$ and $n \ge 1$, we have

(3.4.2) $$w_j^{(n)} = 2^{\alpha+\beta}\,\frac{(2n+\alpha+\beta)\,\Gamma(n+\alpha)\,\Gamma(n+\beta)}{n!\,\Gamma(n+\alpha+\beta+1)}$$

$$\times\left[P_{n-1}^{(\alpha,\beta)}(\xi_j^{(n)})\,\frac{d}{dx}P_n^{(\alpha,\beta)}(\xi_j^{(n)})\right]^{-1}, \qquad 1\le j\le n.$$

For the proof we argue as follows. By equating the leading coefficients, (1.3.5) yields

(3.4.3) $$\frac{u_n(x)}{x-\xi_j^{(n)}} = \frac{(2n+\alpha+\beta)(2n+\alpha+\beta-1)}{2n\,(n+\alpha+\beta)}\,u_{n-1}(x) + q_j(x), \qquad 1\le j\le n,$$

where $q_j \in \mathbf{P}_{n-2}$, $1\le j\le n$, and $u_n = P_n^{(\alpha,\beta)}$. Therefore, recalling (3.4.1) and (3.2.4),

(3.4.4) $$w_j^{(n)} = \frac{1}{u_{n-1}(\xi_j^{(n)})}\sum_{i=1}^{n}(l_j^{(n)}u_{n-1})(\xi_i^{(n)})\,w_i^{(n)}$$

$$= \frac{1}{u_{n-1}(\xi_j^{(n)})}\int_{-1}^{1} l_j^{(n)}u_{n-1}w\,dx = \frac{1}{(u_{n-1}u_n')(\xi_j^{(n)})}\int_{-1}^{1}\frac{u_n}{x-\xi_j^{(n)}}\,u_{n-1}w\,dx$$

$$= \frac{1}{(u_{n-1}u_n')(\xi_j^{(n)})}\,\frac{(2n+\alpha+\beta)(2n+\alpha+\beta-1)}{2n\,(n+\alpha+\beta)}\,\|u_{n-1}\|_w^2, \qquad 1\le j\le n.$$

Finally, (3.4.2) is obtained with the help of (2.2.10).

Legendre case - To obtain an expression for the weights, we simply set $\alpha=\beta=0$ in (3.4.2). This gives

(3.4.5) $$w_j^{(n)} = \frac{2}{n}\left[P_{n-1}(\xi_j^{(n)})P_n'(\xi_j^{(n)})\right]^{-1}, \quad 1\le j\le n.$$

Chebyshev case - This is the most remarkable case. We recall (1.5.1), (3.1.16) and the relation $T_{n-1}(\xi_j^{(n)}) = (-1)^{j+n}\sqrt{1-[\xi_j^{(n)}]^2}$, $1\le j\le n$. By taking $\alpha=\beta=-\frac{1}{2}$ in (3.4.2), one obtains

(3.4.6) $$w_j^{(n)} = \frac{\pi}{n}, \qquad 1\le j\le n.$$

The set of the Chebyshev zeroes is the only distribution of Jacobi nodes for which the corresponding Gaussian weights are all the same. This further sets Chebyshev polynomials apart from the other families of orthogonal polynomials.

Laguerre case - For $\alpha > -1$ and $n \geq 1$, we have

$$(3.4.7) \qquad w_j^{(n)} = -\frac{\Gamma(n+\alpha)}{n!}\left[L_{n-1}^{(\alpha)}(\xi_j^{(n)})\,\frac{d}{dx}L_n^{(\alpha)}(\xi_j^{(n)})\right]^{-1}, \quad 1 \leq j \leq n.$$

This time the counterpart of (3.4.3) is

$$(3.4.8) \qquad \frac{u_n(x)}{x-\xi_j^{(n)}} = -\frac{1}{n}\,u_{n-1}(x) + q_j(x), \quad 1 \leq j \leq n,$$

where $q_j \in \mathbf{P}_{n-2}$, $1 \leq j \leq n$, and $u_n = L_n^{(\alpha)}$. The proof proceedes like that of the Jacobi case after recalling (2.2.13).

Hermite case - We have for $n \geq 1$

$$(3.4.9) \qquad w_j^{(n)} = \sqrt{\pi}\,2^{n+1}\,n!\left[H_n'(\xi_j^{(n)})\right]^{-2}, \quad 1 \leq j \leq n.$$

For the proof we can argue as in the previous cases. We only note that, by (1.7.8) one has $H_n'(\xi_j^{(n)}) = 2nH_{n-1}(\xi_j^{(n)})$, $1 \leq j \leq n$.

We observe that for all the cases the weights are strictly positive. This is clear if we consider that

$$w_j^{(n)} = \sum_{i=1}^{n} [l_j^{(n)}]^2(\xi_i^{(n)})\,w_i^{(n)} = \int_I [l_j^{(n)}]^2 w\,dx > 0, \quad 1 \leq j \leq n.$$

Once the zeroes have been computed, the weights can be evaluated using the recurrence formula (1.1.2) to recover the values of u_{n-1} and u_n'.

Due to their high accuracy, Gauss formulas play a fundamental role in the theoretical analysis of spectral methods. The possibility of integrating polynomials of degree $2n-1$ just by knowing their values at n points will be widely used. The reader interested in collecting more results may consult SZEGÖ(1939), STROUD and SECREST(1966), GHIZZETTI and OSSICINI(1970), DAVIS and RABINOWITZ(1984), where other explicit expressions of the weights are also given.

3.5 Gauss-Lobatto integration formulas

In this section, we study an integration formula based on the nodes obtained from the derivatives of Jacobi polynomials. Integrating (3.2.6), we get for $n \geq 1$ and any polynomial $p \in \mathbf{P}_n$, the so called *Gauss-Lobatto formula*

$$\int_{-1}^{1} pw \, dx = \sum_{j=0}^{n} p(\eta_j^{(n)}) \, \tilde{w}_j^{(n)}, \tag{3.5.1}$$

where w is the Jacobi weight function and $\tilde{w}_j^{(n)} := \int_{-1}^{1} \tilde{l}_j^{(n)} w dx$, $0 \leq j \leq n$, are the *Gauss-Lobatto weights*.

The equation (3.5.1) represents a high degree integration formula by virtue of the following theorem.

Theorem 3.5.1 - *Formula* (3.5.1) *is true for any polynomial* $p \in \mathbf{P}_{2n-1}$.

Proof - Here we define $q := \tilde{I}_{w,n}p$. Thus $p - q = (1 - x^2)u_n' r$, where $r \in \mathbf{P}_{n-2}$. Moreover, r is the derivative of some polynomial $s \in \mathbf{P}_{n-1}$. Then, recalling (2.2.8) and that $a = (1 - x^2)w$, one finally obtains

$$\int_{-1}^{1} pw \, dx = \int_{-1}^{1} qw \, dx + \int_{-1}^{1} au_n' s' \, dx = \int_{-1}^{1} qw \, dx$$

$$= \sum_{j=0}^{n} q(\eta_j^{(n)}) \, \tilde{w}_j^{(n)} = \sum_{j=0}^{n} p(\eta_j^{(n)}) \, \tilde{w}_j^{(n)}.$$

■

Formula (3.5.1) does not hold in general when $p \in \mathbf{P}_{2n}$.

One is now left with the determination of the weights. To this end we prove the following proposition.

Theorem 3.5.2 - *For any $n \geq 2$, we have*

(3.5.2)

$$\tilde{w}_j^{(n)} = \begin{cases} 2^{\alpha+\beta}\frac{(\beta+1)\,\Gamma^2(\beta+1)(2n+\alpha+\beta)(n-2)!\,\Gamma(n+\alpha)}{\Gamma(n+\alpha+\beta+2)\,\Gamma(n+\beta+1)} \displaystyle\sum_{m=1}^{n-1}(1-\eta_m^{(n)}) & \text{if } j=0, \\ -2^{\alpha+\beta}\frac{(2n+\alpha+\beta)\Gamma(n+\alpha)\Gamma(n+\beta)}{(n+\alpha+\beta+1)\; n!\;\Gamma(n+\alpha+\beta)}\left[\left(P_n^{(\alpha,\beta)}\frac{d}{dx}P_{n-1}^{(\alpha,\beta)}\right)(\eta_j^{(n)})\right]^{-1} & \text{if } 1\leq j\leq n-1, \\ 2^{\alpha+\beta}\frac{(\alpha+1)\,\Gamma^2(\alpha+1)(2n+\alpha+\beta)(n-2)!\,\Gamma(n+\beta)}{\Gamma(n+\alpha+\beta+2)\,\Gamma(n+\alpha+1)} \displaystyle\sum_{m=1}^{n-1}(1+\eta_m^{(n)}) & \text{if } j=n. \end{cases}$$

Proof - As in section 3.4 for the Gauss weights, we start with the relation

$$\text{(3.5.3)} \quad \frac{u_n'(x)}{x-\eta_j^{(n)}} = \frac{(2n+\alpha+\beta)(2n+\alpha+\beta-1)}{2(n-1)(n+\alpha+\beta)}\, u_{n-1}'(x) + q_j'(x), \quad 1\leq j\leq n-1,$$

where $q_j \in \mathbf{P}_{n-2}$ and $u_n = P_n^{(\alpha,\beta)}$. To check (3.5.3), it is sufficient to compare the highest degree terms and recall (1.3.5). Thus, if $1 \leq j \leq n-1$, by (3.2.8) and (2.2.15) we get

(3.5.4)

$$\tilde{w}_j^{(n)} = \frac{1}{u_{n-1}'(\eta_j^{(n)})}\sum_{i=0}^{n}(\tilde{l}_j^{(n)}u_{n-1}')(\eta_i^{(n)})\,\tilde{w}_i^{(n)} = \frac{1}{u_{n-1}'(\eta_j^{(n)})}\int_{-1}^{1}\tilde{l}_j^{(n)}u_{n-1}'\, w dx$$

$$= \frac{-1}{n(n+\alpha+\beta+1)\left[(u_{n-1}'u_n)(\eta_j^{(n)})\right]}\int_{-1}^{1} a\,\frac{u_n'}{x-\eta_j^{(n)}}\,u_{n-1}'\,dx$$

$$= -\frac{(2n+\alpha+\beta)(2n+\alpha+\beta-1)}{2n(n-1)(n+\alpha+\beta)(n+\alpha+\beta+1)\left[(u_{n-1}'u_n)(\eta_j^{(n)})\right]}\,\|u_{n-1}'\|_a^2$$

$$= -\frac{(2n+\alpha+\beta)(2n+\alpha+\beta-1)}{2n(n+\alpha+\beta+1)\left[(u_{n-1}'u_n)(\eta_j^{(n)})\right]}\,\|u_{n-1}\|_w^2.$$

The proof is completed using (2.2.10). For the weights relative to the endpoints, we write

$$(3.5.5)\qquad 0 = \int_{-1}^{1} u_n u'_{n-1}(1-x)w\,dx = \sum_{m=0}^{n} (u_n u'_{n-1})(\eta_m^{(n)})(1-\eta_m^{(n)})\,\tilde{w}_m^{(n)}$$

$$= 2u_n(-1)u'_{n-1}(-1)\tilde{w}_0^{(n)} - 2^{\alpha+\beta}\frac{(2n+\alpha+\beta)\Gamma(n+\alpha)\Gamma(n+\beta)}{(n+\alpha+\beta+1)\,n!\,\Gamma(n+\alpha+\beta)}\sum_{m=1}^{n-1}(1-\eta_m^{(n)}).$$

The first equality in (3.5.5) is due to orthogonality, while in the last equality we used the expression of the weights obtained above for $1 \le m \le n-1$. Then, by (1.3.3) and (3.1.18) we deduce the value of $\tilde{w}_0^{(n)}$. In the same way one gets the value of $\tilde{w}_n^{(n)}$.

■

In the ultraspherical case, since the polynomials are even or odd, the nodes are symmetrically distributed in $[-1,1]$. Therefore, when $\alpha=\beta$ we have $\sum_{j=m}^{n-1}\eta_m^{(n)} = 0$. Hence, in the Legendre case we obtain

$$(3.5.6)\qquad \tilde{w}_j^{(n)} = \begin{cases} \dfrac{2}{n(n+1)} & \text{if } j=0 \text{ or } j=n,\\[2ex] \dfrac{-2}{n+1}\left[P_n(\eta_j^{(n)})P'_{n-1}(\eta_j^{(n)})\right]^{-1} & \text{if } 1\le j\le n-1. \end{cases}$$

For the Chebyshev case, relations (3.5.2) are drastically simplified and give

$$(3.5.7)\qquad \tilde{w}_j^{(n)} = \begin{cases} \dfrac{\pi}{2n} & \text{if } j=0 \text{ or } j=n,\\[2ex] \dfrac{\pi}{n} & \text{if } 1\le j\le n-1. \end{cases}$$

This is a consequence of (3.1.15) and the relation $T'_{n-1}(\eta_j^{(n)}) = -(n-1)(-1)^{j+n}$, $1 \le j \le n-1$.

Also for Gauss-Lobatto formulas it is possible to prove that the weights are positive. More about the subject can be found in ENGELS(1980) or DAVIS and RABINOWITZ(1984). Integration formulas of Gauss-Lobatto type allow the determination of the weighted integral of polynomials of degree $2n-1$, knowing their values only at $n+1$ points. With respect to Gauss formulas there is a slight loss in accuracy. Nevertheless, the fact that the points $x=-1$ and $x=1$ are included in the nodes is important when imposing boundary conditions in the approximation of differential equations (see section 7.4).

3.6 Gauss-Radau integration formulas

Other integration formulas, known as *Gauss-Radau formulas*, are based on the nodes given in (3.1.12), (3.1.13) and (3.1.14) respectively. The first case actually corresponds to a Gauss type integration formula since, thanks to (1.3.6), the nodes given in (3.1.12) are the zeroes of $P_{n-1}^{(3/2,3/2)}$. Therefore, the relative formula is exact for polynomials up to degree $2n-3$. The other two sets of n nodes include the points $x=1$ or $x=-1$ respectively. Appropriate weights can be defined so that the associated integration formulas are valid for polynomials of degree up to $2n-2$. Some details are discussed for instance in RALSTON(1965) or DAVIS and RABINOWITZ(1984).

In this section we are concerned with an additional formula, based on the zeroes of $\frac{d}{dx}L_n^{(\alpha)}$ plus the point $x=0$ (the only boundary point of the interval $[0,+\infty[$). This is

$$\int_0^{+\infty} pw\,dx = \sum_{j=0}^{n-1} p(\eta_j^{(n)})\,\tilde{w}_j^{(n)}. \tag{3.6.1}$$

Here w is the Laguerre weight function and $\tilde{w}_j^{(n)} := \int_0^{+\infty} \tilde{l}_j^{(n)} w dx$, where the Lagrange polynomials are defined in (3.2.11). With a proof similar to that of theorem 3.5.1, equation (3.6.1) turns out to be true for any polynomial p of degree at most $2n-2$. For the weights, we obtain

$$\tilde{w}_j^{(n)} = \begin{cases} \dfrac{(\alpha+1)\,\Gamma^2(\alpha+1)\,(n-1)!}{\Gamma(n+\alpha+1)} & \text{if } j=0, \\[2ex] \dfrac{\Gamma(n+\alpha)}{n!}\left[L_n^{(\alpha)}(\eta_j^{(n)})\dfrac{d}{dx}L_{n-1}^{(\alpha)}(\eta_j^{(n)})\right]^{-1} & \text{if } 1\le j\le n-1. \end{cases} \tag{3.6.2}$$

These formulas are proven with an argument similar to that used in theorem 3.5.2.

3.7 Clenshaw-Curtis integration formulas

The formulas analyzed in the previous sections are very accurate when the weighted integral of a polynomial is needed. The weight function is strictly related to the set of nodes in which the polynomial is known. Nevertheless, very often in applications, one is concerned with evaluating the integral corresponding to a different weight function. For example, we would like to compute $\int_{-1}^{1} p dx$, where p is determined at some Jacobi Gauss-Lobatto nodes. This time, after integrating (3.2.7), we end up with the formula

$$\text{(3.7.1)} \qquad \int_{-1}^{1} p \, dx = \sum_{j=0}^{n} p(\eta_j^{(n)}) \, \chi_j^{(n)},$$

where $\chi_j^{(n)} := \int_{-1}^{1} \tilde{l}_j^{(n)} dx, \quad 0 \le j \le n.$

Of course, relation (3.7.1) is exact for all polynomials in the space $\mathbf{P}_n$. Unfortunately, if the nodes are not related to zeroes of derivatives of Legendre polynomials (in this case we would obtain: $\chi_j^{(n)} = \tilde{w}_j^{(n)}, \ 0 \le j \le n$), we cannot expect the formula to be exact for polynomials of degree higher than n. On the other hand, (3.7.1) is possibly interesting for future applications. For this reason, we give the expression of the weights in the case of Chebyshev nodes (which are presented in (3.1.11)), when $n \ge 4$ is even. In practice, we have

$$\text{(3.7.2)} \qquad \chi_j^{(n)} = \begin{cases} \dfrac{1}{n^2-1} & \text{if } j = 0, \\[2ex] \dfrac{2}{n}\left[1 - \dfrac{(-1)^j}{n^2-1} + \displaystyle\sum_{k=1}^{n/2-1} \dfrac{2}{1-4k^2} \cos\left(\dfrac{2kj\pi}{n}\right)\right] & \text{if } 1 \le j \le \frac{n}{2}, \\[2ex] \chi_{n-j}^{(n)} & \text{if } \frac{n}{2}+1 \le j \le n. \end{cases}$$

With this choice of nodes and weights, relation (3.7.1) is known as *Clenshaw-Curtis formula* (see DAVIS and RABINOWITZ(1984)). This is exact for polynomials of degree not larger than n.

3.8 Discrete norms

Two polynomials $p, q \in \mathbf{P}_{n-1}$ are uniquely characterized by the values they take at the Gauss nodes $\xi_j^{(n)}$, $1 \le j \le n$. Therefore, since pq is a polynomial of degree at most $2n-2$, the scalar product $\int_I pqw\, dx$ can be determined with the help of (3.4.1). The situation is different when the polynomials $p, q \in \mathbf{P}_n$ are distinguished by their values at the points $\eta_j^{(n)}$, $0 \le j \le n$. This time $pq \in \mathbf{P}_{2n}$ and the Gauss-Lobatto formula (3.5.1) becomes invalid. This justifies the definition of an inner product $(\cdot,\cdot)_{w,n} : \mathbf{P}_n \times \mathbf{P}_n \to \mathbf{R}$, and a norm $\|\cdot\|_{w,n} : \mathbf{P}_n \to \mathbf{R}^+$, given by

$$(3.8.1) \qquad (p,q)_{w,n} := \sum_{j=0}^{n} p(\eta_j^{(n)})\, q(\eta_j^{(n)})\, \tilde{w}_j^{(n)}, \qquad \forall p, q \in \mathbf{P}_n,$$

$$(3.8.2) \qquad \|p\|_{w,n} := \left(\sum_{j=0}^{n} p^2(\eta_j^{(n)})\, \tilde{w}_j^{(n)} \right)^{\frac{1}{2}}, \qquad \forall p \in \mathbf{P}_n.$$

These expressions are called *discrete inner product* and *discrete norm* respectively. We first compute the discrete norm of u_n. For the sake of simplicity we restrict ourselves to the ultraspherical case.

Theorem 3.8.1 - *For any $n \ge 1$, we have*

$$(3.8.3) \qquad \|u_n\|_{w,n}^2 = 2^{2\alpha+1} \frac{\Gamma^2(n+\alpha+1)}{n\; n!\; \Gamma(n+2\alpha+1)},$$

where $u_n = P_n^{(\alpha,\alpha)}$, $\alpha > -1$.

Proof - The key point is the relation

$$(3.8.4) \qquad u'_{n-1}(\eta_j^{(n)}) = -\frac{n(n+2\alpha)}{n+\alpha}\, u_n(\eta_j^{(n)}), \qquad 1 \le j \le n-1.$$

To obtain (3.8.4), we refer to SZEGÖ(1939), p.71, formula (4.5.7). We substitute $n-1$ for n in the second part of that formula, and then we eliminate u_{n-1} with the help of the first part. Substituting $x = \eta_j^{(n)}$, $1 \leq j \leq n-1$, the terms containing u'_n vanish. When (3.8.4) is achieved, the proof is straightforward with the help of (3.5.2), (3.1.17) and (3.1.18).

■

In particular, when $\alpha = -\frac{1}{2}$, (3.8.4) yields

(3.8.5) $$\|T_n\|_{w,n}^2 = \pi, \qquad \forall n \geq 1.$$

For any fixed $n \geq 1$, the two norms $\|\cdot\|_w$ and $\|\cdot\|_{w,n}$ defined in $\mathbf{P}_n$ are equivalent. This means that it is possible to find two constants $\gamma_1 > 0, \gamma_2 > 0$, such that

(3.8.6) $$\gamma_1 \|p\|_{w,n} \leq \|p\|_w \leq \gamma_2 \|p\|_{w,n}, \qquad \forall p \in \mathbf{P}_n.$$

This is not surprising, since in a finite dimensional space (such as $\mathbf{P}_n$) any two norms are always equivalent. More important is the following statement, which we prove in the ultraspherical case.

Theorem 3.8.2 - *The constants* γ_1 *and* γ_2 *in* (3.8.6) *do not depend on* n.

Proof - Expanding p with respect to the orthogonal basis $\{u_k\}_{0\leq k\leq n}$, we can write $p = c_n u_n + r$, where $r \in \mathbf{P}_{n-1}$. Using that (3.5.1) is true in $\mathbf{P}_{2n-1}$, one obtains

(3.8.7) $$\|p\|_{w,n}^2 = c_n^2 \|u_n\|_{w,n}^2 + 2c_n (u_n, r)_{w,n} + \|r\|_{w,n}^2$$

$$= c_n^2 \|u_n\|_{w,n}^2 + 2c_n(u_n, r)_w + \|r\|_w^2 = c_n^2 \left(\|u_n\|_{w,n}^2 - \|u_n\|_w^2\right) + \|p\|_w^2, \quad \forall p \in \mathbf{P}_n.$$

By direct comparison of (2.2.10) and (3.8.3) we get $\|u_n\|_{w,n}^2 \geq \|u_n\|_w^2$, $\forall n \geq 1$. This implies $\gamma_2 = 1$. On the other hand by (2.3.7) and by the Schwarz inequality (2.1.7), we get

(3.8.8) $$c_n^2 = \left[\frac{(p, u_n)_w}{\|u_n\|_w^2}\right]^2 \leq \frac{\|p\|_w^2}{\|u_n\|_w^2}, \qquad \forall p \in \mathbf{P}_n.$$

Thus, substituting in (3.8.7),

(3.8.9)
$$\|p\|_{w,n}^2 \le \frac{\|u_n\|_{w,n}^2}{\|u_n\|_w^2}\,\|p\|_w^2, \qquad \forall p \in \mathbf{P}_n.$$

Again, from a direct computation, we have that the ratio on the right-hand side of (3.8.9) is bounded by $\max\{2, 3+2\alpha\} = \gamma_1^{-2}$.

■

It is possible to compute the norm $\|p\|_w$ of a polynomial $p \in \mathbf{P}_n$ as a function of the values it attains at the points $\eta_j^{(n)}$, $0 \le j \le n$. This is shown by the following proposition.

Theorem 3.8.3 - *For any $n \ge 1$, we have*

(3.8.10)
$$\|p\|_w^2 = \|p\|_{w,n}^2 - \frac{\|u_n\|_{w,n}^2 - \|u_n\|_w^2}{\|u_n\|_{w,n}^4}\,[(p,u_n)_{w,n}]^2, \qquad \forall p \in \mathbf{P}_n,$$

where $u_n = P_n^{(\alpha,\beta)}$, $\alpha > -1$, $\beta > -1$.

Proof - By (2.3.7) and the exactness of (3.5.1) in $\mathbf{P}_{2n-1}$, one has

(3.8.11)
$$c_n\,\|u_n\|_w^2 = (p,u_n)_w = (p,u_n)_{w,n} + c_n(\|u_n\|_w^2 - \|u_n\|_{w,n}^2), \qquad \forall p \in \mathbf{P}_n.$$

Recovering c_n, one gets the interesting property

(3.8.12)
$$c_n = \frac{(p,u_n)_{w,n}}{\|u_n\|_{w,n}^2}, \qquad \forall p \in \mathbf{P}_n.$$

The proof is concluded after substitution of c_n in (3.8.7).

■

The right-hand side of (3.8.10) actually depends only on the values of p at the nodes. Using (2.2.10) and (3.8.3), equation (3.8.10) becomes

(3.8.13)
$$\|p\|_w = \left(\|p\|_{w,n}^2 - \frac{n(n+1)}{2(2n+1)}\,[(p,P_n)_{w,n}]^2\right)^{\frac{1}{2}}, \qquad \forall p \in \mathbf{P}_n,$$

(3.8.14) $$\|p\|_w = \left(\|p\|^2_{w,n} - \frac{1}{2\pi}\,[(p,T_n)_{w,n}]^2\right)^{\frac{1}{2}}, \quad \forall p \in \mathbf{P}_n,$$

when specialized to the Legendre and Chebyshev cases respectively.

The result of theorem 3.8.3 can be generalized as follows:

(3.8.15) $$(p,q)_w = (p,q)_{w,n} - \frac{\|u_n\|^2_{w,n} - \|u_n\|^2_w}{\|u_n\|^4_{w,n}}\,(p,u_n)_{w,n}\,(q,u_n)_{w,n}, \quad \forall p,q \in \mathbf{P}_n.$$

We leave the proof as exercise.

Another quantity we would like to compute is the integral $\int_{-1}^1 p^2 dx$, in terms of the values of $p \in \mathbf{P}_n$ at some nodes. When the nodes are the Legendre Gauss-Lobatto points, we can use (3.8.13). On the contrary, we are in a situation similar to that of section 3.7. We only solve the problem for Chebyshev Gauss-Lobatto nodes. Here we have $p^2 \in \mathbf{P}_{2n}$, while (3.7.1) is satisfied only in $\mathbf{P}_n$. The idea is to duplicate the number of points and still use (3.7.1). Recalling (3.1.11) and (3.1.4), in the Chebyshev case we have the remarkable result that

(3.8.16) $$\{\xi_i^{(n)}\}_{1\leq i\leq n} \cup \{\eta_j^{(n)}\}_{0\leq j\leq n} = \{\eta_k^{(2n)}\}_{0\leq k\leq 2n}.$$

Since p is known at the points (3.1.11) and $p(\eta_j^{(n)}) = p(\eta_{2j}^{(2n)})$, $0 \leq j \leq n$, we can extrapolate its value at the points (3.1.4) with the help of (3.2.7):

(3.8.17) $$p(\eta_{2i-1}^{(2n)}) = p(\xi_i^{(n)}) = \sum_{j=0}^{n} p(\eta_j^{(n)})\,\tilde{l}_j^{(n)}(\xi_i^{(n)}), \qquad 1 \leq i \leq n.$$

Now, we can use (3.7.1), obtaining $\forall p \in \mathbf{P}_n$

(3.8.18) $$\int_{-1}^{1} p^2\,dx = \sum_{m=0}^{2n} p^2(\eta_m^{(2n)})\,\chi_m^{(2n)}$$

$$= \sum_{j=0}^{n} p^2(\eta_j^{(n)})\,\chi_{2j}^{(2n)} + \sum_{i=1}^{n}\left(\sum_{k=0}^{n} p(\eta_k^{(n)})\,\tilde{l}_k^{(n)}(\xi_i^{(n)})\right)^2 \chi_{2i-1}^{(2n)}.$$

Finally, the weights can be determined from (3.7.2), and the coefficients $\tilde{l}_k^{(n)}(\xi_i^{(n)})$ from (3.2.10) and (3.1.16). This procedure has a computational cost proportional to n^2, versus the cost of implementing (3.8.10), which is proportional to n.

A discrete inner product with its relative norm, can also be conveniently defined for polynomials in $\mathbf{P}_{n-1}$ distinguished by their values at the n nodes including the $n-1$ zeroes of $\frac{d}{dx}L_n^{(\alpha)}$ and the point $x = 0$. However, this definition would be redundant. Indeed, when $r, s \in \mathbf{P}_{n-1}$ we have $p := rs \in \mathbf{P}_{2n-2}$; therefore, by virtue of (3.6.1), the new inner product coincides with the usual one, i.e. $\|\cdot\|_w$, where w is the Laguerre weight function.

3.9 Discrete maximum norms

Discrete equivalents of the maximum norm, introduced in section 2.5, can be defined. We study two cases. For any $n \geq 1$, we first consider $|||\cdot|||_{w,n} : \mathbf{P}_{n-1} \to \mathbf{R}^+$ given by

$$|||p|||_{w,n} := \max_{1\leq j\leq n} |p(\xi_j^{(n)})|, \quad p \in \mathbf{P}_{n-1}, \tag{3.9.1}$$

where the $\xi_j^{(n)}$'s are the zeroes of $P_n^{(\alpha,\beta)}$.
Similarly, for any $n \geq 1$, we define $|||\cdot|||_{w,n}^* : \mathbf{P}_n \to \mathbf{R}^+$ as

$$|||p|||_{w,n}^* := \max_{0\leq j\leq n} |p(\eta_j^{(n)})|, \quad p \in \mathbf{P}_n, \tag{3.9.2}$$

where the $\eta_j^{(n)}$ are the zeroes of $\frac{d}{dx}P_n^{(\alpha,\beta)}$ plus the endpoints of the interval $\bar{I} = [-1,1]$.
The following two results are easy to derive:

$$|||p|||_{w,n} \leq \|p\|_\infty, \qquad \forall p \in \mathbf{P}_{n-1}, \tag{3.9.3}$$

$$|||p|||_{w,n}^* \leq \|p\|_\infty, \qquad \forall p \in \mathbf{P}_n. \tag{3.9.4}$$

Of course, working within finite dimensional spaces, the inequalities (3.9.3) and (3.9.4) can be reversed. For instance, $\forall n \geq 1$, we can find γ such that

(3.9.5) $$\|p\|_\infty \leq \gamma\ |||p|||_{w,n}, \qquad \forall p \in \mathbf{P}_{n-1}.$$

This time, a statement like that of theorem 3.8.2 cannot be proven. In fact, $\gamma \equiv \gamma(n)$ actually grows with n. Following ERDÖS(1961), it is possible to find a constant $\delta > 0$ such that

(3.9.6) $$\gamma(n) > \frac{2}{\pi} \log(n) - \delta.$$

However, using Chebyshev nodes, we get the relation (see NATANSON(1965))

(3.9.7) $$\gamma(n) < \frac{2}{\pi} \log(n) + 1.$$

Further inequalities of this type will be covered in section 6.3.

Within the context of the never ending properties of Chebyshev polynomials, we state a result which generalizes theorem 2.5.1. The proof is in DUFFIN and SCHAEFFER(1941).

Theorem 3.9.1 - *Let* $w(x) = 1/\sqrt{1-x^2}$, *then for any* $n \geq 1$, *we have*

(3.9.8) $$\|p'\|_\infty \leq n^2\ |||p|||^*_{w,n}, \qquad \forall p \in \mathbf{P}_n.$$

■

Another theorem, closely related to theorem 2.5.2, is stated in RIVLIN(1974), p.104.

Theorem 3.9.2 - *Let* $w(x) = 1/\sqrt{1-x^2}$, *then for any* $n \geq 1$, *we have*

(3.9.9) $$|p'(\xi_j^{(n)})| \leq \frac{n}{\sqrt{1-[\xi_j^{(n)}]^2}}\ |||p|||^*_{w,n}, \quad 1 \leq j \leq n, \quad \forall p \in \mathbf{P}_n.$$

■

3.10 Scaled weights

For large n, the determination of the Laguerre weights can sometimes lead to severe numerical problems. In fact, for the nodes largest in magnitude, the corresponding weights are very small and decay quite fast to zero for increasing values of n. Following the hints in section 1.6, it is advisable to introduce some *scaled weights*.

First, using (1.6.12) and (1.6.13), we note that, for $n \geq 1$,

$$\frac{d}{dx}L_n^{(\alpha)}(\xi_j^{(n)}) = \binom{n+\alpha}{n} \frac{d}{dx}\hat{L}_n^{(\alpha)}(\xi_j^{(n)}) \prod_{k=1}^{n}\left(1+\frac{\xi_j^{(n)}}{4k}\right), \qquad 1 \leq j \leq n. \tag{3.10.1}$$

Next, starting from (3.4.7), we define a new set of Gauss type weights

$$\omega_j^{(n)} := w_j^{(n)} \left[S_n^{(\alpha)}(\xi_j^{(n)})\right]^{-2} \tag{3.10.2}$$

$$= -\frac{\Gamma(n+\alpha+1)}{4n^2\, n!}(4n+\xi_j^{(n)}) \left[\hat{L}_{n-1}^{(\alpha)}(\xi_j^{(n)})\frac{d}{dx}\hat{L}_n^{(\alpha)}(\xi_j^{(n)})\right]^{-1}, \qquad 1 \leq j \leq n.$$

The values of the scaled Laguerre functions at the nodes are evaluated by (1.6.14) and (1.6.15). Recalling (3.4.1), we obtain the formula

$$\int_0^{+\infty} p^2(x)\, x^\alpha e^{-x}\, dx = \sum_{j=1}^{n} \hat{p}^2(\xi_j^{(n)})\, \omega_j^{(n)}, \qquad \forall p \in \mathbf{P}_{n-1}, \tag{3.10.3}$$

where $\hat{p} := pS_n^{(\alpha)}$ are used in place of p in numerical computations. The advantages to this approach consist in a better numerical treatment. More details about the implementation are examined later in section 7.5.

A similar definition is given for Gauss-Radau type weights (see (3.6.2)), i.e.

$$\tilde{\omega}_0^{(n)} := \tilde{w}_0^{(n)} \left[S_n^{(\alpha)}(0)\right]^{-2} = \frac{(\alpha+1)\,\Gamma(n+\alpha+1)}{n\, n!}, \tag{3.10.4}$$

(3.10.5)
$$\tilde{\omega}_j^{(n)} := \tilde{w}_j^{(n)} \left[S_n^{(\alpha)}(\eta_j^{(n)})\right]^{-2}$$

$$= \frac{\Gamma(n+\alpha+1)}{4n^2\ n!} \frac{4n+\eta_j^{(n)}}{\hat{L}_n^{(\alpha)}(\eta_j^{(n)})} \left(\frac{d}{dx}\hat{L}_{n-1}^{(\alpha)}(\eta_j^{(n)}) + \hat{L}_{n-1}^{(\alpha)}(\eta_j^{(n)}) \sum_{m=1}^{n-1} \frac{1}{4m+\eta_j^{(n)}}\right)^{-1},$$
$$1 \le j \le n-1,$$

where we noted that, for $n \geq 2$

(3.10.6)
$$\frac{d}{dx}\left[S_{n-1}^{(\alpha)}(x)\right]^{-1} = \left[S_{n-1}^{(\alpha)}(x)\right]^{-1} \sum_{m=1}^{n-1} \frac{1}{4m+x}, \quad x \in [0,+\infty[.$$

Again we have

(3.10.7)
$$\int_0^{+\infty} p^2(x)\ x^\alpha e^{-x}\ dx = \sum_{j=0}^{n-1} \hat{p}^2(\eta_j^{(n)})\ \tilde{\omega}_j^{(n)}, \quad \forall p \in \mathbf{P}_{n-1},$$

where $\hat{p} := pS_n^{(\alpha)}$.

The same procedure applied to the Hermite weights in (3.4.9), yields

(3.10.8)
$$\omega_j^{(n)} := w_j^{(n)} \left[S_{n/2}^{(-1/2)}([\xi_j^{(n)}]^2)\right]^{-2}$$

$$= \frac{\sqrt{\pi}\ n!}{2^{n-1}\ [(n/2)!]^2} \left[\hat{H}'_n(\xi_j^{(n)})\right]^{-2}, \quad 1 \le j \le n, \quad \text{if } n \text{ is even},$$

(3.10.9)
$$\omega_j^{(n)} := w_j^{(n)} \left[S_{(n-1)/2}^{(1/2)}([\xi_j^{(n)}]^2)\right]^{-2}$$

$$= \frac{\sqrt{\pi}\ n!}{2^{n-1}\ [((n-1)/2)!]^2} \left[\hat{H}'_n(\xi_j^{(n)})\right]^{-2}, \quad 1 \le j \le n, \quad \text{if } n \text{ is odd}.$$

Here the $\xi_j^{(n)}$'s are the Hermite zeroes. Relations (1.7.9), (1.7.10), (1.7.11) and (1.7.12) have been taken into account.

4
TRANSFORMS

The aim of the previous chapters was to show that a polynomial admits a double representation. One is given by its Fourier coefficients with respect to set of orthogonal basis functions, the other by the values at the nodes associated with a suitable high accuracy integration formula. Clearly, it is possible to switch from one representation to the other. The way to do it is the subject of this chapter.

4.1 Fourier transforms

One of the main results of chapter two was to show that the polynomials introduced in chapter one were orthogonal in relation to a suitable weight function. In addition, other polynomials, i.e., the Lagrange polynomials with respect to a certain set of nodes, have been presented in chapter three. These too are in general orthogonal. Actually, due to (3.4.1) and theorem 3.4.1, we have for any fixed $n \geq 1$,

$$\int_I l_i^{(n)} l_j^{(n)} \, w dx \; = \; \sum_{k=1}^{n} l_i^{(n)}(\xi_k^{(n)}) \, l_j^{(n)}(\xi_k^{(n)}) \, w_k^{(n)} \; = \; \begin{cases} 0 & \text{if } i \neq j, \\ w_i^{(n)} & \text{if } i = j. \end{cases} \tag{4.1.1}$$

A similar relation holds when Gauss-Radau nodes are considered. When Gauss-Lobatto points are used, we loose orthogonality. However, in terms of the discrete inner product introduced in section 3.8, we still get orthogonality. In fact, it is trivial to show that, for any $n \geq 1$:

(4.1.2) $$(\tilde{l}_i^{(n)}, \tilde{l}_j^{(n)})_{w,n} = \delta_{ij}\, \tilde{w}_i^{(n)}, \qquad 0 \le i \le n,\ 0 \le j \le n.$$

This is not true anymore when using the inner product given by (2.1.8). In fact, using (3.8.15), one obtains

(4.1.3) $$\int_{-1}^{1} \tilde{l}_i^{(n)} \tilde{l}_j^{(n)}\, w dx = \delta_{ij}\, \tilde{w}_i^{(n)} - \frac{\|u_n\|_{w,n}^2 - \|u_n\|_w^2}{\|u_n\|_{w,n}^4}\, u_n(\eta_i^{(n)}) u_n(\eta_j^{(n)})\, \tilde{w}_i^{(n)} \tilde{w}_j^{(n)},$$
$$0 \le i \le n,\ 0 \le j \le n,$$

where $u_n = P_n^{(\alpha,\beta)}$, $\alpha > -1$, $\beta > -1$.
In general, the last term on the right-hand side of (4.1.3) tends to zero (when n diverges) faster than the first term, so that the polynomials considered are almost orthogonal.

Thus, for a fixed $n \ge 1$, two orthogonal bases are available in $\mathbf{P}_{n-1}$, i.e., $\{u_k\}_{0 \le k \le n-1}$ and $\{l_j^{(n)}\}_{1 \le j \le n}$. Therefore, we can define a linear transformation $K_n : \mathbf{P}_{n-1} \to \mathbf{P}_{n-1}$, mapping the vector at the point values of a polynomial p into the vector of its Fourier coefficients. This application is called *discrete Fourier transform*, and is represented by a $n \times n$ matrix. Each Fourier coefficient is explicitly determined recalling (2.3.7) and (3.4.1):

(4.1.4) $$c_i = \frac{1}{\|u_i\|_w^2} \sum_{j=1}^{n} p(\xi_j^{(n)})\, u_i(\xi_j^{(n)})\, w_j^{(n)}, \qquad 0 \le i \le n-1.$$

Therefore, we have the matrix relation

(4.1.5) $$K_n := \{k_{ij}\}_{\substack{0 \le i \le n-1 \\ 0 \le j \le n-1}}, \quad \text{where } k_{ij} := \frac{u_i(\xi_{j+1}^{(n)})\, w_{j+1}^{(n)}}{\|u_i\|_w^2}.$$

Obviously K_n is invertible. We express its inverse using (2.3.1):

(4.1.6) $$p(\xi_i^{(n)}) = \sum_{j=0}^{n-1} c_j\, u_j(\xi_i^{(n)}), \qquad 1 \le i \le n,$$

and obtain

(4.1.7) $$K_n^{-1} = \{u_j(\xi_{i+1}^{(n)})\}_{\substack{0 \le i \le n-1 \\ 0 \le j \le n-1}}.$$

It is evident that $K_n^{-1} = D_n^{(1)} K_n^t D_n^{(2)}$, where $D^{(1)}$ and $D^{(2)}$ are the diagonal matrices given respectively by $\mathrm{diag}\{\|u_j\|_w^2\}_{0\le j\le n-1}$ and $\mathrm{diag}\{1/w_{i+1}^{(n)}\}_{0\le i\le n-1}$. This shows that, after an appropriate normalization of the elements of the two bases, the inverse of K_n is given by its transpose.

Similar arguments can be applied to the analysis of transforms based on Gauss-Lobatto nodes. Here the space $\mathbf{P}_n$ is generated either by $\{u_k\}_{0\le k\le n}$ or by $\{\tilde{l}_j^{(n)}\}_{0\le j\le n}$. Hence, we introduce a discrete Fourier transform $\tilde{K}_n : \mathbf{P}_n \to \mathbf{P}_n$, which corresponds to an $(n+1)\times(n+1)$ matrix $\{\tilde{k}_{ij}\}_{\substack{0\le i\le n\\ 0\le j\le n}}$. To get the first $n-1$ coefficients we use (2.3.7) and (3.5.1):

$$\text{(4.1.8)} \qquad c_i = \frac{1}{\|u_i\|_w^2} \sum_{j=0}^{n} p(\eta_j^{(n)})\, u_i(\eta_j^{(n)})\, \tilde{w}_j^{(n)}, \qquad 0 \le i \le n-1.$$

For the last coefficient we recall (3.8.12)

$$\text{(4.1.9)} \qquad c_n = \frac{1}{\|u_n\|_{w,n}^2} \sum_{j=0}^{n} p(\eta_j^{(n)})\, u_n(\eta_j^{(n)})\, \tilde{w}_j^{(n)}.$$

Conversely, thanks to (2.3.1), one has

$$\text{(4.1.10)} \qquad p(\eta_i^{(n)}) = \sum_{j=0}^{n} c_j\, u_j(\eta_i^{(n)}), \qquad 0 \le i \le n.$$

Finally, for the Laguerre Gauss-Radau case, we define $\tilde{K}_n : \mathbf{P}_{n-1} \to \mathbf{P}_{n-1}$ as described above. Then, we have by virtue of (2.3.7) and (3.6.1)

$$\text{(4.1.11)} \qquad c_i = \frac{1}{\|L_i^{(\alpha)}\|_w^2} \sum_{j=0}^{n-1} p(\eta_j^{(n)})\, L_i^{(\alpha)}(\eta_j^{(n)})\, \tilde{w}_j^{(n)}, \quad 0 \le i \le n-1,$$

$$\text{(4.1.12)} \qquad p(\eta_i^{(n)}) = \sum_{j=0}^{n-1} c_j\, L_j^{(\alpha)}(\eta_i^{(n)}), \qquad 0 \le i \le n-1.$$

The scaled weights introduced in section 3.10 can be taken into account in the evaluation of (4.1.11).

Henceforth, depending on the basis used for the representation of a polynomial, we distinguish among the two cases with the help of the following terminology. When the space of polynomials is intended to be the span of the u_k's, then it is isomorph to the set of Fourier coefficients. In this situation we call it the *frequency space*. Otherwise, when the isomorphism is naturally established with the set of the point values, it will be denoted by ***physical space***. There are situations in which it is more convenient to work with one of the two representations. The discrete Fourier transform allows passage from the physical to the frequency space, thus both representations can be used to their fullest advantage.

The cost of implementing the transform or its inverse is proportional to n^2, since it corresponds to a matrix-vector multiplication. Most of this chapter is dedicated to the analysis of a special case which occurs when Chebyshev polynomials are used. For this choice of polynomials, fast algorithms have been developed to perform the discrete Fourier transform, considerably reducing the computational expense.

4.2 Aliasing

Two operators $\Pi_{w,n}$ and $I_{w,n}$ have been respectively introduced in sections 2.4 and 3.3. For a continuous function f, both the images $\Pi_{w,n-1}f$ and $I_{w,n}f$ are in $\mathbf{P}_{n-1}$, which is interpreted as frequency or physical space respectively. Unless f is a polynomial in $\mathbf{P}_{n-1}$, the projection and the interpolant do not coincide. We show two examples. In figures 4.2.1 and 4.2.2, we plot in $[-1,1]$ the projection and the interpolant of the same degree, corresponding to two different functions. In the first figure we have set $f(x) = \big|4|x| - 2\big| - 1$, $w(x) \equiv 1$ (Legendre weight function) and $n = 8$. In the second we have set $f(x) = \frac{1}{2} - \sqrt{|\sin \pi x|}$, $w(x) = 1/\sqrt{1-x^2}$ (Chebyshev weight function) and $n = 13$.

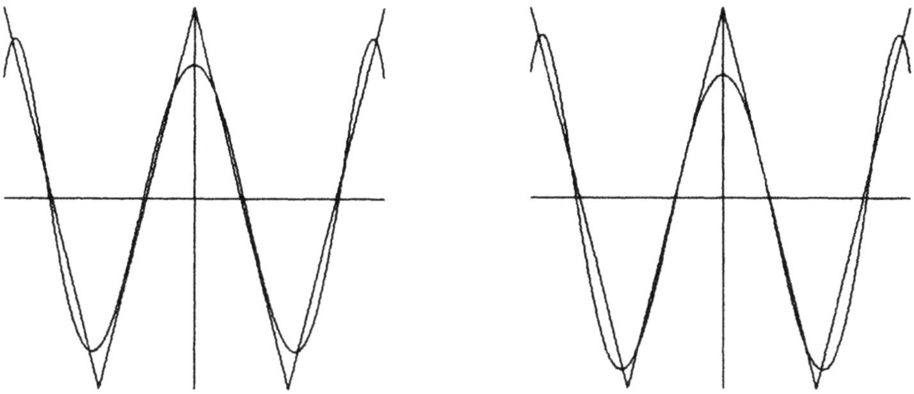

Figure 4.2.1 - Legendre projection and interpolant of degree 7 for the function $f(x) = \Big|4|x| - 2\Big| - 1, \quad x \in [-1, 1].$

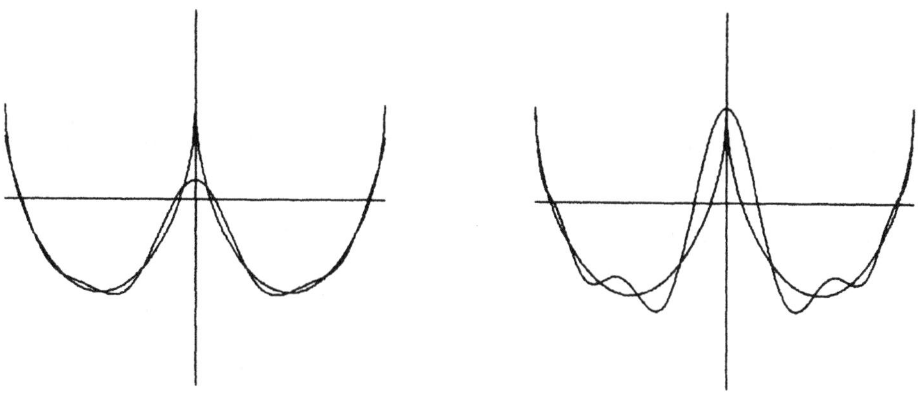

Figure 4.2.2 - Chebyshev projection and interpolant of degree 12 for the function $f(x) = \frac{1}{2} - \sqrt{|\sin \pi x|}, \quad x \in [-1, 1].$

Thus, it is natural to define, for $n \geq 1$, the operator $A_{w,n} = I_{w,n} - \Pi_{w,n-1}$. The polynomial in $\mathbf{P}_{n-1}$

$$A_{w,n}f = I_{w,n}f - \Pi_{w,n-1}f, \quad n \geq 1, \ f \in C^0([-1,1]), \tag{4.2.1}$$

takes the name of *aliasing error*.

Depending on the smoothness of f, this error tends to zero when $n \to +\infty$. The proof of this fact and the characterization of the aliasing error is given in section 6.6.

Another type of aliasing error is obtained by setting

$$\tilde{A}_{w,n}f = \tilde{I}_{w,n}f - \Pi_{w,n}f, \quad f \in C^0([-1,1]), \tag{4.2.2}$$

where $n \geq 1$. This is also studied later on.

4.3 Fast Fourier transform

Let us examine what happens by applying the discrete Fourier transform in the special case of the Chebyshev basis. By virtue of (1.5.6), (3.1.4), (3.4.6) and (2.2.12), the entries of K_n in (4.1.5) take the form

$$k_{ij} = \left(\frac{(-1)^i}{n}\cos\frac{i(2j+1)\pi}{2n}\right) \times \begin{cases} 1 & \text{if } i=0,\ 0 \leq j \leq n-1 \\ 2 & \text{if } 1 \leq i \leq n-1,\ 0 \leq j \leq n-1 \end{cases} . \tag{4.3.1}$$

For the same reason, due to (3.1.11), (3.5.7) and (3.8.5), the entries of $\tilde{K}_n$ are

$$\tilde{k}_{ij} = \left(\frac{(-1)^i}{n}\cos\frac{ij\pi}{n}\right) \times \begin{cases} \frac{1}{2} & \text{if } i=0,\ j=0 \text{ or } j=n \\ 1 & \text{if } i=0,\ 1 \leq j \leq n-1 \\ 1 & \text{if } 1 \leq i \leq n-1,\ j=0 \text{ or } j=n \\ 2 & \text{if } 1 \leq i \leq n-1,\ 1 \leq j \leq n-1 \\ \frac{1}{2} & \text{if } i=n,\ j=0 \text{ or } j=n \\ 1 & \text{if } i=n,\ 1 \leq j \leq n-1 \end{cases} . \tag{4.3.2}$$

The main ingredient in expressions (4.3.1) and (4.3.2) is a term of the form $\cos \frac{m_1 m_2}{n}\pi$, where m_1 and m_2 are integers. Due to the periodicity of the cosine function, the entries of K_n and $\tilde{K}_n$ may attain the same values for different choices of $i, j \in \mathbf{N}$. The most remarkable case is when n is of the form 2^k for some $k \in \mathbf{N}$. In this situation a lot of coefficients are identical and regularly distributed in the matrix. The poorest case is when n is a prime number. The entries generated then attain a larger number of distinct values and substructures are not easily recognized. These observations are the starting point of an efficient algorithm for the evaluation of the Fourier transform and its inverse, with a computational cost less than that needed for a matrix-vector multiplication. This procedure is known as *Fast Fourier Transform* (abbreviated by FFT).

The original version of the FFT was officially introduced in COOLEY and TUKEY(1965), although the concept had already been published. Fast Fourier transforms are best defined in complex space. On the other hand, the transform involving the cosine can be interpreted as a type of projection into the set of the real numbers. A lot of variants and improvements have been added by various authors. At the same time, computer programs have been coded, with the aim of giving a fast and reliable product. High quality versions of the FFT based on modern parallel architectures are also available. Applications in the framework of spectral computations for parallel processors are examined in PELZ(1990).

The FFT algorithm is a little bit tricky and needs a certain amount of perseverance in order to be fully understood. Many books explain, using numerous examples, the basic ideas of the FFT as well as series of generalizations. Usually, they start with the complex formulation restricted to the case $n = 2^k$. Other generalizations (known as multi-radix FFT algorithms) apply when n is not a power of 2, giving however less efficient performances. Among the texts we suggest for instance BRIGHAM(1974), BINI, CAPOVANI, LOTTI and ROMANI(1981), NUSSBAUMER(1981). Since a full discussion of the FFT implementation is beyond the scope of this book, we only present an elementary exposition to bring out its essential features.

The first problem we have in mind is to determine a fast way to compute matrix-vector multiplications with (4.3.1). As we mentioned before, it is convenient to operate with complex numbers. Thus, let $\mathbf{i}$ denote the complex unity, i.e., $\mathbf{i}^2 = -1$. Then, we consider the transform

$$(4.3.3) \qquad \gamma_m = \sum_{j=0}^{n-1} \delta_j \, e^{2\mathbf{i}mj\pi/n}, \qquad 0 \le m \le n-1.$$

Here γ_m, $0 \le m \le n-1$, and δ_j, $0 \le j \le n-1$, are complex variables, related by the standard complex FFT. We study later how to efficiently compute (4.3.3). We now establish a relation between (4.3.3) and (4.1.4) in the Chebyshev case. To this end, the papers of COOLEY, LEWIS and WELSH(1970), and BRACHET *et al.*(1983) give useful indications. Assume that n is even. We first define the input data as follows:

$$(4.3.4) \qquad \delta_j := \begin{cases} p(\xi_{2j+1}^{(n)}) & 0 \le j \le \frac{n}{2} - 1, \\ p(\xi_{2n-2j}^{(n)}) & \frac{n}{2} \le j \le n-1. \end{cases}$$

With this choice, we compute γ_m, $0 \le m \le n-1$. Finally, one verifies that

$$(4.3.5) \qquad c_m = \frac{(-1)^m}{n} \cdot \begin{cases} \gamma_0 & \text{if } m = 0, \\ \gamma_m \, e^{\mathbf{i}m\pi/2n} + \gamma_{n-m} \, e^{-\mathbf{i}m\pi/2n} & \text{if } 1 \le m \le n-1. \end{cases}$$

Of course, this final computation has a cost only proportional to n. The procedure can be further improved by virtue of the analysis presented in SWARZTRAUBER(1986). Since the δ_j's are real, one can split the data in a suitable way, to allow the computation of two complex FFTs of length $\frac{n}{2}$ instead of n. We skip the details for simplicity.
The inverse transform (see (4.1.6)) is clearly treated in a very similar way.

The next problem is related to matrix-vector multiplications involving (4.3.2). We now define

$$(4.3.6) \qquad \delta_j := \begin{cases} p(\eta_0^{(n)}) & j=0, \\ p(\eta_{2j}^{(n)}) + \mathbf{i}\left[p(\eta_{2j+1}^{(n)}) - p(\eta_{2j-1}^{(n)})\right] & 1 \le j \le \frac{n}{2} - 1, \\ p(\eta_n^{(n)}) & j = \frac{n}{2}, \\ p(\eta_{2n-2j}^{(n)}) + \mathbf{i}\left[p(\eta_{2n-2j-1}^{(n)}) - p(\eta_{2n-2j+1}^{(n)})\right] & \frac{n}{2} + 1 \le j \le n-1. \end{cases}$$

After evaluating the γ_m's by (4.3.3), the output data can be recovered by

(4.3.7)

$$c_m = \frac{(-1)^m}{2n} \cdot \begin{cases} p(\eta_0^{(n)}) + p(\eta_n^{(n)}) + 2\sum_{j=1}^{n-1} p(\eta_j^{(n)}) & m = 0, \\ \gamma_m \left[1 + \left(2\sin\frac{\pi m}{n}\right)^{-1}\right] + \overline{\gamma_{n-m}}\left[1 - \left(2\sin\frac{\pi m}{n}\right)^{-1}\right] & 1 \le m \le n-1, \\ p(\eta_0^{(n)}) + p(\eta_n^{(n)}) + 2\sum_{j=1}^{n-1}(-1)^j p(\eta_j^{(n)}) & m = n. \end{cases}$$

Here the overbar denotes complex conjugate. The cumbersome verification of (4.3.7) is left to the reader. We note that the output coefficients in (4.3.7) are real. Other suggestions and improvements are contained for instance in SWARZTRAUBER(1986). Again, once the γ_m's have been determined, the cost to implement (4.3.7) is proportional to n.

We proceed now with the analysis of relation (4.3.3) and we consider the special case when $n = 8 = 2^3$. The matrix corresponding to the linear transformation takes the form

(4.3.8)

$$\Phi = \begin{bmatrix} \phi^0 & \phi^0 & \phi^0 & \phi^0 & \phi^0 & \phi^0 & \phi^0 & \phi^0 \\ \phi^0 & \phi^1 & \phi^2 & \phi^3 & \phi^4 & \phi^5 & \phi^6 & \phi^7 \\ \phi^0 & \phi^2 & \phi^4 & \phi^6 & \phi^0 & \phi^2 & \phi^4 & \phi^6 \\ \phi^0 & \phi^3 & \phi^6 & \phi^1 & \phi^4 & \phi^7 & \phi^2 & \phi^5 \\ \phi^0 & \phi^4 & \phi^0 & \phi^4 & \phi^0 & \phi^4 & \phi^0 & \phi^4 \\ \phi^0 & \phi^5 & \phi^2 & \phi^7 & \phi^4 & \phi^1 & \phi^6 & \phi^3 \\ \phi^0 & \phi^6 & \phi^4 & \phi^2 & \phi^0 & \phi^6 & \phi^4 & \phi^2 \\ \phi^0 & \phi^7 & \phi^6 & \phi^5 & \phi^4 & \phi^3 & \phi^2 & \phi^1 \end{bmatrix},$$

where $\phi := e^{i\pi/4}$ and we noted that $\phi^{k+8} = \phi^k, \quad \forall k \in \mathbf{N}$.
Moreover, the linear system can be also written as

(4.3.9)

$$\begin{bmatrix} \gamma_0 \\ \gamma_4 \\ \gamma_2 \\ \gamma_6 \\ \gamma_1 \\ \gamma_5 \\ \gamma_3 \\ \gamma_7 \end{bmatrix} = \Phi_1 \; \Phi_2 \; \Phi_3 \begin{bmatrix} \delta_0 \\ \delta_1 \\ \delta_2 \\ \delta_3 \\ \delta_4 \\ \delta_5 \\ \delta_6 \\ \delta_7 \end{bmatrix},$$

where the matrices $\Phi_k, \quad 1 \le k \le 3$, are given by

$$(4.3.10)\qquad \Phi_1 = \begin{bmatrix} \phi^0 & \phi^0 & 0 & 0 & 0 & 0 & 0 & 0 \\ \phi^0 & \phi^4 & 0 & 0 & 0 & 0 & 0 & 0 \\ 0 & 0 & \phi^0 & \phi^2 & 0 & 0 & 0 & 0 \\ 0 & 0 & \phi^0 & \phi^6 & 0 & 0 & 0 & 0 \\ 0 & 0 & 0 & 0 & \phi^0 & \phi^1 & 0 & 0 \\ 0 & 0 & 0 & 0 & \phi^0 & \phi^5 & 0 & 0 \\ 0 & 0 & 0 & 0 & 0 & 0 & \phi^0 & \phi^3 \\ 0 & 0 & 0 & 0 & 0 & 0 & \phi^0 & \phi^7 \end{bmatrix},$$

$$(4.3.11)\qquad \Phi_2 = \begin{bmatrix} \phi^0 & 0 & \phi^0 & 0 & 0 & 0 & 0 & 0 \\ 0 & \phi^0 & 0 & \phi^0 & 0 & 0 & 0 & 0 \\ \phi^0 & 0 & \phi^4 & 0 & 0 & 0 & 0 & 0 \\ 0 & \phi^0 & 0 & \phi^4 & 0 & 0 & 0 & 0 \\ 0 & 0 & 0 & 0 & \phi^0 & 0 & \phi^2 & 0 \\ 0 & 0 & 0 & 0 & 0 & \phi^0 & 0 & \phi^2 \\ 0 & 0 & 0 & 0 & \phi^0 & 0 & \phi^6 & 0 \\ 0 & 0 & 0 & 0 & 0 & \phi^0 & 0 & \phi^6 \end{bmatrix},$$

$$(4.3.12)\qquad \Phi_3 = \begin{bmatrix} \phi^0 & 0 & 0 & 0 & \phi^0 & 0 & 0 & 0 \\ 0 & \phi^0 & 0 & 0 & 0 & \phi^0 & 0 & 0 \\ 0 & 0 & \phi^0 & 0 & 0 & 0 & \phi^0 & 0 \\ 0 & 0 & 0 & \phi^0 & 0 & 0 & 0 & \phi^0 \\ \phi^0 & 0 & 0 & 0 & \phi^4 & 0 & 0 & 0 \\ 0 & \phi^0 & 0 & 0 & 0 & \phi^4 & 0 & 0 \\ 0 & 0 & \phi^0 & 0 & 0 & 0 & \phi^4 & 0 \\ 0 & 0 & 0 & \phi^0 & 0 & 0 & 0 & \phi^4 \end{bmatrix}.$$

This splitting is the kernel of the FFT algorithm. In general, starting from the $n \times n$ full matrix Φ (where n is a power of 2), we can find a suitable sequence of $n \times n$ matrices Φ_k, $1 \le k \le \log_2 n$, whose product $\Phi^* := \prod_{k=1}^{\log_2 n} \Phi_k$ allows the recovery of Φ. As the reader can observe, the data entries in the left-hand side of (4.3.9) have been permuted. Therefore, Φ is obtained after a suitable recombination of the rows of Φ^*.

The new matrices are very sparse. Actually, a vector multiplication just involves two nonzero entries for each row. The FFT algorithm takes advantage of the regular structure of the Φ_k's. This reduces the matrix-vector multiplication for any

Φ_k, $1 \le k \le \log_2 n$, to a procedure whose cost is only proportional to n. Consequently, the complete FFT can be performed with a cost proportional to $n\log_2 n$, provided n is a power of 2. This shows that, for large values of n, the procedure is efficient. Whether this approach is competitive for small values of n, depends strongly on code implementation. Performance is also affected by different computer architectures.
A clear description, both intuitive and rigorous, of the steps to be followed to minimize the operations, is given in BRIGHAM(1974), where a flow chart is presented. Generalizations to the case where n is not a power of 2, are discussed e.g. in SINGLETON(1969) and TEMPERTON(1983).

Finally, we touch on the problem of reordering the output data, known as *unscrambling procedure*. Consider for example equation (4.3.9). We associate to any index $0 \le m \le 7$, its 3-digit binary representation, i.e.

0	000	0
1	001	4
2	010	2
3	011	6
4	100	1
5	101	5
6	110	3
7	111	7

The last column shows the indices obtained by reading the binary numbers from right to left. These exactly match the indices of the left-hand side vector in (4.3.9). Reordering this last set results in the reordering of the output data. The rough outline above can be easily generalized. Recent variants of the FFT algorithm (*self-sorting* FFT) do not require unscrambling of the data after applying the transform. We refer the reader to specialized publications for further information.

4.4 Other fast methods

There are situations in which a polynomial, defined by the values assumed at the nodes of some Gauss integration formula, must be determined at the points corresponding to the relative Gauss-Lobatto formula. More precisely, from (3.2.2), we get for $p \in \mathbf{P}_{n-1}$

$$(4.4.1) \qquad p(\eta_i^{(n)}) = \sum_{j=1}^{n} p(\xi_j^{(n)})\, l_j^{(n)}(\eta_i^{(n)}), \qquad 0 \le i \le n.$$

Conversely, if $p \in \mathbf{P}_n$, we obtain by (3.2.7)

$$(4.4.2) \qquad p(\xi_i^{(n)}) = \sum_{j=0}^{n} p(\eta_j^{(n)})\, \tilde{l}_j^{(n)}(\xi_i^{(n)}), \qquad 1 \le i \le n.$$

As usual, particularly interesting is the Chebyshev case, where this interpolation process can be optimized. By (4.1.4) or (4.1.8)-(4.1.9), we recover the Fourier coefficients using the FFT. Then, in place of (4.4.1), one computes

$$(4.4.3) \qquad p(\eta_i^{(n)}) = \sum_{j=0}^{n-1} c_j\, T_j(\eta_i^{(n)}) = -\sum_{j=0}^{n-1} c_j \cos\frac{ij}{n}\pi, \quad 0 \le i \le n.$$

Furthermore, in place of (4.4.2), we can write

$$(4.4.4) \qquad p(\xi_i^{(n)}) = \sum_{j=0}^{n} c_j\, T_j(\xi_i^{(n)}) = -\sum_{j=0}^{n} c_j \cos\frac{j(2i-1)}{2n}\pi, \quad 1 \le i \le n,$$

so that an FFT-like algorithm can still be used.
The advantages of using two sets of points, like those considered above, in the discretization of certain classes of partial differential equations, such as the ***incompressible Navier-Stokes*** equations (see section 13.4), have been investigated in MALIK, ZANG and HUSSAINI(1985), and theoretically analyzed and extended in BERNARDI and MADAY (1988). The need for efficient transfer of data between the two grids, could favor the Chebyshev polynomials for these kind of computations.

Additional algorithms, based on the cosine FFT described in the previous section, have been adapted for other Jacobi polynomials (see ORSZAG(1986)). Unfortunately, they are deduced from suitable asymptotic expansions. Therefore, they are fast and accurate only when n is quite large.

5
FUNCTIONAL SPACES

Before studying the convergence properties of orthogonal polynomials, it is necessary to specify the classes of functions to be approximated. This leads to the definition of new functional spaces. An elementary introduction of the so called Sobolev spaces is given in this chapter. Beginners, with a weak mathematical background, can skip over this part in the first reading.

5.1 The Lebesgue integral

The *Lebesgue integral* extends the classical Riemann integral presented in the introductory courses of calculus. This generalization allows the definition of integrals of functions that belong to very large classes. This improvement does not only offer the possibility of dealing with pathological functions (most of which are not of interest to us), but provides new tools to help prove sophisticated results within the framework of a simple and elegant theory.

We start by reviewing the main properties of the Lebesgue measure theory. For a deeper analysis, we refer to the books by HALMOS(1971), KOLMOGOROV and FOMIN(1961), HARTMAN and MIKUSIŃSKI(1961), HILDEBRANDT(1963), TEMPLE(1971), RAO(1987), etc..

We only discuss the one dimensional case. Let $I =]a,b[$, $a < b$, be a bounded interval. The Lebesgue measure μ associated to a subset $J \subset \bar{I}$ is a non negative real number. However, not all the subsets are measurable, though these cases are quite involved. Let us begin by considering very simple sets. Briefly, we review the most significant cases.

The empty set $J = \emptyset$ has measure zero. If $J \subset \bar{I}$ contains a single point or a finite number of points, then $\mu(J) := 0$. When J is an interval with endpoints c and d, $a \leq c < d \leq b$, it is natural to require that

$$\mu(]c,d[) = \mu([c,d[) = \mu(]c,d]) = \mu([c,d]) := d - c.$$

When J is the union of a finite number m of disjoint subintervals K_i, $1 \leq i \leq m$, then its measure is given by $\mu(J) := \sum_{i=1}^{m} \mu(K_i)$. Sets of this form are called *elementary sets*. Union, intersection and difference of two elementary sets is still an elementary set. The measure of an elementary set is easily reduced to the sum of lengths of one or more segments.

A further generalization is to assume that J is the union of a countable collection of disjoint intervals $K_i \subset \bar{I}$, $i \in \mathbf{N}$; we then set $\mu(J) = \sum_{i=0}^{\infty} \mu(K_i)$. This series converges since it is bounded by the value $b-a$. We note that, if the intervals K_i, $i \in \mathbf{N}$, are not all disjoint (i.e., there exist $i, j \in \mathbf{N}$ such that $K_i \cap K_j \neq \emptyset$), then we can replace the family $\{K_i\}_{i\in\mathbf{N}}$ by the family $\{K_i^*\}_{i\in\mathbf{N}}$ such that $J = \cup_{i=0}^{\infty} K_i^*$ and the K_i^*'s are disjoint. In this case, we have the inequality

$$\mu(J) = \sum_{i=0}^{\infty} \mu(K_i^*) \leq \sum_{i=0}^{\infty} \mu(K_i). \tag{5.1.1}$$

Now, let $J \subset \bar{I}$ be a generic subset. Inspired by (5.1.1), we can always define the *outer measure* μ^* of J. This is given by

$$\mu^*(J) := \inf\left\{ \sum_{i=0}^{\infty} \mu(K_i) \;\middle|\; J \subset \bigcup_{i=0}^{\infty} K_i \right\}, \tag{5.1.2}$$

where the lower bound is taken over all the coverings of J by countable families of intervals K_i, $i \in \mathbf{N}$. For elementary sets the outer measure μ^* coincides with the actual measure μ.

We are now ready to give the definition of a Lebesgue measurable set. The subset $J \subset \bar{I}$ is measurable if and only if

$$\mu^*(J) \quad = \quad \mu(\bar{I}) \; - \; \mu^*(\bar{I} - J). \tag{5.1.3}$$

Then we set $\mu(J) := \mu^*(J)$. The idea is the following. The term on the left-hand side of (5.1.3) is obtained by approximating J from outside with families of intervals. The term on the right-hand side represents in some sense an approximation from inside. When both the estimates coincide, then J is measurable.

It is possible to show that the union and intersection of a finite number of measurable sets are still measurable. Moreover, and this is the main improvement with respect to Riemann measure theory, union and intersection of a countable family of measurable sets are still measurable. If the sets J_i, $i \in \mathbf{N}$, are measurable and disjoint, then

$$\mu\left(\bigcup_{i=0}^{\infty} J_i\right) = \sum_{i=0}^{\infty} \mu(J_i). \tag{5.1.4}$$

For instance, the set of rational numbers $J = \mathbf{Q} \cap \bar{I}$, is not Riemann measurable. Being the union of a countable collection of points, J is Lebesgue measurable and by (5.1.4) $\mu(J) = 0$. If I is an unbounded interval, then $J \subset \bar{I}$ is measurable when $J \cap K$ is measurable for any bounded interval $K \subset \bar{I}$. In this situation we assume that μ can also take the value $+\infty$.

We can now introduce the concept of *measurable function*. A function $f : \bar{I} \to \mathbf{R}$ is measurable when the set $\{x \in \bar{I} |\; f(x) < \gamma\}$ is measurable for any $\gamma \in \mathbf{R}$. It is possible to prove that the sum and the product of two measurable functions are still measurable.

A measurable function that assumes a finite or countable set of values in $\bar{I}$, is called *simple function*. Any measurable function f can be represented as a uniform limit of simple functions f_n, $n \in \mathbf{N}$ (this means that $\lim_{n\to\infty} \sup_{x\in I} |f_n(x) - f(x)| = 0$).

Two measurable functions f and g coincide *almost everywhere* (abbreviated by *a.e.*), when the measure of the set $\{x \in \bar{I}|\ f(x) \neq g(x)\}$ is zero. In this situation f and g are called *equivalent*.

We have now collected enough ingredients to define the Lebesgue integral of a measurable function. Let us start with the integral of a simple function f. In this case, f attains at most a countable set of values γ_i, $i \in \mathbf{N}$, and the sets $A_i := \{x \in \bar{I}|\ f(x) = \gamma_i\}$, $i \in \mathbf{N}$, are measurable. Then we have

$$\int_I f\, dx := \sum_{i=0}^{\infty} \gamma_i\, \mu(A_i), \tag{5.1.5}$$

provided the series is absolutely convergent.
In general, let f be a measurable function; then there exists a sequence of simple functions f_n, $n \in \mathbf{N}$, uniformly convergent to f. Thus we set

$$\int_I f\, dx := \lim_{n\to\infty} \int_I f_n\, dx, \tag{5.1.6}$$

provided the limit exists. In this case, the limit does not depend on the approximating sequence $\{f_n\}_{n\in\mathbf{N}}$.
The sum of two integrable functions is an integrable function. Moreover, for continuous or monotone functions, the Lebesgue and Riemann integrals coincide.
We have the implication

$$\int_I |f|\, dx = 0 \qquad \Longleftrightarrow \qquad f \equiv \mathbf{0} \quad a.e. \tag{5.1.7}$$

This shows that point values, achieved on sets whose measure is zero, are not meaningful for the evaluation of the integral. This observation is crucial. In the following, besides a given integrable function f, we shall not distinguish among functions equivalent to f. This is because we shall work with norms of f based on the Lebesgue integral, and these are not affected by modifications of f on a set of measure zero.

5.2 Spaces of measurable functions

In section 2.1, we introduced an inner product and a norm in the space of continuous functions (see (2.1.8) and (2.1.9)). As we shall see in the sequel, $C^0(\bar{I})$ is not a suitable space for this kind of norm. Besides, in view of the chosen applications, we must to enrich our space with new functions. Using the approach of section 2.1, this is quite cumbersome. For instance, including discontinuous functions is not straightforward. One of the reasons is that property (2.1.3) could not be satisfied in an extended space (for example, take u vanishing in I, with the exception of one point).
Here, we would like to make this extension in the very large family of measurable functions. To this purpose, for any measurable function u, we shall denote by $\mathcal{C}_u$ the class of all the functions equivalent to u. Let $w : I \to \mathbf{R}$, be a positive continuous weight function (we prefer to avoid generalizations concerning w). Then, we define the following functional space:

$$(5.2.1) \qquad L^2_w(I) := \left\{ \mathcal{C}_u \middle| \ u \text{ is measurable and } \int_I u^2 \ w dx < +\infty \right\}.$$

Although the elements of $L^2_w(I)$ are not functions, we shall not distinguish between u and $\mathcal{C}_u$, since the integral in (5.2.1) does not recognize the different representatives of a certain class. So that, due to (5.1.7), we get the implication

$$(5.2.2) \qquad u \in L^2_w(I) \text{ and } \int_I u^2 \ w dx = 0 \quad \Longleftrightarrow \quad u \equiv \mathbf{0}.$$

The correct interpretation of the right-hand side in (5.2.2) is that a generic representative of the class of functions associated with u is an element of the class of functions vanishing almost everywhere. When I is bounded, the inclusion $C^0(\bar{I}) \subset L^2_w(I)$ is similarly justified.
In $L^2_w(I)$ we can define an inner product and a norm which are the natural extensions of those given in section 2.1, i.e.

$$(5.2.3) \qquad (u,v)_{L^2_w(I)} := \int_I uv \ w dx, \qquad \forall u, v \in L^2_w(I),$$

(5.2.4) $$\|u\|_{L^2_w(I)} := \left(\int_I u^2 \, w dx\right)^{\frac{1}{2}}, \qquad \forall u \in L^2_w(I).$$

The property (2.1.3) is now exactly expressed by (5.2.2).

Another functional space is related to the norm introduced in section 2.5. We first give some definitions. We say that a measurable function u is *essentially bounded* when we can find a constant $\gamma \geq 0$ such that $|u| \leq \gamma$ almost everywhere. An essentially bounded function u behaves like a bounded function, except on sets of measure zero. Thus, we set

(5.2.5) $$L^\infty(I) := \left\{ \mathcal{C}_u \middle| \; u \text{ is essentially bounded} \right\}.$$

The corresponding norm is given by

(5.2.6) $$\|u\|_{L^\infty(I)} := \inf \left\{ \gamma \geq 0 \middle| \; |u| \leq \gamma \;\; a.e. \right\}.$$

When I is bounded, continuous functions in $\bar{I}$ belong to $L^\infty(I)$. In this case, the norm (5.2.6) is equal to the norm $\|\cdot\|_\infty$, presented in section 2.5.
Introductory discussions on the spaces analyzed herein, and their generalizations, are contained e.g. in VULIKH(1963), GOFFMAN and PEDRICK(1965), OKIKIOLU (1971), and WOUK(1979). Other references are given in the following sections.

5.3 Completeness

We point out an important property of the functional spaces introduced above. We first recall that a sequence $\{u_n\}_{n\in\mathbf{N}}$ in a normed space $\mathbf{X}$ is a *Cauchy sequence* when, for any $\epsilon > 0$, we can find $N \in \mathbf{N}$ such that

(5.3.1) $$\|u_{n_1} - u_{n_2}\| < \epsilon, \qquad \forall n_1, n_2 > N.$$

Every convergent sequence in $\mathbf{X}$ is a Cauchy sequence. The converse depends on the structure of $\mathbf{X}$. This justifies the following definition. A space $\mathbf{X}$, where every Cauchy sequence converges to an element of $\mathbf{X}$, is called *complete space*.
Let I be a bounded interval. The space $\mathbf{X} = C^0(\bar{I})$ with the norm $\|\cdot\|_w$ given in (2.1.9), is not complete. For instance, let $\bar{I} = [0,1]$ and $u_n(x) = x^n$, $x \in \bar{I}$, $n \in \mathbf{N}$. Therefore, we obtain a Cauchy sequence in $C^0(\bar{I})$ converging in the norm $\|\cdot\|_w$, $w \equiv 1$, to a discontinuous function.
The correct norm to be used in $C^0(\bar{I})$ is defined by (2.5.1). This leads to a complete space, i.e., Cauchy sequences in $C^0(\bar{I})$ with the norm $\|\cdot\|_\infty$ always converge to a continuous function in that norm (the well-known *uniform convergence*). Moreover, $L^\infty(I)$ is also complete with the norm given in (5.2.6). In the following we set (see (5.2.6))

$$(5.3.2) \qquad \|u\|_{C^0(\bar{I})} := \|u\|_{L^\infty(I)}, \qquad \forall u \in C^0(\bar{I}),$$

provided either I or u are bounded.

Another interesting result is that $L^2_w(I)$ is complete with the norm given in (5.2.4). When I is bounded, $L^2_w(I)$ is the *completion* of $C^0(\bar{I})$ with the norm (2.1.9). This means that any function in $L^2_w(I)$ can be approximated in the norm of $L^2_w(I)$ by a sequence of continuous functions. In this case, not only does $C^0(\bar{I}) \subset L^2_w(I)$, but there exists a constant $K > 0$ (depending on $\mu(I)$ and w) such that

$$(5.3.3) \qquad \|u\|_{L^2_w(I)} \leq K\, \|u\|_{C^0(\bar{I})}, \qquad \forall u \in C^0(\bar{I}).$$

A complete space, whose norm is related to an inner product, is called a *Hilbert space*. In particular $L^2_w(I)$ is a Hilbert space, $C^0(\bar{I})$ is not a Hilbert space. More results are presented in the classical texts of functional analysis.

5.4 Weak derivatives

We recall that the space $C^k(\bar{I})$, where $k \geq 1$ is an integer, is given by the functions whose derivatives, of order less or equal to k, exist and are continuous in $\bar{I}$. Following section 5.3, when I is bounded, a norm for $C^k(\bar{I})$ is obtained by setting

$$(5.4.1) \qquad \|u\|_{C^k(\bar{I})} := \|u\|_{C^0(\bar{I})} + \sum_{m=1}^{k} \left\| \frac{d^m u}{dx^m} \right\|_{C^0(\bar{I})}, \qquad u \in C^k(\bar{I}), \quad k \geq 1.$$

With this norm $C^k(\bar{I})$, $k \geq 1$, is a complete space. In particular, when $u \in C^k(\bar{I})$, $\forall k \in \mathbf{N}$, we say that $u \in C^\infty(\bar{I})$. We are mainly concerned with a particular subspace of $C^\infty(\bar{I})$. This is denoted by $C_0^\infty(I)$ and consists of all the functions $\phi \in C^\infty(\bar{I})$ such that there exists a closed and bounded interval strictly included in I, outside of which ϕ vanishes. Therefore, since I is open, any function in $C_0^\infty(I)$ is zero in a neighborhood of the possible boundary points of I. Therefore, all the derivatives of a function in $C_0^\infty(I)$ still belong to $C_0^\infty(I)$.

Up to this point, we only considered subspaces of the set of continuous functions. This is against the policy of expanding our functional spaces. We now give meaning to the differential calculus of measurable functions by introducing the concept of ***weak derivative***. A Lebesgue integrable function u is differentiable in a weak sense, if there exists another integrable function v such that

$$(5.4.2) \qquad \int_I u\phi' \, dx = -\int_I v\phi \, dx, \qquad \forall \phi \in C_0^\infty(I).$$

The function v is the weak derivative of u which is indicated using classical notation. This definition is not in conflict with the usual one. Actually, we claim that, if $u \in C^1(\bar{I})$, then $v \in C^0(\bar{I})$ is exactly the standard derivative of u. To prove this statement, we just integrate the left-hand side of (5.4.2) by parts. Recalling the vanishing conditions on ϕ, we obtain as requested

$$(5.4.3) \qquad \int_I (u' - v)\phi \, dx = 0, \quad \forall \phi \in C_0^\infty(I) \quad \Longrightarrow \quad v \equiv u'.$$

The implication in (5.4.3) can be proven with a little knowledge of basic calculus.

Higher-order derivatives are similarly defined. The measurable function v is the k-th derivative of u if

$$\int_I u \frac{d^k\phi}{dx^k}\,dx = (-1)^k \int_I v\phi\,dx, \qquad \forall\phi \in C_0^\infty(I), \quad k \geq 1. \tag{5.4.4}$$

The weak derivative of a function u is not uniquely determined. However, two different derivatives are equivalent, hence they are in the same class $\mathcal{C}_v$.

In general, not all integrable functions admit a weak derivative, unless we enlarge the space of possible derivatives. This extension leads to the definition of the space of ***distributions***, whose elements are often not even representable as functions (the typical example is the ***Dirac delta***). For simplicity, we skip this subject and refer the interested reader to the following authors: SCHWARTZ(1966), TREVES(1967), GUELFAND, GRAEV and VILENKIN(1970), and GARNIR, DE WILDE and SCHMETS(1973).

In the following sections, we study how to use weak derivatives to define additional functional spaces.

5.5 Transformation of measurable functions in R

Let $\mathbf{C}$ denote the set of complex numbers. A function $u : \mathbf{R} \to \mathbf{C}$ is measurable when both the real part $\mathrm{Re}(u)$ and imaginary part $\mathrm{Im}(u)$ are measurable. Then we define

$$L^2(\mathbf{R};\mathbf{C}) := \left\{ \mathcal{C}_u \,\middle|\, u \text{ measurable and } \int_{\mathbf{R}} [(\mathrm{Re}(u))^2 + (\mathrm{Im}(u))^2]dx < +\infty \right\}. \tag{5.5.1}$$

The space in (5.5.1) is a Hilbert space with the (complex) inner product

$$(u,v)_{L^2(\mathbf{R};\mathbf{C})} := \int_{\mathbf{R}} u\bar{v}\,dx, \qquad u,v \in L^2(\mathbf{R};\mathbf{C}), \tag{5.5.2}$$

where $\bar{v}$ denotes the complex conjugate of v.

Let us now consider the operator $\mathcal{F} : L^2(\mathbf{R};\mathbf{C}) \to L^2(\mathbf{R};\mathbf{C})$. This maps $u \in L^2(\mathbf{R};\mathbf{C})$ to the function defined by:

(5.5.3) $$[\mathcal{F}u](t) := \frac{1}{\sqrt{2\pi}} \int_{\mathbf{R}} u(x)\, e^{-\mathbf{i}tx} dx \quad \in L^2(\mathbf{R};\mathbf{C}),$$

where $\mathbf{i}$ is the imaginary unity and the equality holds in $\mathbf{R}$ almost everywhere. Since $u \in L^2(\mathbf{R};\mathbf{C})$ does not in general imply that u is integrable in $\mathbf{R}$, a little care is necessary when manipulating expression (5.5.3). With some technical reasoning, one verifies that the setting is correct after appropriate interpretation (see for instance RUDIN(1966)). The function $\mathcal{F}u$ is called *Fourier transform* of u. The operator $\mathcal{F}$ suitably generalizes the complex discrete Fourier transform (4.3.3).
The following relation is well-known:

(5.5.4) $$\|\mathcal{F}u\|_{L^2(\mathbf{R};\mathbf{C})} = \|u\|_{L^2(\mathbf{R};\mathbf{C})}, \qquad \forall u \in L^2(\mathbf{R};\mathbf{C}).$$

In other words, $\mathcal{F}$ is an *isometry*.
Moreover, the inverse operator $\mathcal{F}^{-1} : L^2(\mathbf{R};\mathbf{C}) \rightarrow L^2(\mathbf{R};\mathbf{C})$, admits the representation

(5.5.5) $$[\mathcal{F}^{-1}u](t) = \frac{1}{\sqrt{2\pi}} \int_{\mathbf{R}} u(x)\, e^{\mathbf{i}tx} dx, \qquad a.e. \in \mathbf{R}, \ \ \forall u \in L^2(\mathbf{R};\mathbf{C}).$$

Many other properties characterize the transformation $\mathcal{F}$ (see RUDIN(1966)). We just mention some of those useful in the following sections.
If both the real and the imaginary part of $u \in L^2(\mathbf{R};\mathbf{C})$ have a weak derivative (see previous section) and $u' \in L^2(\mathbf{R};\mathbf{C})$, then the following relation can be established for almost every $t \in \mathbf{R}$:

(5.5.6) $$[\mathcal{F}u'](t) = -\frac{\mathbf{i}\,t}{\sqrt{2\pi}}\, [\mathcal{F}u](t).$$

Therefore, setting $v(t) := t[\mathcal{F}u](t)$, $t \in \mathbf{R}$, we have

(5.5.7) $$u' \in L^2(\mathbf{R};\mathbf{C}) \quad \Longleftrightarrow \quad v \in L^2(\mathbf{R};\mathbf{C}).$$

Since $t \in \mathbf{R}$ diverges at infinity at the same rate as $\rho(t) := (1+t^2)^{1/2}$, $t \in \mathbf{R}$, (5.5.7) can be rewritten as

(5.5.8) $$u' \in L^2(\mathbf{R};\mathbf{C}) \quad \Longleftrightarrow \quad \rho\mathcal{F}u \in L^2(\mathbf{R};\mathbf{C}).$$

The use of $\rho(t)$ in place of t will be understood later on. Now, due to (5.5.8) and a proposition proven in ESCOBEDO and KAVIAN(1987), if $w(x) = e^{x^2}$, $x \in \mathbf{R}$, one can deduce that

$$(5.5.9) \qquad u \in L^2_w(\mathbf{R}), \;\; u' \in L^2_w(\mathbf{R}) \qquad \Longleftrightarrow \qquad \rho\mathcal{F}(u\sqrt{w}) \in L^2(\mathbf{R};\mathbf{C}).$$

More generally, for any integer $k \geq 1$, when u is differentiable k times, we have

$$(5.5.10) \qquad \frac{d^m u}{dx^m} \in L^2_w(\mathbf{R}), \quad 0 \leq m \leq k \qquad \Longleftrightarrow \qquad \rho^k\mathcal{F}(u\sqrt{w}) \in L^2(\mathbf{R};\mathbf{C}).$$

5.6 Sobolev spaces in R

Let $k \geq 1$ be an integer. We introduce the *weighted Sobolev space* of order k in $\mathbf{R}$ as follows (see ADAMS(1975), BERGH and LÖFSTRÖM(1976), TRIEBEL(1978)):

$$(5.6.1) \qquad H^k_w(\mathbf{R}) := \left\{ \mathcal{C}_u \;\middle|\; u \text{ differentiable } k \text{ times and } \frac{d^m u}{dx^m} \in L^2_w(\mathbf{R}), \text{ for } 0 \leq m \leq k \right\}.$$

The derivative is intended in the weak sense and $\frac{d^m u}{dx^m} = u$ when $m = 0$. We also set $H^0_w(\mathbf{R}) := L^2_w(\mathbf{R})$. Recalling (5.2.3) and (5.2.4), an inner product and a norm are introduced in $H^k_w(\mathbf{R})$, $k \in \mathbf{N}$, namely

$$(5.6.2) \qquad (u,v)_{H^k_w(\mathbf{R})} := \sum_{m=0}^{k} \left(\frac{d^m u}{dx^m}, \frac{d^m v}{dx^m} \right)_{L^2_w(\mathbf{R})}, \qquad \forall u,v \in H^k_w(\mathbf{R}),$$

$$(5.6.3) \qquad \|u\|_{H^k_w(\mathbf{R})} := \left(\sum_{m=0}^{k} \left\| \frac{d^m u}{dx^m} \right\|^2_{L^2_w(\mathbf{R})} \right)^{\frac{1}{2}}, \qquad \forall u \in H^k_w(\mathbf{R}).$$

With this norm, $H^k_w(\mathbf{R})$, $k \in \mathbf{N}$, is a Hilbert space. The higher is the index k and the smoother the functions belonging to $H^k_w(\mathbf{R})$ will be.

From now on, let us assume that $w(x) = e^{x^2}$, $x \in \mathbf{R}$, i.e. the inverse of the Hermite weight function. A first characterization is immediately obtained by virtue of relation (5.5.10)

$$(5.6.4) \qquad H^k_w(\mathbf{R}) \; = \; \left\{ \mathcal{C}_u \;\middle|\; \rho^k \mathcal{F}(u\sqrt{w}) \in L^2(\mathbf{R};\mathbf{C}) \right\},$$

where $\rho(t) = (1+t^2)^{1/2}$, $t \in \mathbf{R}$. This new point of view suggests the introduction of other Sobolev spaces. Namely, let $s \geq 0$ be a real number, we define

$$(5.6.5) \qquad H^s_w(\mathbf{R}) \; := \; \left\{ \mathcal{C}_u \;\middle|\; \rho^s \mathcal{F}(u\sqrt{w}) \in L^2(\mathbf{R};\mathbf{C}) \right\}.$$

We note that ρ^s, $s \geq 0$ is a smooth function in $\mathbf{R}$. This justifies the use of $\rho(t)$ in place of t.

Of course, when s is integer, the two definitions (5.6.1) and (5.6.5) coincide. It is clear that we have the inclusion $H^{s_1}_w(\mathbf{R}) \subset H^{s_2}_w(\mathbf{R})$, if $s_1 > s_2 \geq 0$. In this way, although we cannot speak of derivatives of order s when s is not an integer, we can instead decide when a function u belongs to $H^s_w(\mathbf{R})$. In this case, u will be more regular than a function in $H^{[s]}_w(\mathbf{R})$ and less regular than a function in $H^{[s]+1}_w(\mathbf{R})$, where $[s]$ denotes the integer part of s. For instance, let $u(x) = \sqrt{|x|}$, $x \in \mathbf{R}$. Then, one can easily check that $u \notin H^1_w(\mathbf{R})$. On the other hand, for any s such that $0 \leq s < 1$, we have $u \in H^s_w(\mathbf{R})$. By (5.5.2), a norm in $H^s_w(\mathbf{R})$, $s \geq 0$, is defined by

$$(5.6.6). \qquad |||u|||_{H^s_w(\mathbf{R})} \; := \; \|\rho^s \mathcal{F}(u\sqrt{w})\|_{L^2(\mathbf{R};\mathbf{C})}, \qquad \forall u \in H^s_w(\mathbf{R}).$$

One can prove that, when s is an integer, the norms (5.6.3) and (5.6.6) are equivalent (see TRIEBEL(1978), p.177). Actually, for any $k \in \mathbf{N}$, we can find two constants $C_1, C_2 > 0$, such that

$$(5.6.7) \qquad C_1 \, \|u\|_{H^k_w(\mathbf{R})} \; \leq \; |||u|||_{H^k_w(\mathbf{R})} \; \leq \; C_2 \|u\|_{H^k_w(\mathbf{R})}, \qquad \forall u \in H^k_w(\mathbf{R}).$$

When s is not integer, it is possible to show (see TRIEBEL(1978), p.189) that the norm $|||\cdot|||_{H^s_w(\mathbf{R})}$ is equivalent to:

$$(5.6.8)\quad \|u\|_{H^s_w(\mathbf{R})} := \|u\|_{H^{[s]}_w(\mathbf{R})} + \left(\int_{\mathbf{R}\times\mathbf{R}} \frac{|v(x)-v(y)|^2}{|x-y|^{1+2\sigma}}\,dxdy\right)^{\frac{1}{2}}, \qquad \forall u \in H^s_w(\mathbf{R}),$$

where $v := \frac{d^{[s]}}{dx^{[s]}}(u\sqrt{w})$ and $\sigma := s - [s]$.

Depending on the context, we shall make use either of the representations (5.6.3) and (5.6.8) or of the expression (5.6.6). In all three cases, $H^s_w(\mathbf{R})$ turns out to be a complete space.

We can easily check that any function of the form pe^{-x^2}, where p is a polynomial, belongs to $H^s_w(\mathbf{R})$, $\forall s \geq 0$. At the same time, if $u \in C^k(\mathbf{R})$, $k \in \mathbf{N}$, and u and its derivatives of order less or equal to k decay faster than $e^{-x^2/2}$ at infinity, then we obtain $u \in H^k_w(\mathbf{R})$. The converse of this statement leads to the *Sobolev embedding theorem* (see TRIEBEL(1978), p.206), which can be stated as follows. For any $k \in \mathbf{N}$, if $u \in H^s_w(\mathbf{R})$, with $s > k + \frac{1}{2}$, then $u \in C^k(\mathbf{R})$. Furthermore, a constant $K > 0$ exists such that

$$(5.6.9)\qquad \|u\sqrt{w}\|_{C^k(\mathbf{R})} \leq K\ \|u\|_{H^s_w(\mathbf{R})}, \qquad \forall u \in H^s_w(\mathbf{R}),\ s > k + \tfrac{1}{2}.$$

This result is significant since it relates abstract Sobolev spaces with the classical spaces of continuous and differentiable functions. To be precise, the inclusion $H^s_w(\mathbf{R}) \subset C^0(\mathbf{R})$, for $s > \frac{1}{2}$, means that any class $C_u \in H^s_w(\mathbf{R})$ contains a continuous representative u. We also have (see FUNARO and KAVIAN(1988))

$$(5.6.10)\quad \|\rho^\mu u\sqrt{w}\|_{C^0(\mathbf{R})} \leq K\ \|u\|_{H^s_w(\mathbf{R})}, \qquad \forall u \in H^s_w(\mathbf{R}),\ s > \tfrac{1}{2} + \mu \ \text{ and } \ \mu \geq 0.$$

An interesting property is that any function in $H^s_w(\mathbf{R})$, $s \geq 0$, can be approximated in the norm of $H^s_w(\mathbf{R})$ by a sequence of functions in $C^\infty_0(\mathbf{R})$.

Finally, we mention the following inequality. Let $\sigma, \tau \geq 0$ and $u \in H^\sigma_w(\mathbf{R}) \cap H^\tau_w(\mathbf{R})$, then we have $u \in H^s_w(\mathbf{R})$, where $s = (1-\theta)\sigma + \theta\tau$, $\theta \in [0,1]$, and we can find a constant $K > 0$ such that

$$(5.6.11)\qquad \|u\|_{H^s_w(\mathbf{R})} \leq K\ \|u\|^{1-\theta}_{H^\sigma_w(\mathbf{R})}\ \|u\|^{\theta}_{H^\tau_w(\mathbf{R})}, \qquad \forall u \in H^\sigma_w(\mathbf{R}) \cap H^\tau_w(\mathbf{R}).$$

5.7 Sobolev spaces in intervals

For any integer $k \geq 1$, the weighted Sobolev space of order k in $I \subset \mathbf{R}$ is defined by

(5.7.1) $$H_w^k(I) := \left\{ \mathcal{C}_u \;\middle|\; u \text{ differentiable } k \text{ times and } \frac{d^m u}{dx^m} \in L_w^2(I), \text{ for } 0 \leq m \leq k \right\}.$$

Here, $w : I \to \mathbf{R}$ is a weight function. Moreover $H_w^0(I) := L_w^2(I)$. For every $k \in \mathbf{N}$, $H_w^k(I)$ is a Hilbert space with the inner product and the norm

(5.7.2) $$(u,v)_{H_w^k(I)} := \sum_{m=0}^{k} \left(\frac{d^m u}{dx^m}, \frac{d^m v}{dx^m} \right)_{L_w^2(I)}, \qquad \forall u, v \in H_w^k(I),$$

(5.7.3) $$\|u\|_{H_w^k(I)} := \left(\sum_{m=0}^{k} \left\| \frac{d^m u}{dx^m} \right\|_{L_w^2(I)}^2 \right)^{\frac{1}{2}}, \qquad \forall u \in H_w^k(I).$$

Many fundamental properties about these spaces are given in KUFNER(1980). We recall that, for a function $u \in H_w^k(I)$, there exists a sequence of regular functions $\{v_n\}_{n\in\mathbf{N}}$ (we can assume for example $v_n \in C^\infty(\bar{I})$, $\forall n \in \mathbf{N}$), converging to u in the norm of $H_w^k(I)$, i.e., $\lim_{n\to+\infty} \|v_n - u\|_{H_w^k(I)} = 0$. In the following, using a standard functional analysis technique, many theorems will be proven with the help of this property. The statement to be proven is checked for regular functions and successively extended to Sobolev spaces with a limit procedure. For simplicity, we shall often omit the details when applying this kind of technique.

Let us examine the Jacobi weight function $w(x) = (1-x)^\alpha(1+x)^\beta$, $x \in I =]-1,1[$, where the parameters satisfy the conditions $-1 < \alpha < 1$ and $-1 < \beta < 1$. In this case, one verifies that $H_w^1(I) \subset C^0(\bar{I})$. Moreover, we can find two constants K_1, $K_2 > 0$, such that

(5.7.4) $$\|u\|_{L_w^2(I)} \leq K_1 \|u\|_{C^0(\bar{I})} \leq K_2 \|u\|_{H_w^1(I)}, \qquad \forall u \in H_w^1(I).$$

As a byproduct of Hardy's inequality (see for instance LIONS and MAGENES(1972), p.49), one obtains the relation

$$(5.7.5) \qquad \left\| \frac{u}{1-x^2} \right\|_{L^2_w(I)} \leq K \, \|u'\|_{L^2_w(I)}, \qquad \forall u \in H^1_w(I) \text{ with } u(\pm 1) = 0.$$

Now, we introduce another space, namely

$$(5.7.6) \qquad H^1_{0,w}(I) := \left\{ u \in H^1_w(I) \;\middle|\; u(\pm 1) = 0 \right\}.$$

The following norm will be adopted in $H^1_{0,w}(I)$:

$$(5.7.7) \qquad \|u\|_{H^1_{0,w}(I)} := \|u'\|_{L^2_w(I)}, \qquad \forall u \in H^1_{0,w}(I).$$

To show that this is a norm we have to check (2.1.3). Actually, by (5.7.5), the condition $\|u\|_{H^1_{0,w}(I)} = 0$ implies $u \equiv \mathbf{0}$. In practice, in $H^1_{0,w}(I)$, it is sufficient to bound the derivative of u in order to control the function itself.

As in the previous section, we would like to define Sobolev spaces $H^s_w(I)$, where $s \geq 0$ is a real number. There are several approaches. A first approach consists in constructing a suitable prolongation of the weight w and functions in $L^2_w(I)$ to the whole domain $\mathbf{R}$. Then we can provide a definition involving the transformation $\mathcal{F}$ following the guideline of section 5.6. Another definition can be based on a formula similar to (5.6.8). However, the common practice is to define $H^s_w(I)$ as an *intermediate space* between $H^{[s]}_w(I)$ and $H^{[s]+1}_w(I)$, via a technique called *interpolation* (see for instance LIONS and MAGENES(1972)). For the sake of simplicity, we shall not enter into the details of this procedure.

The case $w \equiv 1$ has been extensively investigated by many authors. For a general weight function, we cannot expect the various definitions to match. The literature on this subject does not cover all the possible cases and the theory is often quite complicated. Results have recently been published in MADAY(1990) on the interpolation of Sobolev spaces with the Chebyshev weight function. Though the use of spaces obtained by interpolation seems to be a promising approach in view of the applications, with a few exceptions, one can easily end up with abstract intermediate spaces, without a satisfactory characterization of their elements.

* * * * * * * * * * * * * *

From the point of view of functional analysis, the brief presentation set forth in these pages is far from exhaustive. In our opinion, however, it contains sufficient material for an in depth understanding of the applications that follow. We refer the readers that are more concerned with theoretical issues, to the numerous publications in this field. Very attractive to applied mathematicians are, for instance, the text of BREZIS(1983) and the introductory essay of BREZZI and GILARDI(1987), where the functional spaces are rigorously treated, but without the tiresome formalism of the more specialized books.

6 RESULTS IN APPROXIMATION THEORY

According to a theorem of Weierstrass, any continuous function in a bounded closed interval can be uniformly approximated by polynomials. From this celebrated statement, a large variety of sophisticated results have emerged. We review those that are most relevant to the analysis of spectral methods.

6.1 The problem of best approximation

We begin with a classical theorem in approximation theory.

Theorem 6.1.1 *(Weierstrass) - Let I be a bounded interval and let $f \in C^0(\bar{I})$. Then, for any $\epsilon > 0$, we can find $n \in \mathbf{N}$ and $p_n \in \mathbf{P}_n$ such that*

$$|f(x) - p_n(x)| < \epsilon, \qquad \forall x \in \bar{I}. \tag{6.1.1}$$

Hints for the proof - We just give a brief sketch of the proof in the interval $\bar{I} = [0,1]$. We follow the guideline of the original approach presented in BERNSTEIN(1912b). Other techniques are discussed for instance in TIMAN(1963), TODD(1963) and CHENEY(1966). For sufficiently large n, the approximating polynomial is required to be of the form

$$p_n(x) := \sum_{j=0}^{n} f\left(\frac{j}{n}\right) \binom{n}{j} x^j (1-x)^{n-j}, \qquad x \in [0,1], \tag{6.1.2}$$

which leads to

$$(6.1.3) \qquad f(x) - p_n(x) = \sum_{j=0}^{n} \left[f(x) - f\left(\frac{j}{n}\right)\right] \binom{n}{j} x^j (1-x)^{n-j}, \qquad x \in [0,1].$$

The proof that this error tends to zero uniformly is now a technical exercise. One argues as follows. For any $x \in [0,1]$, the sum in (6.1.2) is decomposed into two parts. In the first part, the indices j are such that $|x - j/n|$ is suitably small. Therefore, from the uniform continuity of f, the term $|f(x) - f(j/n)|$ will also be small. To handle the remaining part of the summation, we observe that the function $\binom{n}{j} x^j (1-x)^{n-j}$ achieves its maximum for $x = j/n$, and far away from this value it decays quite fast. By appropriately applying the above information, one can, after some manipulation, conclude the proof.

■

At this point, the following question arises. Among all the polynomials of degree less or equal to a fixed integer n, find the one which best approximates, uniformly in $\bar{I}$, a given continuous function f. In practice, with the notations of section 2.5, for any $n \in \mathbf{N}$, we would like to study the existence of $\Psi_{\infty,n}(f) \in \mathbf{P}_n$ such that

$$(6.1.4) \qquad \|f - \Psi_{\infty,n}(f)\|_\infty = \inf_{\psi \in \mathbf{P}_n} \|f - \psi\|_\infty .$$

This problem admits a unique solution, though the proof is very involved. An extensive and general treatise on this subject is given for instance in TIMAN(1963). The n-degree polynomial $\Psi_{\infty,n}(f)$ is called the polynomial of *best uniform approximation* of f in $\bar{I}$. As a consequence of theorem 6.1.1, one immediately obtains

$$(6.1.5) \qquad \lim_{n \to +\infty} \|f - \Psi_{\infty,n}(f)\|_\infty = 0.$$

One can try to characterize the polynomial of best approximation of a certain degree. Results in this direction are considered in TIMAN(1963) and RIVLIN(1969) where Chebyshev polynomials play a fundamental role. As an example, theorem 2.5.3 provides the polynomial that best approximates the zero function in $[-1,1]$, in the subset of $\mathbf{P}_n$ given by the polynomials of the form $x^n + \{\textit{lower degree terms}\}$.

We now collect some result on the problem of best approximation. We begin with the celebrated theorem by Jackson (see for instance FEINERMAN and NEWMAN(1974)).

Theorem 6.1.2 *(Jackson) - Let $I =]-1,1[$, then it is possible to find a constant $C > 0$ such that, for any $f \in C^0(\bar{I})$, we have*

$$\|f - \Psi_{\infty,n}(f)\|_\infty \leq C \sup_{|x_1 - x_2| < 2/n} |f(x_1) - f(x_2)|, \qquad \forall n \geq 1. \tag{6.1.6}$$

■

The original proof can be found in JACKSON(1911). The above proposition can be generalized as follows.

Theorem 6.1.3 *(Jackson) - Let $I =]-1,1[$ and $k \in \mathbf{N}$. One can then find a constant $C > 0$ such that, for any $f \in C^k(\bar{I})$, we have*

$$\|f - \Psi_{\infty,n}(f)\|_\infty \leq C \left(\frac{1}{n}\right)^k \sup_{|x_1 - x_2| < 2/n} \left|\frac{d^k f}{dx^k}(x_1) - \frac{d^k f}{dx^k}(x_2)\right|, \qquad \forall n \geq k. \tag{6.1.7}$$

■

This shows that the speed of convergence to zero of the error depends on the regularity of the function f. An exponential decay to zero is obtained when approximating analytic functions (see TIMAN(1963), p.280).

To extend theorem 6.1.1 to the case of unbounded domains, we present two propositions (see TIMAN(1963), p.16).

Theorem 6.1.4 - *Let $I =]0,+\infty[$ and let $f \in C^0(\bar{I})$ satisfy $\lim_{x \to +\infty} f(x) e^{-\delta x} = 0$, for a given $\delta > 0$. Then, for any $\epsilon > 0$, we can find $n \in \mathbf{N}$ and $p_n \in \mathbf{P}_n$ such that*

$$|f(x) - p_n(x)| \, e^{-\delta x} < \epsilon, \qquad \forall x \in \bar{I}. \tag{6.1.8}$$

■

Theorem 6.1.5 - *Let $I = \mathbf{R}$ and let $f \in C^0(\mathbf{R})$ satisfy $\lim_{x \to \pm\infty} f(x) e^{-\delta x^2} = 0$, for a given $\delta > 0$. Then, for any $\epsilon > 0$, we can find $n \in \mathbf{N}$ and $p_n \in \mathbf{P}_n$ such that*

$$|f(x) - p_n(x)| \, e^{-\delta x^2} < \epsilon, \qquad \forall x \in \mathbf{R}. \tag{6.1.9}$$

■

6.2 Estimates for the projection operator

As usual, we denote by $\{u_n\}_{n\in\mathbf{N}}$ the sequence of orthogonal polynomials in $\bar{I}$ (where $\bar{I}=[-1,1]$, $\bar{I}=[0,+\infty[$ or $\bar{I}\equiv\mathbf{R}$), with respect to the weight function w (we refer in particular to those introduced in chapter one).
Another best approximation problem can be formulated in terms of the norm $\|\cdot\|_w$ introduced in (2.1.9). In this case, we wish to find $\Psi_{w,n}(f)\in\mathbf{P}_n$ such that

$$\|f-\Psi_{w,n}(f)\|_w = \inf_{\psi\in\mathbf{P}_n} \|f-\psi\|_w, \tag{6.2.1}$$

where f is a given function in $\bar{I}$.
For the reader acquainted with the results of chapter five, the natural space to develop the theory is $L^2_w(I)$. Under this assumption, we can still define the Fourier coefficients of f as

$$c_k := \frac{(f,u_k)_{L^2_w(I)}}{\|u_k\|^2_{L^2_w(I)}}, \qquad k\in\mathbf{N}. \tag{6.2.2}$$

We note that the integral $\int_I f u_k w dx$ is finite in virtue of the Schwarz inequality (2.1.7). In this formulation we can also include the cases when I is not bounded, corresponding respectively to Laguerre and Hermite polynomials. Now, even if the functions in $L^2_w(I)$ are not required to be bounded, the exponential decay of the weight w is sufficient to insure the existence of the coefficients in (6.2.2). The reader unfamiliar with these functional spaces, can draw conclusions by thinking in terms of Riemann integrals and continuous functions, and by replacing the norms $\|\cdot\|_{L^2_w(I)}$ and $\|\cdot\|_{C^0(\bar{I})}$, by $\|\cdot\|_w$ and $\|\cdot\|_\infty$ respectively.
The projector $\Pi_{w,n}: L^2_w(I)\to\mathbf{P}_n$, $n\in\mathbf{N}$, is defined in the usual way (see section 2.4). The next proposition fully characterizes the solution of problem (6.2.1).

Theorem 6.2.1 - *For any* $f\in L^2_w(I)$ *and any* $n\in\mathbf{N}$, *there exists a unique polynomial* $\Psi_{w,n}(f)\in\mathbf{P}_n$ *that satisfies* (6.2.1). *Moreover* $\Psi_{w,n}(f)=\Pi_{w,n}f$.

Proof - A polynomial $\psi \in \mathbf{P}_n$ can be written in the form $\psi = \sum_{k=0}^{n} d_k u_k$, for some real coefficients d_k, $0 \le k \le n$. Minimizing $\|f-\psi\|_{L^2_w(I)}$, or equivalently $\|f-\psi\|^2_{L^2_w(I)}$, requires the derivatives

(6.2.3) $$\frac{\partial}{\partial d_j}\|f-\psi\|^2_{L^2_w(I)} = \frac{\partial}{\partial d_j}\left(\|f\|^2_{L^2_w(I)} - 2\sum_{k=0}^{n} d_k (f,u_k)_{L^2_w(I)} + \sum_{k=0}^{n} d_k^2 \|u_k\|^2_{L^2_w(I)}\right) = -2\,(f,u_j)_{L^2_w(I)} + 2d_j\,\|u_j\|^2_{L^2_w(I)}, \qquad 0 \le j \le n.$$

We deduce that the unique minimum is attained when $d_j = c_j$, $0 \le j \le n$, where the c_j's are the Fourier coefficients of f (see (6.2.2)). This ends the proof.

■

In short, we can write

(6.2.4) $$\|f - \Pi_{w,n} f\|_{L^2_w(I)} = \inf_{\psi \in \mathbf{P}_n} \|f-\psi\|_{L^2_w(I)}.$$

Another interesting characterization is given in the following theorem.

Theorem 6.2.2 - *For any* $f \in L^2_w(I)$ *and* $n \in \mathbf{N}$, *we have*

(6.2.5) $$\int_I (f - \Pi_{w,n} f)\phi w \, dx = 0, \qquad \forall \phi \in \mathbf{P}_n.$$

Proof - We fix $\phi \in \mathbf{P}_n$ and define $G : \mathbf{R} \to \mathbf{R}$ by

$$G(\nu) := \|f - \Pi_{w,n} f + \nu\phi\|^2_{L^2_w(I)}.$$

We know from theorem 6.2.1 that $\nu = 0$ is a minimum for G. Therefore

(6.2.6) $$G'(\nu) = 2\int_I (f - \Pi_{w,n} f)\phi w \, dx + 2\nu\|\phi\|^2_{L^2_w(I)}, \qquad \forall \nu \in \mathbf{R}.$$

Imposing that $G'(0) = 0$, we get (6.2.5).

■

By virtue of this theorem, the operator $\Pi_{w,n}$ takes the name of *orthogonal projector*, since the error $f - \Pi_{w,n}f$ is orthogonal to the space $\mathbf{P}_n$. Choosing in particular $\phi = \Pi_{w,n}f$ in (6.2.5), application of the Schwarz inequality leads to

(6.2.7) $$\|\Pi_{w,n}f\|_{L^2_w(I)} \leq \|f\|_{L^2_w(I)}, \qquad \forall n \in \mathbf{N}, \ \ \forall f \in L^2_w(I).$$

The counterpart of relation (6.1.5) is deduced from the next result.

Theorem 6.2.3 - *Let I be bounded, then for any $f \in L^2_w(I)$, we have*

(6.2.8) $$\lim_{n\to+\infty} \|f - \Pi_{w,n}f\|_{L^2_w(I)} = 0.$$

Proof - If $f \in C^0(\bar{I})$, (6.2.4) and (5.3.3) together imply that

(6.2.9) $$\|f-\Pi_{w,n}f\|_{L^2_w(I)} \leq \|f-\Psi_{\infty,n}(f)\|_{L^2_w(I)} \leq \|f-\Psi_{\infty,n}(f)\|_{C^0(\bar{I})} \left(\int_I w\ dx\right)^{\frac{1}{2}}.$$

The last term in (6.2.9) tends to zero in view of (6.1.5).
Now, we prove the statement for a general $f \in L^2_w(I)$. We can find a sequence of functions $\{f_m\}_{m\in\mathbf{N}}$, with $f_m \in C^0(\bar{I})$, $m \in \mathbf{N}$, converging to f in the $L^2_w(I)$ norm. Therefore, using the triangle inequality, we first note that

(6.2.10) $$\|f - \Pi_{w,n}f\|_{L^2_w(I)} \leq \|f - f_m\|_{L^2_w(I)}$$

$$+\|f_m - \Pi_{w,n}f_m\|_{L^2_w(I)} + \|\Pi_{w,n}(f_m - f)\|_{L^2_w(I)}, \qquad \forall m \in \mathbf{N}.$$

Finally, each one of the three terms on the right-hand side of (6.2.10) tends to zero when m and n grow (for the last term we can recall (6.2.7)).

■

When I is not bounded, the proof of the previous theorem does not apply anymore. Nevertheless, the expression (6.2.8) holds also in the case of Laguerre and Hermite polynomials. The proof of this fact is more delicate and we refer to COURANT and HILBERT(1953), Vol.1, page 95, for the details. This means that, for all the orthogonal systems of polynomials here considered, one is allowed to write

$$f = \sum_{k=0}^{\infty} c_k u_k, \qquad \forall f \in L^2_w(I), \tag{6.2.11}$$

where the c_k's are the Fourier coefficients of f and the equality has to be intended almost everywhere. By orthogonality, one also obtains the *Parseval identity*

$$\|f\|^2_{L^2_w(I)} = \sum_{k=0}^{\infty} c_k^2 \, \|u_k\|^2_{L^2_w(I)}, \qquad \forall f \in L^2_w(I). \tag{6.2.12}$$

From (6.2.11) and (6.2.12), one easily deduces that

$$\|f - \Pi_{w,n} f\|_{L^2_w(I)} \le \|f\|_{L^2_w(I)}, \qquad \forall f \in L^2_w(I). \tag{6.2.13}$$

We remark that the convergence of the series (6.2.11) is not in general uniform in I. On the other hand, discontinuous functions can be also approximated. For instance, when f is continuous with the exception of the point $x_0 \in]-1,1[$, then for Legendre expansions one has

$$\sum_{k=0}^{\infty} c_k u_k(x_0) = \frac{1}{2}\left[\lim_{x\to x_0^-} f(x) + \lim_{x\to x_0^+} f(x)\right], \tag{6.2.14}$$

provided the two limits exist.

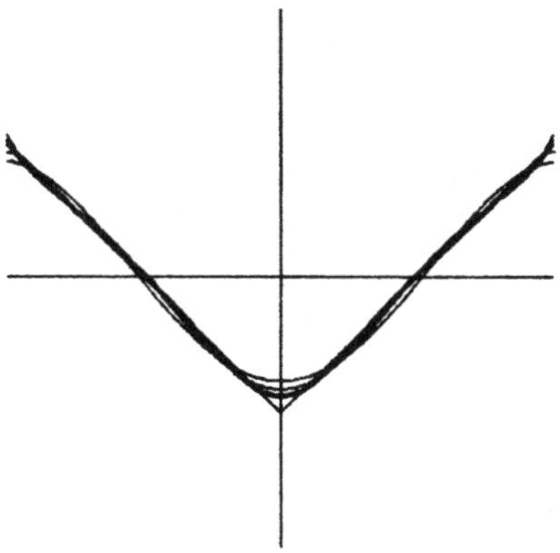

Figure 6.2.1 - Legendre orthogonal projections for $f(x) = |x| - 1/2$.

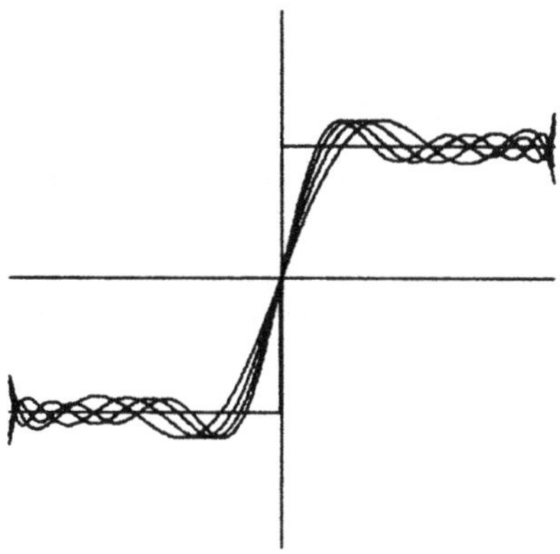

Figure 6.2.2 - Legendre orthogonal projections for $f(x) = -1/2,\ x < 0$, $f(x) = 1/2,\ x \ge 0$.

Figure 6.2.1 shows the behavior of the approximating polynomials $\Pi_{w,n}f$, when w is the Legendre weight function, $n = 4,\ 6,\ 8,\ 10$, and f is defined by $f(x) = |x| - \frac{1}{2}$. A slower rate of convergence is measured at the point $x = 0$, where f presents a singularity in the derivative. In figure 6.2.2, a typical discontinuous function, i.e., $f(x) = \begin{cases} -1/2 & \text{if } x < 0 \\ 1/2 & \text{if } x \geq 0 \end{cases}$, is approximated by $\Pi_{w,n}f$, with $w \equiv 1$ and $n = 7,\ 9,\ 11,\ 13$. In situations like this one, the *Gibbs phenomenon* develops (see COURANT and HILBERT(1953), Vol.1). As a consequence, the oscillations of the approximating polynomials, in the neighborhood of the point $x = 0$, stay bounded but do not damp for increasing n.

Next, we evaluate the rate of convergence of the limit in (6.2.8). This depends on the *regularity* of the function f. The smoother the function, the faster is the convergence. This problem has been widely studied by many authors. The classical approach relates the speed of convergence to the so called *modulus of continuity* of f (see TIMAN(1963)). A more recent approach expresses the smoothness of f in terms of its norm in appropriate Sobolev spaces. We prefer the latter, since it is more suitable for the analysis of differential equations. Results in this direction are given for instance in BABUŠKA, SZABO and KATZ(1981) for the Legendre case. Extensions are considered in CANUTO and QUARTERONI(1980) and CANUTO and QUARTERONI(1982a). Other references are listed in CANUTO, HUSSAINI, QUARTERONI and ZANG(1988). The next theorem summarizes the results for Jacobi type approximations. Thus, $\{\lambda_j = j(j + \alpha + \beta + 1)\}_{j \in \mathbf{N}}$ denotes the sequence of eigenvalues corresponding to the Sturm-Liouville problem (1.3.1). We recall that a and w are related by the formula $a(x) = (1 - x^2)w(x), \quad x \in I =]-1,1[$. According to definition (5.7.1), $f \in H^k_w(I)$ when f and its derivatives of order up to k belong to $L^2_w(I)$.

Theorem 6.2.4 - *Let $k \in \mathbf{N}$, then there exists a constant $C > 0$ such that, for any $f \in H^k_w(I)$, one has*

$$(6.2.15) \qquad \|f - \Pi_{w,n}f\|_{L^2_w(I)} \leq C \left(\frac{1}{n}\right)^k \left\|(1 - x^2)^{k/2} \frac{d^k f}{dx^k}\right\|_{L^2_w(I)}, \quad \forall n > k.$$

Proof - For $k = 0$, (6.2.15) is a trivial consequence of (6.2.13). For $k \geq 1$, let us set $v := (1-x^2)^k w$, $x \in I$, and observe that $L^2_w(I) \subset L^2_v(I)$, since $v \leq w$. This implies that $f' \in L^2_w(I) \subset L^2_a(I)$ admits the representation

(6.2.16)
$$f' = \sum_{j=1}^{\infty} c_j \, u'_j, \qquad a.e. \text{ in } I,$$

where the c_j's are the Fourier coefficients of f. To show (6.2.16), we recall that $\{u'_j\}_{j\geq 1}$, up to normalizing factors, is an orthogonal basis of polynomials with respect to the weight function a (see (1.3.6) and (2.2.8)). Therefore, the Fourier coefficients d_j, $j \geq 1$, of f' with respect to this new basis, are

(6.2.17)
$$d_j := \frac{\int_I f' u'_j a \, dx}{\|u'_j\|^2_{L^2_a(I)}} = - \frac{\int_I f (u'_j a)' \, dx}{\|u'_j\|^2_{L^2_a(I)}}$$

$$= \lambda_j \frac{\int_I f u_j w \, dx}{\|u'_j\|^2_{L^2_a(I)}} = \frac{\int_I f u_j w \, dx}{\|u_j\|^2_{L^2_w(I)}} = c_j, \qquad j \geq 1.$$

Hence we get (6.2.16). In (6.2.17) we first integrated by parts (recalling that $a(\pm 1) = 0$), then we used the fact that $\{u_j\}_{j\in\mathbf{N}}$ are the solutions of a Sturm-Liouville problem, and finally we used relation (2.2.15). For a correct justification of the equalities in (6.2.17), one should first argue with $f \in C^1(\bar{I})$ and then approximate a general $f \in H^1_w(I)$ by a sequence of functions in $C^1(\bar{I})$.
In a similar way, we expand $\frac{d^k f}{dx^k} \in L^2_v(I)$ in series of the polynomial basis $\left\{\frac{d^k u_j}{dx^k}\right\}_{j\geq k}$, orthogonal with respect to the weight function v. The elements of this basis are solutions of a Sturm-Liouville problem with eigenvalues $(j-k)(j+\alpha+\beta+1+k)$, $j \geq k$ (see (1.3.6)). This yields

(6.2.18)
$$\frac{d^k f}{dx^k} = \sum_{j=k}^{\infty} c_j \frac{d^k u_j}{dx^k}, \qquad a.e. \text{ in } I.$$

Finally, repeated application of (2.2.15) to successive derivatives of u_j, $j \geq 1$, gives for $n > k$:

$$\text{(6.2.19)} \qquad \|f - \Pi_{w,n}f\|^2_{L^2_w(I)} = \sum_{j=n+1}^{\infty} c_j^2 \, \|u_j\|^2_{L^2_w(I)}$$

$$= \sum_{j=n+1}^{\infty} c_j^2 \left(\prod_{i=0}^{k-1} [(j-i)(j+\alpha+\beta+1+i)]^{-1} \right) \left\| \frac{d^k u_j}{dx^k} \right\|^2_{L^2_v(I)}$$

$$\leq \left(\max_{j \geq n+1} \left\{ \prod_{i=0}^{k-1} [(j-i)(j+\alpha+\beta+1+i)]^{-1} \right\} \right) \sum_{j=k}^{\infty} c_j^2 \left\| \frac{d^k u_j}{dx^k} \right\|^2_{L^2_v(I)}$$

$$\leq \left(\prod_{i=0}^{k-1} [(n+1-i)(n+\alpha+\beta+2+i)]^{-1} \right) \left\| \frac{d^k f}{dx^k} \right\|^2_{L^2_v(I)}$$

$$\leq C^2 \, n^{-2k} \left\| (1-x^2)^{k/2} \frac{d^k f}{dx^k} \right\|^2_{L^2_w(I)},$$

where C depends on α, β and k. Thus, we obtained (6.2.15).

■

The hypotheses of the previous theorem can be weakened by requiring only that f satisfies $\frac{d^m f}{dx^m}(1-x^2)^{m/2} \in L^2_w(I)$, $0 \leq m \leq k$.
A proof based on similar arguments has been provided in BERNARDI and MADAY(1989) for ultraspherical polynomials. As pointed out in GOTTLIEB and ORSZAG(1977), p.33, a strict requirement to obtain estimates like (6.2.15), is that the polynomial basis considered derives from a Sturm-Liouville problem, which is singular at the boundary points of the domain (see section 1.1).
Equations (6.2.16) and (6.2.17) lead to the following remarkable relation, $\forall n \geq 1$:

$$\text{(6.2.20)} \qquad (\Pi_{w,n}f)' = \Pi_{a,n-1}f', \qquad \forall f \in L^2_w(I) \cap H^1_a(I).$$

Inequality (6.2.15) is a typical example of a *spectral error estimate*. The rate of convergence depends only on the smoothness of the function f. In contrast to other approximation techniques, for instance those based on *splines* where the rate of convergence is controlled by the regularity of the approximating functions in the domain,

the approximating polynomials, being analytic, do not affect the convergence behavior. In comparison with other methods, smaller values of the parameter n are in general sufficient to guarantee a satisfactory approximation error.
Derivatives of the error $f - \Pi_{w,n} f$ can be also estimated, which result in a similar convergence behavior. Nevertheless, the rate of convergence is slower than that exhibited by other projection operators which are introduced later on. The reader interested in this field is referred to CANUTO and QUARTERONI(1982a), where other error estimates concerning Chebyshev and Legendre weights are taken into account. With techniques related to interpolation spaces (see section 5.7), formula (6.2.15) can be suitably generalized to Sobolev spaces with real exponent. For other best approximation polynomials in more general functional spaces (for instance $L^p_w(I)$ spaces, $p > 1$) the reader is addressed to DITZIAN and TOTIK(1987) and the references therein. Chebyshev expansions for functions with singularities at the endpoints of the interval $[-1, 1]$ are studied in BOYD(1981).

The same kind of result can be established for other systems of orthogonal polynomials. Laguerre polynomials require a little care. Here $\lambda_j = j$, $j \in \mathbf{N}$, and $a(x) = xw(x) = x^{\alpha+1}e^{-x}$, $\alpha > -1$, $x \in I =]0, +\infty[$. The inequality $a(x) \leq w(x)$ only holds for $x \in]0, 1] \subset I$, which implies that $L^2_w(I) \not\subset L^2_a(I)$. Nevertheless, the proof of theorem 6.2.4 is still applicable and gives the next result.

Theorem 6.2.5 - *Let $k \in \mathbf{N}$. Then there exists a constant $C > 0$ such that, for any f satisfying $\frac{d^m f}{dx^m} x^{m/2} \in L^2_w(I)$, $0 \leq m \leq k$, one has*

$$\|f - \Pi_{w,n} f\|_{L^2_w(I)} \leq C \left(\frac{1}{\sqrt{n}}\right)^k \left\| x^{k/2} \frac{d^k f}{dx^k} \right\|_{L^2_w(I)}, \qquad \forall n > k. \tag{6.2.21}$$

■

Further estimates are given in MADAY, PERNAUD-THOMAS and VANDEVEN(1985) for the case $\alpha = 0$. In that paper, f is assumed to belong to the weighted space $H^k_\omega(I)$, with $\omega := e^{-(1-\epsilon)x}$, $\epsilon > 0$ arbitrarily small. In this situation, we obtain the inclusion $H^1_\omega(I) \subset L^2_w(I) \cap H^1_a(I)$.

A proof identical to that of theorem 6.2.4 holds for Hermite polynomials, where $\lambda_j = 2j$, $j \in \mathbf{N}$, and $a(x) = w(x) = e^{-x^2}$, $x \in \mathbf{R}$.

Theorem 6.2.6 - *Let $k \in \mathbf{N}$. Then there exists a constant $C > 0$ such that, for any $f \in H_w^k(\mathbf{R})$, one has*

$$(6.2.22) \qquad \|f - \Pi_{w,n}f\|_{L_w^2(\mathbf{R})} \leq C \left(\frac{1}{\sqrt{n}}\right)^k \left\|\frac{d^k f}{dx^k}\right\|_{L_w^2(\mathbf{R})}, \qquad \forall n > k.$$

■

In particular, (6.2.20) takes the form

$$(6.2.23) \qquad (\Pi_{w,n}f)' = \Pi_{w,n-1}f', \qquad \forall f \in H_w^1(\mathbf{R}),\ \forall n \geq 1.$$

From the examples illustrated here, it is clear that the speed of convergence to zero of the error is strictly related to the eigenvalues λ_j, $j \in \mathbf{N}$ (i.e., the *spectrum* of the Sturm-Liouville problem (1.1.1)). Somehow, this justifies the adoption of the adjective *spectral*, commonly used in the frame of these approximation methods.

6.3 Inverse inequalities

Inverse inequalities are an effective tool for the analysis of convergence in approximation theory. They are based on the fact that, in finite dimensional spaces, two given norms are always equivalent. Thus, we are allowed to give a bound to *strong* norms by using *weaker* norms. The drawback is that the constants in the equivalence relation, depend on the dimension of the space considered, so that it is not possible to extrapolate the same inequalities at the limit. Classical examples in finite element method are given for instance in CIARLET(1978), p.140.
Let us start with the Jacobi case (i.e., $a(x) = (1 - x^2)w(x)$, $x \in I =]-1, 1[$).

Lemma 6.3.1 - *We can find a constant $C > 0$ such that, for any $n \geq 1$*

$$\|p\|_{L^2_w(I)} \leq Cn\, \|p\|_{L^2_a(I)}, \qquad \forall p \in \mathbf{P}_n. \tag{6.3.1}$$

Proof - Let $m = n + 2$. Since $(1 - x^2)p^2 \in \mathbf{P}_{2m-2}$, formula (3.4.1) and theorem 3.4.1 lead to

$$\int_I p^2 w\, dx = \sum_{j=1}^{m} p^2(\xi_j^{(m)})\, w_j^{(m)}, \qquad \forall p \in \mathbf{P}_n, \tag{6.3.2}$$

$$\int_I p^2 a\, dx = \sum_{j=1}^{m} p^2(\xi_j^{(m)})(1 - [\xi_j^{(m)}]^2)\, w_j^{(m)}, \qquad \forall p \in \mathbf{P}_n, \tag{6.3.3}$$

where $\xi_j^{(m)}$, $1 \leq j \leq m$, are the zeroes of $P_m^{(\alpha,\beta)}$.

Now, we can easily conclude the proof by noting that there exists a constant $C^* > 0$, depending only on α and β, such that, for any $n \geq 1$

$$1 - [\xi_j^{(m)}]^2 \geq \frac{C^*}{m^2} \geq \frac{C^*}{9n^2}, \qquad 1 \leq j \leq m. \tag{6.3.4}$$

In fact, from (1.3.2) and (3.1.17), the straight-line tangent to $P_m^{(\alpha,\beta)}$ at the point $x = 1$ vanishes at $\hat{x} := 1 - \frac{2(\alpha+1)}{m(m+\alpha+\beta+1)}$. It is easy to see that $P_m^{(\alpha,\beta)}$ is convex in the interval $[\xi_m^{(m)}, 1]$, hence $\hat{x}$ is an upper bound for the zeroes of $P_m^{(\alpha,\beta)}$. A similar argument is valid near the point $x = -1$ which proves (6.3.4).

■

Relation (6.3.1) is an example of inverse inequality. Actually, the norm in $L^2_w(I)$ is in general bigger than the norm in $L^2_a(I)$, because $w \geq a$, but the inequality in the opposite direction holds in the finite dimensional space $\mathbf{P}_n$, for any fixed $n \in \mathbf{N}$.

More interesting cases are obtained for inverse inequalities that involve derivatives of polynomials. Examples have been given in section 2.5. Using Sobolev norms, we can establish similar results.

Theorem 6.3.2 - *We can find a constant $C > 0$ such that, for any $n \geq 1$*

$$\|p'\|_{L^2_a(I)} \leq Cn\, \|p\|_{L^2_w(I)}, \qquad \forall p \in \mathbf{P}_n, \tag{6.3.5}$$

$$(6.3.6) \qquad \|p'\|_{L^2_w(I)} \leq Cn^2 \|p\|_{L^2_w(I)}, \qquad \forall p \in \mathbf{P}_n.$$

Proof - We set $p = \sum_{k=0}^{n} c_k u_k$, where $\{u_k\}_{k\in\mathbf{N}}$ is the sequence of Jacobi polynomials relative to the weight function w. From relations (2.2.15) and (2.2.8), we get

$$(6.3.7) \qquad \|p'\|^2_{L^2_a(I)} = \sum_{k=1}^{n} c_k^2 \|u_k'\|^2_{L^2_a(I)}$$

$$= \sum_{k=1}^{n} c_k^2 \lambda_k \|u_k\|^2_{L^2_w(I)} \leq \lambda_n \sum_{k=0}^{n} c_k^2 \|u_k\|^2_{L^2_w(I)} \leq C^2 n^2 \|p\|^2_{L^2_w(I)},$$

where we observed that the eigenvalue λ_n behaves like n^2. To obtain (6.3.6), we first apply lemma 6.3.1 to the polynomial p' and then use (6.3.5).

■

As a byproduct of the previous theorem, it is not difficult to get, for any $n \geq 1$

$$(6.3.8) \qquad \|p\|_{H^m_w(I)} \leq Cn^{2(m-k)} \|p\|_{H^k_w(I)}, \qquad \forall p \in \mathbf{P}_n, \quad \forall m, k \in \mathbf{N},\ m \geq k.$$

When w is the Legendre weight function, relation (6.3.8) is the Schmidt inequality (see BELLMAN(1944)). Using different arguments, (6.3.8) was established in CANUTO and QUARTERONI (1982a) for Chebyshev weights.

We study now the Laguerre case ($a(x) = xw(x) = x^{\alpha+1}e^{-x}$, $\alpha > -1$, $x \in I =]0, +\infty[$).

Lemma 6.3.3 - *We can find two constants $C_1, C_2 > 0$ such that, for any $n \geq 1$*

$$(6.3.9) \qquad \frac{C_1}{\sqrt{n}} \|p\|_{L^2_a(I)} \leq \|p\|_{L^2_w(I)} \leq C_2\sqrt{n}\, \|p\|_{L^2_a(I)}, \qquad \forall p \in \mathbf{P}_n.$$

Proof - As done in the proof of lemma 6.3.1, let $\xi_j^{(m)}$, $1 \leq j \leq m$, be the zeroes of $L_m^{(\alpha)}$, $m = n+1$. Then one has the inequality $C_1^* \frac{1}{n} \leq \xi_j^{(m)} \leq C_2^* n$, $1 \leq j \leq m$ (see section 3.1), where $C_1^*, C_2^* > 0$ do not depend on n. Now we can easily deduce (6.3.9).

■

Using the same proof of theorem 6.3.2, we also deduce the the following proposition.

Theorem 6.3.4 - *We can find a constant $C > 0$ such that, for any $n \geq 1$*

(6.3.10) $$\|p'\|_{L^2_a(I)} \leq C\sqrt{n}\ \|p\|_{L^2_w(I)}, \qquad \forall p \in \mathbf{P}_n,$$

(6.3.11) $$\|p'\|_{L^2_w(I)} \leq Cn\ \|p\|_{L^2_w(I)}, \qquad \forall p \in \mathbf{P}_n.$$

■

Finally, we are left with the Hermite case (i.e., $a(x) = w(x) = e^{-x^2}$, $x \in \mathbf{R}$). Since we have the identity $a \equiv w$, lemma 6.3.1 is not useful. We also have the counterpart of theorem 6.3.2, which states:

Theorem 6.3.5 - *We can find a constant $C > 0$ such that, for any $n \geq 1$*

(6.3.12) $$\|p'\|_{L^2_w(\mathbf{R})} \leq C\sqrt{n}\ \|p\|_{L^2_w(\mathbf{R})}, \qquad \forall p \in \mathbf{P}_n.$$

Proof - We argue exactly as in (6.3.7).

■

The problem of giving a bound to the norm in $C^0(\bar{I})$, by means of the norm in $L^2_w(I)$, has arisen in section 3.9, where other inverse inequalities are presented.

6.4 Other projection operators

We focus our attention on theorem 6.2.2. As we can see from the next proposition, the expression (6.2.5) can actually be used to define the polynomial $\Pi_{w,n}f$.

Theorem 6.4.1 - *For a given* $f \in L^2_w(I)$, *let* $\psi \in \mathbf{P}_n$ *such that*

(6.4.1) $$(f-\psi,\phi)_{L^2_w(I)} = 0, \qquad \forall \phi \in \mathbf{P}_n.$$

Then we have $\psi = \Pi_{w,n} f$.

Proof - We can write $\psi = \sum_{k=0}^n d_k u_k$ for some coefficients d_k, $0 \le k \le n$. For any $0 \le j \le n$, choosing $\phi = u_j$ as *test function*, (6.4.1) yields

(6.4.2) $$(f,u_j)_{L^2_w(I)} = \sum_{k=0}^{n} d_k \, (u_k,u_j)_{L^2_w(I)} = d_j \, \|u_j\|^2_{L^2_w(I)}, \qquad 0 \le j \le n.$$

Thus, by (6.2.2), the d_k's are the first $n+1$ Fourier coefficients of f. Hence $\psi = \Pi_{w,n} f$.

■

This result is the starting point for the definition of new projection operators. For any $m \in \mathbf{N}$ and $n \ge m$, we introduce the operator $\Pi^m_{w,n} : H^m_w(I) \to \mathbf{P}_n$, such that

(6.4.3) $$(f - \Pi^m_{w,n} f, \phi)_{H^m_w(I)} = 0, \qquad \forall \phi \in \mathbf{P}_n.$$

The inner product in $H^m_w(I)$ has been defined in (5.6.2) and (5.7.2). Of course we have $\Pi^0_{w,n} = \Pi_{w,n}$. One easily verifies that the polynomial $\Pi^m_{w,n} f \in \mathbf{P}_n$ is uniquely determined and represents the orthogonal projection of f in the $H^m_w(I)$ norm. One checks that this projection satisfies a minimization problem like (6.2.4), namely

(6.4.4) $$\|f - \Pi^m_{w,n} f\|_{H^m_w(I)} = \inf_{\psi \in \mathbf{P}_n} \|f - \psi\|_{H^m_w(I)}.$$

Moreover, we have for $n \ge m$

(6.4.5) $$\|\Pi^m_{w,n} f\|_{H^m_w(I)} \le \|f\|_{H^m_w(I)}, \qquad \forall f \in H^m_w(I).$$

As before, the expression in (6.4.4) tends to zero when $n \to +\infty$, with a rate depending on the degree of smoothness of f. For simplicity, we restrict the analysis to the case $m = 1$ and we just consider the Jacobi weight $w(x) = (1-x)^\alpha (1+x)^\beta$, $x \in I =]-1,1[$.

Theorem 6.4.2 - *Let* $-1 < \alpha < 1$, $-1 < \beta < 1$ *and* $k \geq 1$, *then we can find a constant* $C > 0$ *such that, for any* $f \in H^k_w(I)$, *one has*

$$(6.4.6) \qquad \|f - \Pi^1_{w,n} f\|_{H^1_w(I)} \leq C \left(\frac{1}{n}\right)^{k-1} \left\| (1-x^2)^{(k-1)/2} \frac{d^k f}{dx^k} \right\|_{L^2_w(I)}, \qquad \forall n > k.$$

Proof - We first consider $f \in C^\infty(\bar{I})$. Let $x_0 \in I$ and set

$$(6.4.7) \qquad q_n(x) := \left[f(x_0) + \int_{x_0}^{x} (\Pi_{w,n-1} f')(t)\, dt \right] \in \mathbf{P}_n, \quad n \geq 1.$$

From (6.4.4) we can write

$$(6.4.8) \qquad \|f - \Pi^1_{w,n} f\|_{H^1_w(I)} \leq \|f - q_n\|_{H^1_w(I)}$$

$$= \left(\|f - q_n\|^2_{L^2_w(I)} + \|f' - \Pi_{w,n-1} f'\|^2_{L^2_w(I)} \right)^{\frac{1}{2}}.$$

On the other hand, we have by the Schwarz inequality

$$(6.4.9) \qquad \|f - q_n\|^2_{L^2_w(I)} = \int_I \left(f(x) - f(x_0) - \int_{x_0}^{x} (\Pi_{w,n-1} f')(t) dt \right)^2 w(x) dx$$

$$= \int_I \left(\int_{x_0}^{x} (f' - \Pi_{w,n-1} f')(t) \sqrt{w(t)}\, \frac{dt}{\sqrt{w(t)}} \right)^2 w(x)\, dx$$

$$\leq \int_I \left(\left| \int_{x_0}^{x} (f' - \Pi_{w,n-1} f')^2(t)\, w(t)\, dt \right| \cdot \left| \int_{x_0}^{x} w^{-1}(t)\, dt \right| \right) w(x)\, dx$$

$$\leq \|f' - \Pi_{w,n-1} f'\|^2_{L^2_w(I)} \int_I \left| \int_{x_0}^{x} w^{-1}(t) dt \right| w(x)\, dx.$$

We note that the last integral in (6.4.9) is finite when $-1 < \alpha < 1$ and $-1 < \beta < 1$. At this point we can conclude after substituting (6.4.9) in (6.4.8) and recalling theorem 6.2.4.

The proof is completed by approximating $f \in H_w^k(I)$ with a sequence of functions in $C^\infty(\bar{I})$ (see section 5.7).

■

The proof of the above statement was formerly given in MADAY and QUARTERONI(1981) for Legendre and Chebyshev weights. The extension to other ultraspherical weights is provided in BERNARDI and MADAY(1989). In those papers, one also finds estimates of the same error in other norms. For instance, when $-1 < \alpha = \beta < 1$ and $k \geq 1$, there exists a constant $C > 0$ such that, for any $f \in H_w^k(I)$

$$\text{(6.4.10)} \qquad \|f - \Pi_{w,n}^1 f\|_{L_w^2(I)} \leq C \left(\frac{1}{n}\right)^k \|f\|_{H_w^k(I)}, \qquad \forall n > k.$$

Suitable projections can be also defined in $H_{0,w}^1(I) \subset H_w^1(I)$ (see definition (5.7.6)). First of all we introduce the space

$$\text{(6.4.11)} \qquad \mathbf{P}_n^0 := \left\{ p \in \mathbf{P}_n \;\middle|\; p(\pm 1) = 0 \right\}, \qquad n \geq 2.$$

Assuming that $-1 < \alpha < 1$ and $-1 < \beta < 1$, we examine two projection operators. For any $n \geq 2$, we consider $\Pi_{0,w,n}^1 : H_{0,w}^1(I) \to \mathbf{P}_n^0$ and $\hat{\Pi}_{0,w,n}^1 : H_{0,w}^1(I) \to \mathbf{P}_n^0$ such that, for $f \in H_{0,w}^1(I)$, we have

$$\text{(6.4.12)} \qquad \int_I (f - \Pi_{0,w,n}^1 f)' \, \phi' w \, dx = 0, \qquad \forall \phi \in \mathbf{P}_n^0,$$

$$\text{(6.4.13)} \qquad \int_I (f - \hat{\Pi}_{0,w,n}^1 f)' (\phi w)' \, dx = 0, \qquad \forall \phi \in \mathbf{P}_n^0.$$

Existence and uniqueness for $\Pi_{0,w,n}^1 f$ are obtained by recalling definition (5.7.7) and arguing as in theorem 6.2.1. Furthermore, it is easy to verify

$$\text{(6.4.14)} \qquad \|f - \Pi_{0,w,n}^1 f\|_{H_{0,w}^1(I)} = \inf_{\psi \in \mathbf{P}_n^0} \|f - \psi\|_{H_{0,w}^1(I)}.$$

Error estimates for $\Pi_{0,w,n}^1$ can be inferred from estimates of other operators, as illustrated by the following proposition.

Theorem 6.4.3 - *There exists a constant $C > 0$ such that, for any $f \in H^1_{0,w}(I)$, we have*

$$\|f - \Pi^1_{0,w,n} f\|_{H^1_{0,w}(I)} \leq C \, \|f - \Pi^1_{w,n} f\|_{H^1_w(I)}, \qquad \forall n \geq 2. \tag{6.4.15}$$

Proof - We define

$$\psi(x) := (\Pi^1_{w,n} f)(x) + \tfrac{1}{2}\Big[(1+x)(f - \Pi^1_{w,n} f)(1) + (1-x)(f - \Pi^1_{w,n} f)(-1)\Big], \quad x \in I. \tag{6.4.16}$$

Recalling that $f(\pm 1) = 0$, we have $\psi \in \mathbf{P}^0_n$. Hence, substituting in (6.4.14), we get

$$\|f - \Pi^1_{0,w,n} f\|_{H^1_{0,w}(I)} \leq \|f - \Pi^1_{w,n} f\|_{H^1_{0,w}(I)} \tag{6.4.17}$$

$$+ \tfrac{1}{2} \left\| \frac{d}{dx}\Big[(1+x)(f - \Pi^1_{w,n} f)(1) + (1-x)(f - \Pi^1_{w,n} f)(-1)\Big] \right\|_{L^2_w(I)}$$

$$\leq \|f - \Pi^1_{w,n} f\|_{H^1_w(I)} + C^* \, \|f - \Pi^1_{w,n} f\|_{C^0(\bar{I})} \, ,$$

where $C^* > 0$ is a constant. Now, recalling (5.7.4), we immediately conclude the proof. ■

The analysis of the operator $\hat{\Pi}^1_{0,w,n}$ is more involved and will be considered in section 9.4. Of course, when $w \equiv 1$ is the Legendre weight function, the relations (6.4.12) and (6.4.13) are identical.

6.5 Convergence of the Gaussian formulas

High order integration formulas were introduced in sections 3.4, 3.5 and 3.6. Here we consider their asymptotic properties for increasing number of nodes. In the case of Jacobi weights, we begin with the corollary of a convergence result due to Stekloff and Fejér (see SZEGÖ(1939), p.342).

Theorem 6.5.1 - *Let f be such that fw is Riemann integrable in $I=]-1,1[$. Then we have*

(6.5.1) $$\lim_{n\to+\infty}\sum_{j=1}^{n} f(\xi_j^{(n)})\ w_j^{(n)} = \lim_{n\to+\infty}\int_I I_{w,n}f\ w\ dx = \int_I fw\ dx.$$

Proof - We show a simplified proof for $f \in C^0(\bar{I})$. The general case is studied in STEKLOFF(1916). We use the polynomial of best uniform approximation of f (see section 6.1). For any $n \geq 1$, taking into account that $[\Psi_{\infty,n-1}(f) - I_{w,n}f] \in \mathbf{P}_{n-1}$, we have

(6.5.2)
$$\left|\int_I (f - I_{w,n}f)w dx\right| \leq \left|\int_I (f - \Psi_{\infty,n-1}(f))w dx\right| + \left|\sum_{j=1}^{n}(\Psi_{\infty,n-1}(f) - I_{w,n}f)(\xi_j^{(n)})\ w_j^{(n)}\right|$$

$$\leq \|f - \Psi_{\infty,n-1}(f)\|_{C^0(\bar{I})}\left(\int_I w\ dx + \sum_{j=1}^{n} w_j^{(n)}\right) = 2\ \|f - \Psi_{\infty,n-1}(f)\|_{C^0(\bar{I})}\int_I w\ dx,$$

where we made use of $I_{w,n}f(\xi_j^{(n)}) = f(\xi_j^{(n)})$, $1 \leq j \leq n$. Now, because of (6.1.5), the last term in (6.5.2) tends to zero. Moreover, when f is smooth, we can establish the rate of convergence using (6.1.7).

■

For Laguerre and Hermite quadrature formulas we can state similar propositions (see USPENSKY(1928) and DAVIS and RABINOWITZ(1984)).

Theorem 6.5.2 - *Let f be such that fw is Riemann integrable in $I=]0,+\infty[$, where w is the Laguerre weight function. Assume the existence of $x_0 \in I$ and $\epsilon > 0$ such that $|f(x)w(x)| \leq x^{-1-\epsilon}$, $\forall x > x_0$. Under these assumptions* (6.5.1) *holds.*

■

Theorem 6.5.3 - *Let f be such that fw is Riemann integrable in $I = \mathbf{R}$, where w is the Hermite weight function. Assume the existence of $x_0 \in I$ and $\epsilon > 0$ such that $|f(x)w(x)| \leq x^{-1-\epsilon}$, $\forall |x| > x_0$. Under these assumptions* (6.5.1) *holds.*

■

The rate of convergence in (6.5.1) is measured by the degree of smoothness of the function f. If f fulfills certain regularity requirements, we can analytically express the error between the integral and the quadrature. The proof of the following theorem does not present difficulties and is given for instance in TODD(1963).

Theorem 6.5.4 - *Let $n \geq 1$ and $f \in C^{2n}(\bar{I})$. Let us define*

$$E_n(f) := \int_I fw \, dx - \sum_{j=1}^{n} f(\xi_j^{(n)}) \, w_j^{(n)}. \tag{6.5.3}$$

Then we can find $\xi \in I$ such that

(6.5.4) (*Jacobi*)

$$E_n(f) = \left[\frac{2^{2n+\alpha+\beta+1} \, n! \, \Gamma(n+\alpha+1) \, \Gamma(n+\beta+1)}{(2n)! \, \Gamma(2n+\alpha+\beta+1)} \times \frac{\Gamma(n+\alpha+\beta+1)}{\Gamma(2n+\alpha+\beta+2)} \right] \frac{d^{2n}f}{dx^{2n}}(\xi), \quad \alpha > -1, \; \beta > -1,$$

(6.5.5) (*Legendre*)

$$E_n(f) = \frac{2^{2n+1} \, [n!]^4}{(2n+1) \, [(2n)!]^3} \, \frac{d^{2n}f}{dx^{2n}}(\xi),$$

(6.5.6) (*Chebyshev*)

$$E_n(f) = \frac{\pi}{2^{2n-1}(2n)!} \, \frac{d^{2n}f}{dx^{2n}}(\xi),$$

(6.5.7) (*Laguerre*)

$$E_n(f) = \frac{n! \, \Gamma(n+\alpha+1)}{(2n)!} \, \frac{d^{2n}f}{dx^{2n}}(\xi), \quad \alpha > -1,$$

(6.5.8) (*Hermite*)

$$E_n(f) = \frac{n! \, \sqrt{\pi}}{2^n \, (2n)!} \, \frac{d^{2n}f}{dx^{2n}}(\xi).$$

■

Concerning Gauss-Lobatto formulas, we can recover a convergence result for $f \in C^0(\bar{I})$, $I =]-1,1[$, with the same proof given for theorem 6.5.1. A more general theorem, based on weaker assumptions on f, also holds in the Jacobi case:

Theorem 6.5.5 - *Let $f : \bar{I} \to \mathbf{R}$, such that fw is Riemann integrable. Then we have*

$$\lim_{n \to +\infty} \sum_{j=0}^{n} f(\eta_j^{(n)}) \, \tilde{w}_j^{(n)} = \lim_{n \to +\infty} \int_I \tilde{I}_{w,n} f \, w \, dx = \int_I fw \, dx. \tag{6.5.9}$$

Proof - We first recall the equality

(6.5.10) $$\tilde{I}_{w,n}f \;=\; [I_{a,n-1}(fq^{-1})]\,q \;+\; f(-1)\,\tilde{l}_0^{(n)} \;+\; f(1)\,\tilde{l}_n^{(n)}, \qquad n \geq 1,$$

where $q(x) := (1-x^2)$, $x \in I$, $a = qw$, and the Lagrange polynomials are defined in (3.2.8). To check (6.5.10) it is sufficient to note that the relation holds at the points $\eta_j^{(n)}$, $0 \leq j \leq n$. Actually, due to (1.3.6), we have $P_{n-1}^{(\alpha+1,\beta+1)}(\eta_j^{(n)}) = 0$, $1 \leq j \leq n-1$. Hence, the $\eta_j^{(n)}$'s are the nodes of a Gauss formula where α and β are respectively replaced by $\alpha+1$ and $\beta+1$. Therefore $(\tilde{I}_{w,n}g)(\eta_j^{(n)}) = (I_{a,n-1}g)(\eta_j^{(n)}) = g(\eta_j^{(n)})$, $1 \leq j \leq n-1$, for any function g. The verification of (6.5.10) at the points $x = \pm 1$ is trivial. Multiplying both sides of (6.5.10) by w and integrating in I, one gets

(6.5.11) $$\lim_{n\to+\infty} \int_I [I_{a,n-1}(fq^{-1})]\,qw\,dx \;=\; \int_I fq^{-1}a\,dx \;=\; \int_I fw\,dx,$$

(6.5.12) $$\lim_{n\to+\infty} \int_I \tilde{l}_j^{(n)} w\,dx \;=\; \lim_{n\to+\infty} \tilde{w}_j^{(n)} \;=\; 0, \qquad j=0 \text{ or } j=n.$$

The first limit is a consequence of theorem 6.5.1, and (6.5.12) is obtained by studying the behavior for $n \to +\infty$ of the weights in (3.5.2).

■

A statement like that of theorem 6.5.4 also exists for Gauss-Lobatto and Gauss-Radau formulas. We mention DAVIS and RABINOWITZ(1984) for the treatment of the Legendre case and GHIZZETTI and OSSICINI(1970) for a more general discussion. Estimates in the Legendre case are also analyzed in DITZIAN and TOTIK(1987), p.87.

In the case of Laguerre Gauss-Radau formulas (see (3.6.1)), we obtain a proposition like theorem 6.5.2. This time the proof is based on the following relation:

(6.5.13) $$\tilde{I}_{w,n}f \;=\; [I_{a,n-1}(fq^{-1})]\,q \;+\; f(0)\,\tilde{l}_0^{(n)}, \qquad n \geq 1,$$

where $q(x) = x$, $x \in]0,+\infty[$, $a = qw$, and $\tilde{l}_0^{(n)}$ is given in (3.2.11).

6.6 Estimates for the interpolation operator

The interpolation operators $I_{w,n}$ and $\tilde{I}_{w,n}$ have been introduced in section 3.3. In this section, we are concerned with estimating the rate of convergence to zero of the error between a function and its interpolant. We first study the case of ultraspherical polynomials ($\nu := \alpha = \beta$).

Theorem 6.6.1 - *Let* $k \geq 1$ *and* $-1 < \nu \leq 0$. *Then we can find a constant* $C > 0$ *such that, for any* $f \in H_w^k(I)$, *we have*

$$\text{(6.6.1)} \qquad \|f - I_{w,n}f\|_{L_w^2(I)} \leq C \left(\frac{1}{n}\right)^{k-1/2} \left\|(1-x^2)^{-\nu/2} \frac{d^k f}{dx^k}\right\|_{L_w^2(I)}, \qquad \forall n \geq k.$$

The same relation holds for the error $f - \tilde{I}_{w,n}f$.

Proof - We follow the proof of theorem 6.5.1. One has

$$\text{(6.6.2)} \qquad \|f - I_{w,n}f\|_{L_w^2(I)} \leq \|f - \Psi_{\infty,n-1}(f)\|_{L_w^2(I)}$$

$$+ \left(\sum_{j=1}^{n} (\Psi_{\infty,n-1}(f) - I_{w,n}f)^2(\xi_j^{(n)})\, w_j^{(n)}\right)^{\frac{1}{2}} \leq 2\, \|f - \Psi_{\infty,n-1}(f)\|_{C^0(\bar{I})} \left(\int_I w\, dx\right)^{\frac{1}{2}}.$$

Thus, recalling (6.1.7), we can find a constant $C^* > 0$ such that

$$\text{(6.6.3)} \qquad \|f - I_{w,n}f\|_{L_w^2(I)} \leq C^* \left(\frac{1}{n}\right)^{k-1} \sup_{|x_1-x_2|<2/n} \left|\frac{d^{k-1}f}{dx^{k-1}}(x_1) - \frac{d^{k-1}f}{dx^{k-1}}(x_2)\right|$$

$$= C^* \left(\frac{1}{n}\right)^{k-1} \sup_{|x_1-x_2|<2/n} \left|\int_{x_1}^{x_2} \frac{d^k f}{dx^k}\, dx\right|$$

$$\leq C^* \left(\frac{1}{n}\right)^{k-1} \left(\int_I \left[\frac{d^k f}{dx^k}\right]^2 dx\right)^{\frac{1}{2}} \sup_{|x_1-x_2|<2/n} \left(\left|\int_{x_1}^{x_2} dx\right|\right)^{\frac{1}{2}}$$

$$= \sqrt{2}\, C^* \left(\frac{1}{n}\right)^{k-1/2} \left(\int_I (1-x^2)^{-\nu} \left[\frac{d^k f}{dx^k}\right]^2 w\, dx\right)^{\frac{1}{2}},$$

where we used the Schwarz inequality. This proves the statement. Concerning the error $f - \tilde{I}_{w,n}f$, we apply the same arguments. This time, the quadrature formula is not exact for polynomials in $\mathbf{P}_{2n}$. However, we obtain an inequality like (6.6.2), thanks to theorem 3.8.2.

■

For the Chebyshev case ($\nu = -1/2$), a sharper estimate of the error is readily available

$$\|f - I_{w,n}f\|_{L^2_w(I)} \leq C\left(\frac{1}{n}\right)^k \|f\|_{H^k_w(I)}, \tag{6.6.4}$$

where $k \geq 1$ and $w(x) = 1/\sqrt{1-x^2}$, $x \in I$. The proof of this inequality relies on the possibility of using results in approximation theory for trigonometric polynomials (see JACKSON(1930) or ZYGMUND(1988)), related to Chebyshev polynomials via (1.5.6). The same procedure applies for the interpolation operator corresponding to Chebyshev Gauss-Lobatto nodes.

To obtain convergence results for $I_{w,n}$ when $\nu > 0$, we recall the expression

$$I_{w,n}f = [\tilde{I}_{v,n+1}(fq)]\, q^{-1}, \qquad \forall n \in \mathbf{N}, \tag{6.6.5}$$

where $q(x) := (1 - x^2)$, $x \in I$, and $v := q^{-1}w$. Actually, both the terms in (6.6.5) are polynomials in $\mathbf{P}_{n-1}$, coinciding at the nodes $\xi_j^{(n)}$, $1 \leq j \leq n$, as a consequence of relation (1.3.6). Therefore, one gets

$$\|f - I_{w,n}f\|_{L^2_w(I)} = \|(fq) - \tilde{I}_{v,n+1}(fq)\|_{L^2_v(I)}. \tag{6.6.6}$$

Thus, an estimate for $0 < \nu \leq 1$, can be derived from theorem 6.6.1, by noting that v is an ultraspherical weight function, whose exponent is negative.

Convergence results for $\tilde{I}_{w,n}$ in the case $-1 < \nu < 1$ are presented in BERNARDI and MADAY(1989). Proofs are based on techniques borrowed from the theory of interpolation of Sobolev spaces (see section 5.7).

We observe that the inequality (6.6.2) is satisfied for all the Jacobi weights. Therefore, by (6.1.7), the rate of convergence can be estimated in the general case. For instance, for any $f \in C^k(\bar{I})$, $k \geq 1$, the following inequality holds:

$$\|f - I_{w,n}f\|_{L^2_w(I)} \leq C \left(\frac{1}{n}\right)^k \left\|\frac{d^k f}{dx^k}\right\|_{C^0(\bar{I})}, \qquad \forall n \geq k, \tag{6.6.7}$$

where $w(x) = (1-x)^\alpha(1+x)^\beta$, $\alpha > -1$, $\beta > -1$, $x \in I$. The same statement applies for the error $f - \tilde{I}_{w,n}f$. We expect that (6.6.7) can be further refined by weakening the assumptions on f and, with the same rate of convergence, replacing the norm on the right-hand side of (6.6.7) with a weaker one (see also section 6.8).

By means of inverse inequalities, we also give a bound to the derivative of the error as follows:

$$\begin{aligned} \|(f - I_{w,n}f)'\|_{L^2_w(I)} &\leq \|(f - \Pi^1_{w,n}f)'\|_{L^2_w(I)} + \|(\Pi^1_{w,n}f - I_{w,n}f)'\|_{L^2_w(I)} \\ &\leq \|f - \Pi^1_{w,n}f\|_{H^1_w(I)} + Cn^2\,\|\Pi^1_{w,n}f - I_{w,n}f\|_{L^2_w(I)}, \end{aligned} \tag{6.6.8}$$

where $\Pi^1_{w,n}$ is defined in (6.4.3) and where we used the inequality (6.3.6) for the polynomial $p := \Pi^1_{w,n}f - I_{w,n}f \in \mathbf{P}_n$. Assuming a sufficient regularity for f, the last terms in (6.6.8) tend to zero for $n \to +\infty$.

It is clear now that the aliasing errors $A_{w,n}f$ and $\tilde{A}_{w,n}f$, introduced in section 4.2, decay to zero for $n \to +\infty$, with a rate depending on the smoothness of f. From (6.2.4) we know that

$$\|f - \Pi_{w,n-1}f\|_{L^2_w(I)} \leq \|f - I_{w,n}f\|_{L^2_w(I)}, \qquad \forall f \in C^0(\bar{I}), \ \forall n \geq 1. \tag{6.6.9}$$

We can be more precise about this estimate.

Theorem 6.6.2 - *For any* $n \geq 1$ *and any* $f \in C^0(\bar{I})$, *we have*

$$\|f - I_{w,n}f\|^2_{L^2_w(I)} = \|f - \Pi_{w,n-1}f\|^2_{L^2_w(I)} + \|A_{w,n}f\|^2_{L^2_w(I)}, \tag{6.6.10}$$

$$\|f - \tilde{I}_{w,n}f\|^2_{L^2_w(I)} = \|f - \Pi_{w,n}f\|^2_{L^2_w(I)} + \|\tilde{A}_{w,n}f\|^2_{L^2_w(I)}. \tag{6.6.11}$$

Proof - Expand all the squared norms in (6.6.10) (or (6.6.11)) in terms of the inner product in $L^2_w(I)$ and recall (6.2.5).

■

Another characterization of the aliasing error is given by (see (3.3.1) and (6.2.11))

$$(6.6.12) \qquad A_{w,n}f = I_{w,n}(f - \Pi_{w,n-1}f) = I_{w,n}\left(\sum_{k=n}^{\infty} c_k u_k\right) = \sum_{k=n+1}^{\infty} c_k(I_{w,n}u_k),$$

where c_k, $k \in \mathbf{N}$, are the Fourier coefficients of f and $u_k = P_k^{(\alpha,\beta)}$, $k \in \mathbf{N}$. The last equality in (6.6.12) follows by observing that, for any $n \geq 1$, $I_{w,n}$ is a continuous operator in $C^0(\bar{I})$ (see section 6.8). A similar relation holds for $\tilde{A}_{w,n}f$. In the Chebyshev case, the formula (6.6.12) takes a simplified form, thanks to the periodicity of the cosine function, i.e.

$$(6.6.13) \qquad A_{w,n}f = \sum_{m=1}^{n-1}\sum_{j=1}^{\infty}(c_{2jn+m} - c_{2jn-m})T_m.$$

Moreover, we note that $A_{w,n}T_{jn} \equiv \mathbf{0}$, $\forall j \geq 1$. A discussion of the effects of the aliasing error in the computation of solutions of boundary value problems are given in CANUTO, HUSSAINI, QUARTERONI and ZANG(1988).

We analyse now the case when I is not bounded. Let us start with the Laguerre approximations ($w(x) = x^\alpha e^{-x}$, $\alpha > -1$, $x \in I =]0, +\infty[$).

Theorem 6.6.3 - *Let* $\delta \in]0,1[$ *and let* $f \in C^0(\bar{I})$ *satisfy* $\lim_{x\to+\infty} f(x)e^{-\delta x} = 0$. *Then we have*

$$(6.6.14) \qquad \lim_{n\to+\infty} \|f - I_{w,n}f\|_{L^2_w(I)} = 0.$$

The same relation holds for the error $f - \tilde{I}_{w,n}f$, *relative to the Gauss-Radau interpolation operator.*

Proof - For any $p_n \in \mathbf{P}_{n-1}$, we have (compare with (6.6.2))

$$(6.6.15) \qquad \|f - I_{w,n}f\|_{L^2_w(I)} \leq \|f - p_n\|_{L^2_w(I)} + \left(\sum_{j=1}^{n}(p_n - f)^2(\xi_j^{(n)})\, w_j^{(n)}\right)^{\frac{1}{2}} \leq$$

$$\leq \sup_{x\in\bar{I}} \left[|f - p_n|\sqrt{v}\right](x) \; \left\{ \left(\int_I v^{-1} w \; dx\right)^{\frac{1}{2}} + \left(\sum_{j=1}^{n} v^{-1}(\xi_j^{(n)}) \; w_j^{(n)}\right)^{\frac{1}{2}} \right\},$$

where $v(x) := e^{-\delta x}$, $x \in \bar{I}$. Due to theorem 6.5.2, the summation on the right hand side of (6.6.15) converges to $\int_I v^{-1} w dx < +\infty$. Hence, it is bounded by a constant independent of n. According to theorem 6.1.4, we can choose the sequence $\{p_n\}_{n\geq 1}$ to obtain (6.6.14).

■

On the determination of the rate of convergence, we just mention a result given in MADAY, PERNAUD-THOMAS and VANDEVEN(1985), for the case $\alpha = 0$ $(w(x) = e^{-x})$.

Theorem 6.6.4 - *Let* $k \geq 1$ *and* $\delta \in]0,1[$. *Then we can find a constant* $C > 0$ *such that, for any* $f \in H_v^k(I)$, *with* $v(x) := e^{-\delta x}$, *one has*

$$\text{(6.6.16)} \qquad \|f - I_{w,n} f\|_{L_w^2(I)} \leq C \left(\frac{1}{\sqrt{n}}\right)^{k-1} \|f\|_{H_v^k(I)}, \qquad \forall n \geq 1.$$

The same relation holds for the error $f - \tilde{I}_{w,n} f$.

■

We argue similarly for the Hermite weight function $(w(x) = e^{-x^2})$. A theorem like 6.6.3 is derived from theorems 6.1.5 and 6.5.3. Further results are considered in the next section.

6.7 Laguerre and Hermite functions

In the analysis of differential equations in unbounded domains, the solution will be required to decay at infinity with an exponential rate. Therefore, approximation by a polynomial cannot give accurate results (see GOTTLIEB and ORSZAG(1977), p.45). It is thus necessary to introduce new families of approximating functions. We set $v(x) := x^\alpha e^x$, $\alpha > -1$, $x \in I$. For $n \in \mathbf{N}$, we define $\mathbf{S}_n$ to be the space of the functions (named *Laguerre functions*) of the form pe^{-x}, where $p \in \mathbf{P}_n$. By virtue of (2.2.6), this is an orthogonal set of functions where the inner product is weighted by v. We present a preliminary result.

Lemma 6.7.1 - *Let* $f = ge^x$ *then, for any* $k \in \mathbf{N}$, *we have*

$$x^{m/2}\,\frac{d^m f}{dx^m} \in L^2_w(I),\ 0 \le m \le k \quad\Longleftrightarrow\quad x^{k/2}\,\frac{d^k g}{dx^k} \in L^2_v(I). \tag{6.7.1}$$

Proof - This is a direct consequence of the formula

$$\int_I \left(\frac{d^k g}{dx^k}\right)^2 x^{\alpha+k} e^x\, dx \tag{6.7.2}$$

$$= \sum_{m=0}^{k} \binom{k}{m} \frac{\Gamma(\alpha+k+1)}{\Gamma(\alpha+m+1)} \int_I \left(\frac{d^m f}{dx^m}\right)^2 x^{\alpha+m} e^{-x}\, dx,$$

which is proven by induction (see FUNARO(1991)).

■

The operator $\Pi^*_{v,n} : L^2_v(I) \to \mathbf{S}_n$, $n \in \mathbf{N}$, is defined as follows:

$$\Pi^*_{v,n} g := [\Pi_{w,n}(ge^x)]\, e^{-x}, \qquad \forall g \in L^2_v(I). \tag{6.7.3}$$

Combining theorem 6.2.5 and relation (6.7.2), we get the following proposition.

Theorem 6.7.2 - *Let* $k \in \mathbf{N}$. *Then there exists a constant* $C > 0$ *such that, for any* g *satisfying* $\frac{d^k g}{dx^k} x^{k/2} \in L^2_v(I)$, *one has*

$$\|g - \Pi^*_{v,n} g\|_{L^2_v(I)} \le C \left(\frac{1}{\sqrt{n}}\right)^k \left\| x^{k/2}\, \frac{d^k g}{dx^k} \right\|_{L^2_v(I)}, \qquad \forall n > k. \tag{6.7.4}$$

■

Furthermore, we have the following inverse inequality.

Theorem 6.7.3 - *We can find a constant $C > 0$ such that, for any $n \geq 1$*

$$(6.7.5) \qquad \|q'\|_{L^2_v(I)} \leq Cn \, \|q\|_{L^2_v(I)}, \qquad \forall q \in \mathbf{S}_n.$$

Proof - Set $q = pe^{-x}$, $p \in \mathbf{P}_n$, and recall (6.3.11).

■

Next, we define the operator $I^*_{v,n} : C^0(\bar{I}) \to \mathbf{S}_{n-1}$, $n \geq 1$, such that

$$(6.7.6) \qquad I^*_{v,n}g := [I_{w,n}(ge^x)]e^{-x}, \qquad \forall g \in C^0(\bar{I}).$$

In a similar way, one defines the interpolation operator $\tilde{I}^*_{v,n}$, based on the Gauss-Radau nodes. On the basis of theorem 6.6.4, one recovers estimates for the error $g - I^*_{v,n}g$.

Now let us set $v(x) := w^{-1}(x) = e^{x^2}$, $x \in \mathbf{R}$, and define for any $n \in \mathbf{N}$, the space $\mathbf{S}_n$ of the so called *Hermite functions*. Any element of $\mathbf{S}_n$ is of the form pw, where $p \in \mathbf{P}_n$. Furthermore, we introduce the operator $\Pi^*_{v,n} : L^2_v(\mathbf{R}) \to \mathbf{S}_n$, $n \in \mathbf{N}$, such that

$$(6.7.7) \qquad \Pi^*_{v,n}g := [\Pi_{w,n}(gv)]\, w, \qquad \forall g \in L^2_v(\mathbf{R}).$$

As in the previous case, we begin with a basic result.

Lemma 6.7.4 - *The following implication: $f \in H^k_w(\mathbf{R}) \Leftrightarrow g := fw \in H^k_v(\mathbf{R})$, is satisfied for any $k \in \mathbf{N}$. Moreover, the corresponding norms are equivalent.*

Proof - It is possible to find real coefficients $\gamma^{(k)}_m$, $0 \leq k \leq m$ (see FUNARO(1991)), such that

$$(6.7.8) \qquad \int_{\mathbf{R}} \left(\frac{d^k g}{dx^k}\right)^2 v \, dx = \sum_{m=0}^{k} \gamma^{(k)}_m \int_{\mathbf{R}} \left(\frac{d^m f}{dx^m}\right)^2 w \, dx.$$

This completes the proof by virtue of (5.5.10) and (5.6.4).

■

Taking into account lemma 6.7.4, theorems 6.2.6 and 6.3.5 are respectively generalized to the case of Hermite functions as follows.

Theorem 6.7.5 - *Let $k \in \mathbf{N}$, then there exists a constant $C > 0$ such that, for any $g \in H_v^k(\mathbf{R})$, one has*

$$(6.7.9) \qquad \|g - \Pi_{v,n}^* g\|_{L_v^2(\mathbf{R})} \leq C \left(\frac{1}{\sqrt{n}}\right)^k \|g\|_{H_v^k(\mathbf{R})}, \qquad \forall n > k.$$

■

Theorem 6.7.6 - *We can find a constant $C > 0$ such that, for any $n \geq 1$*

$$(6.7.10) \qquad \|q'\|_{L_w^2(\mathbf{R})} \leq C\sqrt{n}\, \|q\|_{L_w^2(\mathbf{R})}, \qquad \forall q \in \mathbf{S}_n.$$

■

We finally define $I_{v,n}^* : C^0(\mathbf{R}) \to \mathbf{S}_{n-1}$, $n \geq 1$, such that

$$(6.7.11) \qquad I_{v,n}^* := [I_{w,n}(gv)]w, \qquad \forall g \in C^0(\mathbf{R}).$$

A theoretical convergence analysis for the operator $I_{v,n}^*$ for $n \to +\infty$ is considered in FUNARO and KAVIAN(1988). Here we state the main result (we recall that the Sobolev spaces with real exponent have been introduced in section 5.6).

Theorem 6.7.7 - *Let $\epsilon > 0$ and $s \geq 1 + \epsilon$. Then there exists a constant $C > 0$ such that, for any $g \in H_v^s(\mathbf{R})$, we have*

$$(6.7.12) \qquad \|g - I_{v,n}^* g\|_{L_v^2(\mathbf{R})} \leq C \left(\frac{1}{\sqrt{n}}\right)^{s-1-\epsilon} \|g\|_{H_v^s(\mathbf{R})}, \qquad \forall n \geq 1.$$

■

Other results, concerning the rate of convergence of the so called *normalized Hermite functions*, are provided in BOYD(1984).

6.8 Refinements

The aim of the previous sections was to present an overview of the fundamental results on the approximation of functions by orthogonal polynomials. Further developments can be taken into account. In general, improvements require a certain skill in dealing with functional spaces and their properties. We therefore confine ourselves to the discussion of some basic facts.

In the book of TIMAN(1963), p.262, the inequality (6.1.7) is generalized as follows:

$$(6.8.1) \qquad |f - \Psi_{\infty,n}(f)|(x) \leq C\, [h_n(x)]^k \sup_{|x_1-x_2|<h_n(x)} \left| \frac{d^k f}{dx^k}(x_1) - \frac{d^k f}{dx^k}(x_2) \right|,$$

where $h_n(x) := \frac{1}{n}\left(\sqrt{1-x^2} + \frac{|x|}{n}\right)$, $n \geq 1$, $x \in [-1,1]$. This shows that a faster convergence is expected for the points x located near the extremes of the interval I, where $h_n(x) \approx 1/n^2$. For example, as noticed in GOTTLIEB and ORSZAG(1977), p.29, in Chebyshev-Fourier expansions the Gibbs phenomenon is reduced when the discontinuity of f is situated near one of the boundary points $x = \pm 1$.

In the approximation of oscillating functions with wavelength λ, polynomials of degree higher than $2\pi/\lambda$ are required for Chebyshev and Legendre expansions, to obtain satisfactory results. A lower degree is sufficient when the oscillations are concentrated within a neighbourhood of the endpoints (see GOTTLIEB and ORSZAG(1977), p.35 and p.38).

Concerning the projection operator, the estimates given in section 6.2 are *optimal*. Actually, it is easy to realize that the rate of convergence cannot be improved upon. A proof of this fact is given in CANUTO and QUARTERONI(1982a), for Chebyshev and Legendre expansions. Similar considerations hold for the inverse inequalities studied in section 6.3. On the contrary, the analysis of the interpolation operator presents complications. The inequalities of section 6.6 have been obtained with the help of the polynomial of best uniform approximation (see for instance relation (6.6.2)). This auxiliary estimate is in general quite rough, especially for Jacobi weight functions with large values of α and β. As a matter of fact, the inequality (6.6.4), which is obtained with different techniques, shows a faster convergence behavior. Although the inequality (6.6.7) ensures the same kind of convergence, it requires f to be more regular. In NEVAI(1976) numerous extensions are provided, and in MADAY(1991) an estimate like (6.6.4) is proven for Legendre Gauss-Lobatto points.

For analytic functions, the error estimates for the interpolation operator are expected to exhibit an exponential decay. Results in this direction are given in TADMOR(1986) for Chebyshev expansions. Various improvements and estimates in different weighted spaces are considered in DITZIAN and TOTIK(1987).

To give an idea of the difficulties one may face when working with the interpolation operators, we note for instance that an inequality like (6.2.7) is not verified, when replacing $\Pi_{w,n}$ by $I_{w,n}$. This shows that the operator $I_{w,n}$ is not continuous in the $L^2_w(I)$ norm, even if $f \in C^0(\bar{I})$. To check this fact, we construct a sequence of functions $f_k \in C^0(\bar{I})$, $k \in \mathbf{N}$, such that $f_k(\xi_j^{(n)}) = 1$, $1 \leq j \leq n$, $\forall k \in \mathbf{N}$. Therefore, one has $I_{w,n} f_k \equiv 1$, $\forall k \in \mathbf{N}$, which implies that $\|I_{w,n} f_k\|_{L^2_w(I)} \neq 0$ is constant with respect to k. On the other hand, every f_k can be defined at the remaining points in such a way that $\lim_{k\to+\infty} \|f_k\|_{L^2_w(I)} = 0$. This is in contrast with the claim that $I_{w,n}$ is continuous. Nevertheless, $I_{w,n}$ is continuous in the norm of $C^0(\bar{I})$, i.e., for any $n \geq 1$, we can find $\gamma \equiv \gamma(n)$ such that

(6.8.2) $$\|I_{w,n} f\|_{C^0(\bar{I})} \leq \gamma \, \|f\|_{C^0(\bar{I})}, \qquad \forall f \in C^0(\bar{I}).$$

In fact, for any $f \in C^0(\bar{I})$, (3.9.5) yields

(6.8.3) $$\|I_{w,n} f\|_{C^0(\bar{I})} \leq \gamma \, |||I_{w,n} f|||_{w,n} = \gamma \max_{1\leq j\leq n} |f(\xi_j^{(n)})| \leq \gamma \, \|f\|_{C^0(\bar{I})}.$$

This is why the natural domain for $I_{w,n}$ is the space of continuous functions. Unfortunately, the constant γ grows with n.

* * * * * * * * * * * *

With the conclusion of this chapter, we have finished the first part of the book. Having established the structure of the approximating functions we intend to use, as well as their properties, we now turn our efforts toward the discretization of derivative operators, and their associated problems.

7
DERIVATIVE MATRICES

The derivative operator is a linear transformation from the space of polynomials into itself. In addition, the exact derivative of a given polynomial can be determined in a finite number of operations. Depending on the basis in which the polynomial is expressed, we will construct the appropriate matrices that allow the computation of its derivative.

7.1 Derivative matrices in the frequency space

Let p be a polynomial in $\mathbf{P}_n$, where $n \geq 1$ is given. From (2.3.1), and $u_0' \equiv \mathbf{0}$, we can write

$$(7.1.1) \qquad p' = \sum_{k=1}^{n} c_k u_k' = \sum_{k=0}^{n} c_k^{(1)} u_k.$$

In (7.1.1), the constants $c_k^{(1)}$, $0 \leq k \leq n$, are the Fourier coefficients of p'. We would like to establish a relation between these new coefficients and the c_k's. Let us begin with the Jacobi case ($u_k = P_k^{(\alpha,\beta)}$, $k \in \mathbf{N}$). For simplicity, we just discuss the ultraspherical case ($\nu := \alpha = \beta$). Thus, we start by considering the following relation (see SZEGÖ(1939), p.84):

$$(7.1.2) \qquad u_n = \frac{d}{dx}\left[\frac{(n+2\nu+1)\, u_{n+1}}{(n+\nu+1)(2n+2\nu+1)} - \frac{(n+\nu)\, u_{n-1}}{(n+2\nu)(2n+2\nu+1)}\right], \quad n \geq 1,$$

with $u_n = P_n^{(\nu,\nu)}$, $\nu > -1$ and $\nu \neq -\frac{1}{2}$.
After tedious calculation, we recover from (7.1.2) the expression of the derivatives of the ultraspherical polynomials, i.e.

$$\frac{d}{dx}u_n = \left\{\frac{\Gamma(n+\nu+1)}{\Gamma(n+2\nu+1)} \times \sum_{m=0}^{[(n-1)/2]} \frac{2n-4m+2\nu-1}{n-2m+2\nu} \frac{\Gamma(n-2m+2\nu+1)}{\Gamma(n-2m+\nu)} u_{n-2m-1}\right\}, \tag{7.1.3}$$

for any $n \geq 1$ and $\nu > -1$ with $\nu \neq -\frac{1}{2}$. The symbol $[\bullet]$ denotes the integer part of $\bullet$. Particularly interesting is the Legendre case ($\nu = 0$), where (7.1.3) takes the form

$$\frac{d}{dx}P_n = \sum_{m=0}^{[(n-1)/2]} (2n-4m-1)P_{n-2m-1}, \qquad n \geq 1. \tag{7.1.4}$$

Using the same arguments, recalling (1.5.10), we get in the Chebyshev case

$$\frac{d}{dx}T_n = \begin{cases} T_0 & \text{if } n = 1, \\ 2n \displaystyle\sum_{m=0}^{n/2-1} T_{2m+1} & \text{if } n \geq 2 \text{ is even}, \\ nT_0 + 2n \displaystyle\sum_{m=1}^{(n-1)/2} T_{2m} & \text{if } n \geq 3 \text{ is odd}. \end{cases} \tag{7.1.5}$$

Substituting formula (7.1.3) into (7.1.1), we can relate the coefficients of p' to those of p. First of all, we note that $c_n^{(1)} = 0$. Then, for $\nu > -1$ with $\nu \neq -\frac{1}{2}$, we have

$$c_i^{(1)} = \frac{2i+2\nu+1}{i+2\nu+1} \frac{\Gamma(i+2\nu+2)}{\Gamma(i+\nu+1)} \sum_{\substack{j=i+1 \\ i+j \text{ odd}}}^{n} \frac{\Gamma(j+\nu+1)}{\Gamma(j+2\nu+1)} c_j, \qquad 0 \leq i \leq n-1. \tag{7.1.6}$$

In the Legendre case ($\nu = 0$), the above formula yields (see also the appendix in the book of GOTTLIEB and ORSZAG(1977))

$$(7.1.7)\qquad c_i^{(1)} = (2i+1) \sum_{\substack{j=i+1 \\ i+j \text{ odd}}}^{n} c_j, \qquad 0 \le i \le n-1.$$

For the Chebyshev basis, we still get $c_n^{(1)} = 0$. Besides, one has

$$(7.1.8)\qquad c_i^{(1)} = \begin{cases} \displaystyle\sum_{\substack{j=1 \\ j \text{ odd}}}^{n} j c_j & \text{if } i = 0, \\[2ex] \displaystyle 2 \sum_{\substack{j=i+1 \\ i+j \text{ odd}}}^{n} j c_j & \text{if } 1 \le i \le n-1. \end{cases}$$

The linear mapping which transforms the vector $\{c_j\}_{0 \le j \le n}$ into the vector $\{c_i^{(1)}\}_{0 \le i \le n}$ is clearly expressed by a $(n+1) \times (n+1)$ matrix which is upper triangular and has a vanishing diagonal . Therefore, all its eigenvalues are zero. By applying k times ($k \ge 1$) the *derivative matrix* to the vector $\{c_j\}_{0 \le j \le n}$ we obtain the vector $\{c_i^{(k)}\}_{0 \le i \le n}$, which represents the Fourier coefficients of the k-th derivative of the polynomial p. This implies that, since $\frac{d^{n+1}}{dx^{n+1}} p \equiv \mathbf{0}$ $\forall p \in \mathbf{P}_n$, the $n+1$ power of the matrix has all its entries equal to zero.

For $2 \le k \le n$, an analytic formula relating the coefficients $c_i^{(k)}$, $0 \le i \le n$, to the coefficients c_j, $0 \le j \le n$, is given in KARAGEORGHIS and PHILLIPS(1989). Here we just present such a formula for $k = 2$, respectively for Legendre and Chebyshev expansions. In these two cases, $c_n^{(2)} = c_{n-1}^{(2)} = 0$, and

$$(7.1.9)\qquad c_i^{(2)} = \left(i + \frac{1}{2}\right) \sum_{\substack{j=i+2 \\ i+j \text{ even}}}^{n} \left(j(j+1) - i(i+1)\right) c_j, \qquad 0 \le i \le n-2,$$

$$(7.1.10)\qquad c_i^{(2)} = \begin{cases} \displaystyle\frac{1}{2} \sum_{\substack{j=2 \\ j \text{ even}}}^{n} j^3 c_j & \text{if } i = 0, \\[2ex] \displaystyle\sum_{\substack{j=i+2 \\ i+j \text{ even}}}^{n} j(j^2 - i^2) c_j & \text{if } 1 \le i \le n-2. \end{cases}$$

As proposed in section 2.3, one can study the relation between the Fourier coefficients of polynomials such as xp' (or more involved expressions) and the coefficients of p. Many examples of this type, for Legendre and Chebyshev expansions, are discussed in the appendix of GOTTLIEB and ORSZAG(1977).

For Laguerre polynomials, relations taken from SZEGÖ(1939), p.98, lead to

$$\frac{d}{dx}L_n^{(\alpha)} = -\sum_{m=0}^{n-1} L_m^{(\alpha)}, \qquad n \geq 1,\ \alpha > -1. \tag{7.1.11}$$

Therefore, one obtains

$$c_i^{(1)} = -\sum_{j=i+1}^{n} c_j, \qquad 0 \leq i \leq n-1, \quad \text{and} \quad c_n^{(1)} = 0. \tag{7.1.12}$$

The Hermite case is very easy to analyze. Actually, from (1.7.8), we have

$$c_i^{(1)} = 2(i+1)\,c_{i+1}, \quad 0 \leq i \leq n-1, \quad \text{and} \quad c_n^{(1)} = 0. \tag{7.1.13}$$

An initial analysis suggests that the evaluation in the frequency space of the derivative of the polynomial $p \in \mathbf{P}_n$ has a computational cost proportional to n^2, since it corresponds to a matrix-vector multiplication. A deeper study reveals an algorithm with a cost only proportional to n. In fact, one easily proves the following recursion formula. First we compute $c_n^{(1)} = 0$ and

$$c_{n-1}^{(1)} = \frac{(2n+2\nu-1)(n+\nu)}{(n+2\nu)}\,c_n, \qquad \nu > -1,\ \nu \neq -\tfrac{1}{2}. \tag{7.1.14}$$

Then, we proceed backwards according to the relations

$$c_i^{(1)} = \frac{(2i+2\nu+1)(i+\nu+1)}{i+2\nu+1}\left[\frac{i+\nu+2}{(2i+2\nu+5)(i+2\nu+2)}\,c_{i+2}^{(1)} + c_{i+1}\right], \tag{7.1.15}$$

$$0 \leq i \leq n-2,\ \nu > -1,\ \nu \neq -\tfrac{1}{2}.$$

Similarly, for the Chebyshev and Laguerre cases, we respectively get

$$(7.1.16)\qquad c_i^{(1)} = \begin{cases} 0 & \text{if } i = n, \\ 2nc_n & \text{if } i = n-1, \\ c_{i+2}^{(1)} + 2(i+1)c_{i+1} & \text{if } 1 \le i \le n-2, \\ \frac{1}{2}c_2^{(1)} + c_1 & \text{if } i = 0, \end{cases}$$

$$(7.1.17)\qquad c_i^{(1)} = \begin{cases} 0 & \text{if } i = n, \\ c_{i+1}^{(1)} - c_{i+1} & \text{if } 0 \le i \le n-1. \end{cases}$$

The Hermite case has already a simplified form given by (7.1.13).

7.2 Derivative matrices in the physical space

Here we consider $\mathbf{P}_{n-1}$, $n \ge 2$, to be generated by the basis of Lagrange polynomials $\{l_j^{(n)}\}_{1\le j\le n}$ introduced in section 3.2. In view of (3.2.2), the derivative of a polynomial $p \in \mathbf{P}_n$ is obtained by evaluating

$$(7.2.1)\qquad \frac{d}{dx}p = \sum_{j=1}^{n} p(\xi_j^{(n)}) \frac{d}{dx} l_j^{(n)}.$$

In particular, we have

$$(7.2.2)\qquad p'(\xi_i^{(n)}) = \sum_{j=1}^{n} d_{ij}^{(1)} \, p(\xi_j^{(n)}), \qquad 1 \le i \le n,$$

where

$$d_{ij}^{(1)} := \left[\frac{d}{dx} l_j^{(n)}\right](\xi_i^{(n)}), \quad 1 \le i \le n, \ 1 \le j \le n. \tag{7.2.3}$$

Since the polynomials p and p' are uniquely determined by their values at the nodes $\xi_i^{(n)}$, $1 \le i \le n$, the linear transformation (7.2.2) is equivalent to performing an exact derivative in the space $\mathbf{P}_{n-1}$. The associated $n \times n$ matrix $D_n := \{d_{ij}^{(1)}\}_{\substack{1\le i\le n \\ 1\le j\le n}}$, allows us to pass from the the vector $\{p(\xi_j^{(n)})\}_{1\le j\le n}$ to the vector $\{p'(\xi_i^{(n)})\}_{1\le i\le n}$, with a computational cost proportional to n^2.

The next step is to give an explicit expression to the entries $d_{ij}^{(1)}$, $1 \le i \le n$, $1 \le j \le n$. Computing the derivatives of the Lagrange polynomials in (3.2.4) yields

$$\left[\frac{d}{dx} l_j^{(n)}\right](x) = \frac{u_n'(x)(x - \xi_j^{(n)}) - u_n(x)}{u_n'(\xi_j^{(n)})\,(x - \xi_j^{(n)})^2}, \qquad x \in I, \ x \ne \xi_j^{(n)}, \ 1 \le j \le n. \tag{7.2.4}$$

After evaluation of (7.2.4) at the nodes, one obtains

$$d_{ij}^{(1)} = \begin{cases} \dfrac{u_n'(\xi_i^{(n)})}{u_n'(\xi_j^{(n)})} \dfrac{1}{\xi_i^{(n)} - \xi_j^{(n)}} & \text{if } i \ne j, \\[3ex] \dfrac{u_n''(\xi_i^{(n)})}{2\,u_n'(\xi_i^{(n)})} & \text{if } i = j. \end{cases} \tag{7.2.5}$$

To get (7.2.5) one recalls that $u_n(\xi_i^{(n)}) = 0$, $1 \le i \le n$. The diagonal entries (i.e., $i = j$) are obtained by noting that

$$\lim_{x \to \xi_i^{(n)}} \left[\frac{d}{dx} l_i^{(n)}\right](x) = \lim_{x \to \xi_i^{(n)}} \frac{u_n''(x)}{2\,u_n'(\xi_i^{(n)})} = \frac{u_n''(\xi_i^{(n)})}{2\,u_n'(\xi_i^{(n)})}, \qquad 1 \le i \le n. \tag{7.2.6}$$

Recalling that u_n is solution of a Sturm-Liouville problem, we can express the values $u_n''(\xi_i^{(n)})$, $1 \le i \le n$, in terms of $u'(\xi_i^{(n)})$, $1 \le i \le n$. This gives

(7.2.7) (*Jacobi*) $$d_{ii}^{(1)} = \frac{(\alpha+\beta+2)\xi_i^{(n)} + \alpha - \beta}{2\,(1-(\xi_i^{(n)})^2)}, \qquad \alpha > -1,\ \beta > -1,$$

(7.2.8) (*Laguerre*) $$d_{ii}^{(1)} = \frac{\xi_i^{(n)} - \alpha - 1}{2\xi_i^{(n)}}, \qquad \alpha > -1,$$

(7.2.9) (*Hermite*) $$d_{ii}^{(1)} = \xi_i^{(n)}.$$

Higher order derivative matrices are obtained by multiplying D_n by itself a suitable number of times. For example, an explicit expression of D_n^2 in the Hermite case is given in FUNARO and KAVIAN(1988).
In general, we denote by $d_{ij}^{(k)}$, $1 \leq i \leq n$, $1 \leq j \leq n$, the entries of the matrix D_n^k, $k \geq 1$. Of course, when $k \geq n$, we have $d_{ij}^{(k)} = 0$, $1 \leq i \leq n$, $1 \leq j \leq n$.

We can argue in the same way using the Gauss-Lobatto points. For any $p \in \mathbf{P}_n$, $n \geq 1$, we have

(7.2.10) $$p'(\eta_i^{(n)}) = \sum_{j=0}^{n} \tilde{d}_{ij}^{(1)}\ p(\eta_j^{(n)}), \qquad 0 \leq i \leq n,$$

where

(7.2.11) $$\tilde{d}_{ij}^{(1)} := \left[\frac{d}{dx}\tilde{l}_j^{(n)}\right](\eta_i^{(n)}), \quad 0 \leq i \leq n,\ 0 \leq j \leq n.$$

Similarly, we define the $(n+1)\times(n+1)$ matrix $\tilde{D}_n := \{\tilde{d}_{ij}^{(1)}\}_{\substack{0\leq i\leq n\\ 0\leq j\leq n}}$. Again, we denote by $\tilde{d}_{ij}^{(k)}$, $0 \leq i \leq n$, $0 \leq j \leq n$, the entries of the $(n+1)\times(n+1)$ matrix $\tilde{D}_n^k$, $k \geq 1$. When $k \geq n+1$, $\tilde{D}_n^k$ is the zero matrix.

The computation of the entries of $\tilde{D}_n$ needs more perseverance, but it is not hard to manage. From (3.2.8), we have

$$
(7.2.12)\qquad \tilde{d}_{ij}^{(1)} = \begin{cases}
\dfrac{\alpha - n(n+\alpha+\beta+1)}{2(\beta+2)} & i=j=0, \\[2ex]
\dfrac{(-1)^n\, n!\, \Gamma(\beta+2)\, u_n(\eta_i^{(n)})}{\Gamma(n+\beta+1)\, (1+\eta_i^{(n)})} & 1 \le i \le n-1, \quad j=0, \\[2ex]
\dfrac{(-1)^n\, \Gamma(\beta+2)\, \Gamma(n+\alpha+1)}{2\, \Gamma(\alpha+2)\, \Gamma(n+\beta+1)} & i=n, \quad j=0, \\[2ex]
\dfrac{-(-1)^n\, \Gamma(n+\beta+1)}{n!\, \Gamma(\beta+2)\, u_n(\eta_j^{(n)})(1+\eta_j^{(n)})} & i=0, \quad 1 \le j \le n-1; \\[2ex]
\dfrac{u_n(\eta_i^{(n)})}{u_n(\eta_j^{(n)})}\, \dfrac{1}{\eta_i^{(n)} - \eta_j^{(n)}} \qquad i \ne j, & 1 \le i \le n-1, \ 1 \le j \le n-1, \\[2ex]
\dfrac{(\alpha+\beta)\eta_i^{(n)} + \alpha - \beta}{2\, (1-(\eta_i^{(n)})^2)} & 1 \le i = j \le n-1, \\[2ex]
\dfrac{\Gamma(n+\alpha+1)}{n!\, \Gamma(\alpha+2)\, u_n(\eta_j^{(n)})(1-\eta_j^{(n)})} & i=n, \quad 1 \le j \le n-1, \\[2ex]
\dfrac{-(-1)^n\, \Gamma(\alpha+2)\, \Gamma(n+\beta+1)}{2\, \Gamma(\beta+2)\, \Gamma(n+\alpha+1)} & i=0, \quad j=n, \\[2ex]
\dfrac{-n!\, \Gamma(\alpha+2)\, u_n(\eta_i^{(n)})}{\Gamma(n+\alpha+1)\, (1-\eta_i^{(n)})} & 1 \le i \le n-1, \quad j=n, \\[2ex]
\dfrac{n(n+\alpha+\beta+1) - \beta}{2(\alpha+2)} & i=j=n.
\end{cases}
$$

To check (7.2.12), the reader must remember that $u_n'(\eta_j^{(n)}) = 0$, $1 \le j \le n-1$, while $u_n'(1)$ and $u_n'(-1)$ are respectively given by (3.1.17) and (3.1.18). Further relations, useful in this computation, are obtained from the differential equation (1.3.1). This can be used for instance to obtain the values of u_n'' at the nodes.

Taking $\nu = 0$ in (7.2.12), we recover the entries of $\tilde{D}_n$ in the Legendre case:

$$
(7.2.13) \qquad \tilde{d}_{ij}^{(1)} = \begin{cases} -\frac{1}{4}n(n+1) & i = j = 0, \\ \dfrac{P_n(\eta_i^{(n)})}{P_n(\eta_j^{(n)})} \dfrac{1}{\eta_i^{(n)} - \eta_j^{(n)}} & 0 \le i \le n,\ 0 \le j \le n,\ i \ne j, \\ 0 & 1 \le i = j \le n-1, \\ \frac{1}{4}n(n+1) & i = j = n. \end{cases}
$$

Similarly, (1.5.1) and (3.1.15) lead to the Chebyshev case ($\nu = -\frac{1}{2}$):

$$
(7.2.14) \qquad \tilde{d}_{ij}^{(1)} = \begin{cases} -\frac{1}{6}(2n^2+1) & i = j = 0, \\ \frac{1}{2}(-1)^i/(1+\eta_i^{(n)}) & 1 \le i \le n-1,\ \ j = 0, \\ \frac{1}{2}(-1)^n & i = n,\ \ j = 0, \\ -2\,(-1)^j/(1+\eta_j^{(n)}) & i = 0,\ \ 1 \le j \le n-1, \\ \dfrac{(-1)^{i+j}}{\eta_i^{(n)} - \eta_j^{(n)}} & 1 \le i \le n-1,\ 1 \le j \le n-1,\ i \ne j, \\ \dfrac{-\eta_i^{(n)}}{2\,(1-(\eta_i^{(n)})^2)} & 1 \le i = j \le n-1, \\ 2\,(-1)^{j+n}/(1-\eta_j^{(n)}) & i = n,\ \ 1 \le j \le n-1, \\ -\frac{1}{2}(-1)^n & i = 0,\ \ j = n, \\ -\frac{1}{2}(-1)^{i+n}/(1-\eta_i^{(n)}) & 1 \le i \le n-1,\ \ j = n, \\ \frac{1}{6}(2n^2+1) & i = j = n. \end{cases}
$$

The matrices (7.2.13) and (7.2.14) have been presented for instance in GOTTLIEB, HUSSAINI and ORSZAG(1984), together with the entries of the matrices corresponding to the second derivatives.

For the Laguerre Gauss-Radau case, the entries of the $n \times n$ matrix $\tilde{D}_n = \{\tilde{d}_{ij}^{(1)}\}_{\substack{0 \le i \le n-1 \\ 0 \le j \le n-1}}$, are obtained by (3.2.11). These are

$$
(7.2.15) \qquad \tilde{d}_{ij}^{(1)} = \begin{cases} -\dfrac{n-1}{\alpha+2} & i = j = 0, \\[2ex] \dfrac{n!\,\Gamma(\alpha+2)\,L_n^{(\alpha)}(\eta_i^{(n)})}{\Gamma(n+\alpha+1)\,\eta_i^{(n)}} & 1 \le i \le n-1, \quad j = 0, \\[2ex] \dfrac{-\Gamma(n+\alpha+1)}{n!\,\Gamma(\alpha+2)\,\eta_j^{(n)}\,L_n^{(\alpha)}(\eta_j^{(n)})} & i = 0, \quad 1 \le j \le n-1, \\[2ex] \dfrac{L_n^{(\alpha)}(\eta_i^{(n)})}{L_n^{(\alpha)}(\eta_j^{(n)})}\,\dfrac{1}{\eta_i^{(n)} - \eta_j^{(n)}} \quad i \ne j & 1 \le i \le n-1, \ 1 \le j \le n-1, \\[2ex] \dfrac{\eta_i^{(n)} - \alpha}{2\,\eta_i^{(n)}} & 1 \le i = j \le n-1. \end{cases}
$$

For the computation one has to recall (3.1.19) and (1.6.1).

In all the examples discussed, the algorithm to perform the derivative in the physical space corresponds to a matrix-vector multiplication. Therefore, its cost is proportional to n^2. The matrices D_n and $\tilde{D}_n$, in the various cases, are full and do not display any particular property. This means that in general we cannot improve the procedure of evaluating a derivative, as suggested in the previous section. Nevertheless, for the Chebyshev case, a faster algorithm exists. In view of the results in section 4.3, when n is a power of 2, we can go from the physical space to the frequency space (and conversely) by applying the FFT with a cost proportional to $n\log_2 n$. Therefore, we can use (7.1.8), the cost of which is only proportional to n. This observation makes the Chebyshev Lagrange basis preferable, when large values of n are taken into account.

We conclude this section with a few additional remarks. Given the function $f \in C^1(\bar{I})$, we can evaluate the quantities $f(\xi_j^{(n)})$, $1 \leq j \leq n$. After applying the matrix D_n, we obtain the vector $\{(I_{w,n}f)'(\xi_i^{(n)})\}_{1\leq i\leq n}$. For n sufficiently large, $(I_{w,n}f)'$ is a fairly good approximation of f' (see (6.6.8)).
Other techniques, such as finite-differences, can be used to compute numerically the derivative of f. Although they generate derivative matrices with quite good structure (for instance, they are banded with a narrow bandwidth), the results are in general very poor when compared with those obtained by the procedure described above. The reason is that finite-differences are *local methods*. To compute the derivative at a given point, these methods use information pertaining to a small neighborhood of the point. In contrast, spectral methods are *global methods*. All n nodes in the domain I are applied in the computation. For example, when f is a polynomial in $\mathbf{P}_{n-1}$, the values at the nodes are representative of the whole function in the domain $\bar{I}$, and the exact derivative is obtained. The particular distribution of the nodes optimizes the process for a general f, and the accuracy increases rapidly with the regularity of the function. The reader may find other persuasive arguments supporting these considerations in BOYD(1989), p.11.

7.3 Boundary conditions in the frequency space

Clearly, the procedure of evaluating the derivative $q := p' \in \mathbf{P}_{n-1}$ of a polynomial $p \in \mathbf{P}_n$ can be inverted except to within an additive constant. Actually, the derivative formulas of section 7.1 do not take into consideration the coefficient c_0. We need an extra condition. The most popular approach is to assume that p satisfies $p(\xi) = \sigma \in \mathbf{R}$, where $\xi \in \bar{I}$. This is a *boundary condition* when $\xi \in \partial I$, i.e., ξ is an element of the boundary of $\bar{I}$. In this case, we can recover the coefficients c_i, $1 \leq i \leq n$, in terms of the coefficients d_j, $0 \leq j \leq n-1$, of the expansion $q = \sum_{j=0}^{n-1} d_j u_j$. Finally, c_0 is determined according to the following relation:

(7.3.1) $$c_0 = \sigma - \sum_{k=1}^{n} c_k\, u_k(\xi).$$

This comes from evaluating (2.3.1) at the point $x = \xi$. Many other conditions are possible. For instance, one can impose $\int_I p dx = \sigma$, $\sigma \in \mathbf{R}$. In the Chebyshev case, by (2.6.4), c_0 is then obtained from

(7.3.2) $$c_0 = \frac{\sigma}{2} - \sum_{\substack{k=2 \\ k \text{ even}}}^{n} \frac{c_k}{1-k^2}.$$

Let us analyze a more general situation. We just consider the ultraspherical case. Let A and σ be real constants and $\xi \in \bar{I}$. Given the polynomial $q \in \mathbf{P}_{n-1}$, we are concerned with finding $p \in \mathbf{P}_n$, such that

(7.3.3) $$\begin{cases} p' + \Pi_{w,n-1}(Ap) = q, \\ p(\xi) = \sigma. \end{cases}$$

The projection operator $\Pi_{w,n}$, $n \in \mathbf{N}$, has been introduced in section 2.4.
For example, when $n = 4$, problem (7.3.3) is equivalent to the following system:

(7.3.4) $$\begin{bmatrix} A & * & 0 & * & 0 \\ 0 & A & * & 0 & * \\ 0 & 0 & A & * & 0 \\ 0 & 0 & 0 & A & * \\ * & * & * & * & * \end{bmatrix} \begin{bmatrix} c_0 \\ c_1 \\ c_2 \\ c_3 \\ c_4 \end{bmatrix} = \begin{bmatrix} d_0 \\ d_1 \\ d_2 \\ d_3 \\ \sigma \end{bmatrix},$$

where the d_k's are the Fourier coefficients of q, and the symbol $*$ denotes the non zero entries. Due to the special structure of the matrix in (7.3.4), the polynomial p in (7.3.3) can be computed by Gauss elimination with a cost proportional to n^2. However, the situation gets far more complicated when $A : I \to \mathbf{R}$ is a non constant function. In this case the matrix associated to the system is full (see section 2.4).

Next, consider second-order derivatives. Now, two additional conditions are required to select a unique primitive function. Since there are several possibilities, we only consider one example. Let A, B, σ_1, σ_2, be real constants and $q \in \mathbf{P}_{n-2}$, $n \geq 2$. In the ultraspherical case, one is concerned with finding $p \in \mathbf{P}_n$ such that

$$(7.3.5) \qquad \begin{cases} -p'' + \Pi_{w,n-2}(Ap' + Bp) = q, \\ p(-1) = \sigma_1, \\ p(1) = \sigma_2. \end{cases}$$

For instance, when $n = 5$, problem (7.3.5) leads to a linear system of the form

$$(7.3.6) \qquad \begin{bmatrix} B & * & * & * & * & * \\ 0 & B & * & * & * & * \\ 0 & 0 & B & * & * & * \\ 0 & 0 & 0 & B & * & * \\ * & * & * & * & * & * \\ * & * & * & * & * & * \end{bmatrix} \begin{bmatrix} c_0 \\ c_1 \\ c_2 \\ c_3 \\ c_4 \\ c_5 \end{bmatrix} = \begin{bmatrix} d_0 \\ d_1 \\ d_2 \\ d_3 \\ \sigma_1 \\ \sigma_2 \end{bmatrix},$$

where the d_k's are the Fourier coefficients of q. Again, this can be solved with a direct procedure with a cost proportional to n^2.
From (1.3.2) and (1.3.3), in the ultraspherical case we have $u_k(1) = (-1)^k u_k(-1)$, $k \in \mathbf{N}$. This implies

$$(7.3.7) \qquad 2 \sum_{\substack{k=0 \\ k \text{ even}}}^{n} c_k\, u_k(1) = \sigma_1 + \sigma_2, \qquad 2 \sum_{\substack{k=1 \\ k \text{ odd}}}^{n} c_k\, u_k(1) = \sigma_1 - \sigma_2.$$

Therefore, when $A = 0$, (7.3.6) can be decoupled into the subsystems

$$(7.3.8) \qquad \begin{bmatrix} B & * & * \\ 0 & B & * \\ * & * & * \end{bmatrix} \begin{bmatrix} c_0 \\ c_2 \\ c_4 \end{bmatrix} = \begin{bmatrix} d_0 \\ d_2 \\ \sigma_1 + \sigma_2 \end{bmatrix}, \qquad \begin{bmatrix} B & * & * \\ 0 & B & * \\ * & * & * \end{bmatrix} \begin{bmatrix} c_1 \\ c_3 \\ c_5 \end{bmatrix} = \begin{bmatrix} d_1 \\ d_3 \\ \sigma_1 - \sigma_2 \end{bmatrix}.$$

This is because $c_i^{(2)}$, $0 \leq i \leq n-2$, is a linear combination of c_j, $i+2 \leq j \leq n$, where $i+j$ is even (see for instance (7.1.9) and (7.1.10)). Following this idea, a fast algorithm for the solution of (7.3.5) is presented in CLENSHAW(1957), for the Chebyshev case. Further hints are given in CANUTO, HUSSAINI, QUARTERONI and ZANG(1988), p.129, and BOYD(1989), p.380.

Different boundary conditions can be examined. For $B > 0$, typical conditions are obtained by imposing $p'(-1) = \sigma_1$, $p'(1) = \sigma_2$. The last two rows in (7.3.6) have to be changed accordingly.

One can also assume that A and B are non constant functions in I. This implies that the matrix relative to the system (7.3.5) is full. Moreover, the expression of the Fourier coefficients of the term $\Pi_{w,n-2}(Ap' + Bp)$, $n \geq 2$, is quite involved (see section 2.4). As we see in the next section, it is more natural to treat these kind of problems in the physical space. Starting from a different approach, further methods to impose boundary conditions in the frequency space will be analyzed in section 9.4. Suggestions for the treatment of the Laguerre and Hermite cases are given in section 9.5.

7.4 Boundary conditions in the physical space

It is standard to work with Gauss-Lobatto nodes in the Jacobi case and Gauss-Radau nodes in the Laguerre case, since these actually include the boundary points of the domain $\bar{I}$. Applications of the Gauss-Radau nodes in (3.1.13) and (3.1.14) have been considered in CANUTO and QUARTERONI(1982b) for boundary conditions on one side of the interval $[-1, 1]$.

We give a survey of the cases that can occur in the discretization of linear boundary-value problems in one dimension. The study of the generating equations and of the behavior of the approximated solutions, when the number of nodes increases, are the subject of chapter nine.

As usual, we first consider the Jacobi case. The problem of inverting the derivative operator in the space of polynomials can be formulated as follows. Let us assume that $\xi = -1$ (similar arguments apply when $\xi = 1$). Let q be a polynomial in $\mathbf{P}_{n-1}$, $n \geq 1$. We are concerned with finding $p \in \mathbf{P}_n$ such that $p' = q$ and $p(\xi) = \sigma$, where $\sigma \in \mathbf{R}$. Evaluating these expressions at the nodes, we end up with the following set of equations:

(7.4.1)
$$\begin{cases} p'(\eta_i^{(n)}) = q(\eta_i^{(n)}) & 1 \le i \le n, \\ p(\eta_0^{(n)}) = \sigma. \end{cases}$$

This is equivalent to a linear system in the unknowns $p(\eta_i^{(n)})$, $0 \le i \le n$. For example, when $n = 3$, with the notations of section 7.2, we get

(7.4.2)
$$\begin{bmatrix} 1 & 0 & 0 & 0 \\ \tilde{d}_{10}^{(1)} & \tilde{d}_{11}^{(1)} & \tilde{d}_{12}^{(1)} & \tilde{d}_{13}^{(1)} \\ \tilde{d}_{20}^{(1)} & \tilde{d}_{21}^{(1)} & \tilde{d}_{22}^{(1)} & \tilde{d}_{23}^{(1)} \\ \tilde{d}_{30}^{(1)} & \tilde{d}_{31}^{(1)} & \tilde{d}_{32}^{(1)} & \tilde{d}_{33}^{(1)} \end{bmatrix} \begin{bmatrix} p(\eta_0^{(n)}) \\ p(\eta_1^{(n)}) \\ p(\eta_2^{(n)}) \\ p(\eta_3^{(n)}) \end{bmatrix} = \begin{bmatrix} \sigma \\ q(\eta_1^{(n)}) \\ q(\eta_2^{(n)}) \\ q(\eta_3^{(n)}) \end{bmatrix}.$$

If $r \in \mathbf{P}_n$ is the solution of (7.4.2) with $\sigma = 0$, we note that r satisfies

(7.4.3)
$$\begin{bmatrix} \tilde{d}_{11}^{(1)} & \tilde{d}_{12}^{(1)} & \tilde{d}_{13}^{(1)} \\ \tilde{d}_{21}^{(1)} & \tilde{d}_{22}^{(1)} & \tilde{d}_{23}^{(1)} \\ \tilde{d}_{31}^{(1)} & \tilde{d}_{32}^{(1)} & \tilde{d}_{33}^{(1)} \end{bmatrix} \begin{bmatrix} r(\eta_1^{(n)}) \\ r(\eta_2^{(n)}) \\ r(\eta_3^{(n)}) \end{bmatrix} = \begin{bmatrix} q(\eta_1^{(n)}) \\ q(\eta_2^{(n)}) \\ q(\eta_3^{(n)}) \end{bmatrix}.$$

Therefore, we reduced the initial $(n+1) \times (n+1)$ system to a $n \times n$ system. The final solution is then obtained by setting $p = r + \sigma$.

Another interesting problem is stated as follows. For $q \in \mathbf{P}_n$ and $\gamma \in \mathbf{R}$, $\gamma \neq 0$, find $p \in \mathbf{P}_n$ such that

(7.4.4)
$$\begin{cases} p'(\eta_i^{(n)}) = q(\eta_i^{(n)}) & 1 \le i \le n, \\ p'(\eta_0^{(n)}) + \gamma p(\eta_0^{(n)}) = q(\eta_0^{(n)}) + \gamma\sigma. \end{cases}$$

We observe that now the boundary constraint is not exactly imposed. Actually, we are trying to enforce both the equation and the boundary condition at the point $\eta_0^{(n)}$. Using the Lagrange polynomials introduced in theorem 3.2.1, an equivalent formulation is:

$$(7.4.5)\qquad p' + \gamma[p(\eta_0^{(n)}) - \sigma]\tilde{l}_0^{(n)} = q.$$

This kind of approach was proposed in FUNARO and GOTTLIEB(1988) for Chebyshev nodes. In terms of matrices, this is equivalent to (we take for instance $n = 3$)

$$(7.4.6)\qquad \begin{bmatrix} \tilde{d}_{00}^{(1)} + \gamma & \tilde{d}_{01}^{(1)} & \tilde{d}_{02}^{(1)} & \tilde{d}_{03}^{(1)} \\ \tilde{d}_{10}^{(1)} & \tilde{d}_{11}^{(1)} & \tilde{d}_{12}^{(1)} & \tilde{d}_{13}^{(1)} \\ \tilde{d}_{20}^{(1)} & \tilde{d}_{21}^{(1)} & \tilde{d}_{22}^{(1)} & \tilde{d}_{23}^{(1)} \\ \tilde{d}_{30}^{(1)} & \tilde{d}_{31}^{(1)} & \tilde{d}_{32}^{(1)} & \tilde{d}_{33}^{(1)} \end{bmatrix} \begin{bmatrix} p(\eta_0^{(n)}) \\ p(\eta_1^{(n)}) \\ p(\eta_2^{(n)}) \\ p(\eta_3^{(n)}) \end{bmatrix} = \begin{bmatrix} q(\eta_0^{(n)}) + \gamma\sigma \\ q(\eta_1^{(n)}) \\ q(\eta_2^{(n)}) \\ q(\eta_3^{(n)}) \end{bmatrix}.$$

If $\gamma = 0$, (7.4.4) is not a ***well-posed*** problem, because the determinant of the corresponding matrix vanishes. Also note that, if $q \in \mathbf{P}_{n-1}$, problems (7.4.1) and (7.4.4) are equivalent. Indeed, in this case, the n conditions $p'(\eta_i^{(n)}) = q(\eta_i^{(n)})$, $1 \le i \le n$, imply that $p' \equiv q$ in $\bar{I}$, hence $p'(\eta_0^{(n)}) = q(\eta_0^{(n)})$. Since $\gamma \neq 0$, we obtain the equivalence.

Let $A : \bar{I} \to \mathbf{R}$ be a continuous function. A generalization of problem (7.4.1) is obtained by setting

$$(7.4.7)\qquad \begin{cases} p'(\eta_i^{(n)}) + (Ap)(\eta_i^{(n)}) = q(\eta_i^{(n)}) & 1 \le i \le n, \\ p(\eta_0^{(n)}) = \sigma, \end{cases}$$

where $q \in \mathbf{P}_{n-1}$ and $\sigma \in \mathbf{R}$.
The corresponding system is obtained by adding to the matrix of system (7.4.1), the $(n+1) \times (n+1)$ diagonal matrix $\mathrm{diag}\{0,\ A(\eta_1^{(n)}), \cdots,\ A(\eta_n^{(n)})\}$.
Problem (7.4.4) is generalized in a similar way. In particular, (7.4.5) becomes

$$(7.4.8)\qquad p' + I_{w,n}(Ap) + \gamma[p(\eta_0^{(n)}) - \sigma]\tilde{l}_0^{(n)} = q,$$

where $q \in \mathbf{P}_n$, $\sigma \in \mathbf{R}$, and $\gamma \in \mathbf{R}$, $\gamma \neq 0$. Now, we must add to the matrix relative to (7.4.4), the $(n+1) \times (n+1)$ diagonal matrix $\mathrm{diag}\{A(\eta_0^{(n)}),\ A(\eta_1^{(n)}), \cdots,\ A(\eta_n^{(n)})\}$.

Let us consider now second-order problems. Let q be a polynomial in $\mathbf{P}_{n-2}$, $n \geq 2$. For $\sigma_1, \sigma_2 \in \mathbf{R}$, one is concerned with finding $p \in \mathbf{P}_n$ such that

$$\text{(7.4.9)} \qquad \begin{cases} -p''(\eta_i^{(n)}) = q(\eta_i^{(n)}) & 1 \leq i \leq n-1, \\ p(\eta_0^{(n)}) = \sigma_1, & \\ p(\eta_n^{(n)}) = \sigma_2. & \end{cases}$$

This corresponds to find the solution of a linear system. We show the case $n = 3$:

$$\text{(7.4.10)} \qquad \begin{bmatrix} 1 & 0 & 0 & 0 \\ -\tilde{d}_{10}^{(2)} & -\tilde{d}_{11}^{(2)} & -\tilde{d}_{12}^{(2)} & -\tilde{d}_{13}^{(2)} \\ -\tilde{d}_{20}^{(2)} & -\tilde{d}_{21}^{(2)} & -\tilde{d}_{22}^{(2)} & -\tilde{d}_{23}^{(2)} \\ 0 & 0 & 0 & 1 \end{bmatrix} \begin{bmatrix} p(\eta_0^{(n)}) \\ p(\eta_1^{(n)}) \\ p(\eta_2^{(n)}) \\ p(\eta_3^{(n)}) \end{bmatrix} = \begin{bmatrix} \sigma_1 \\ q(\eta_1^{(n)}) \\ q(\eta_2^{(n)}) \\ \sigma_2 \end{bmatrix}.$$

Let r be the solution of (7.4.10) with $\sigma_1 = \sigma_2 = 0$. Thus, we can eliminate two unknowns and reduce the system to

$$\text{(7.4.11)} \qquad \begin{bmatrix} -\tilde{d}_{11}^{(2)} & -\tilde{d}_{12}^{(2)} \\ -\tilde{d}_{21}^{(2)} & -\tilde{d}_{22}^{(2)} \end{bmatrix} \begin{bmatrix} r(\eta_1^{(n)}) \\ r(\eta_2^{(n)}) \end{bmatrix} = \begin{bmatrix} q(\eta_1^{(n)}) \\ q(\eta_2^{(n)}) \end{bmatrix}.$$

Finally, we recover p from the relation $p(x) = r(x) + \frac{1}{2}(1-x)\sigma_1 + \frac{1}{2}(1+x)\sigma_2$, $\forall x \in \bar{I}$. Of course, this procedure applies for any $n \geq 2$. We note that the matrix in (7.4.11) is not the square of some first derivative matrix.

Problem (7.4.9) is generalized as follows:

$$\text{(7.4.12)} \qquad \begin{cases} -p''(\eta_i^{(n)}) + (Bp' + Ap)(\eta_i^{(n)}) = q(\eta_i^{(n)}) & 1 \leq i \leq n-1, \\ p(\eta_0^{(n)}) = \sigma_1, & \\ p(\eta_n^{(n)}) = \sigma_2, & \end{cases}$$

where A and B are continuous functions in $\bar{I}$.

Different boundary conditions are possible. A classical problem is to find $p \in \mathbf{P}_n$ such that

$$
(7.4.13) \qquad \begin{cases} -p''(\eta_i^{(n)}) + \mu p(\eta_i^{(n)}) = q(\eta_i^{(n)}) & 1 \le i \le n-1, \\ -p'(\eta_0^{(n)}) = -\sigma_1, \\ p'(\eta_n^{(n)}) = \sigma_2, \end{cases}
$$

where $q \in \mathbf{P}_{n-2}$, $\sigma_1, \sigma_2 \in \mathbf{R}$, and $\mu > 0$.
We changed the sign at relation $p'(\eta_0^{(n)}) = \sigma_1$ for a reason that will be understood in section 8.2. The matrix associated to (7.4.13), for $n = 3$, takes the form

$$
(7.4.14) \qquad \begin{bmatrix} -\tilde{d}_{00}^{(1)} & -\tilde{d}_{01}^{(1)} & -\tilde{d}_{02}^{(1)} & -\tilde{d}_{03}^{(1)} \\ -\tilde{d}_{10}^{(2)} & -\tilde{d}_{11}^{(2)} + \mu & -\tilde{d}_{12}^{(2)} & -\tilde{d}_{13}^{(2)} \\ -\tilde{d}_{20}^{(2)} & -\tilde{d}_{21}^{(2)} & -\tilde{d}_{22}^{(2)} + \mu & -\tilde{d}_{23}^{(2)} \\ \tilde{d}_{30}^{(1)} & \tilde{d}_{31}^{(1)} & \tilde{d}_{32}^{(1)} & \tilde{d}_{33}^{(1)} \end{bmatrix} \begin{bmatrix} p(\eta_0^{(n)}) \\ p(\eta_1^{(n)}) \\ p(\eta_2^{(n)}) \\ p(\eta_3^{(n)}) \end{bmatrix} = \begin{bmatrix} -\sigma_1 \\ q(\eta_1^{(n)}) \\ q(\eta_2^{(n)}) \\ \sigma_2 \end{bmatrix}.
$$

This situation can be easily generalized to mixed type boundary conditions, such as

$$
(7.4.15) \qquad \begin{cases} \tau_1 p'(-1) + \tau_2 p(-1) = \sigma_1 \\ \tau_3 p'(1) + \tau_4 p(1) = \sigma_2 \end{cases} \qquad \tau_i \in \mathbf{R},\ 1 \le i \le 4.
$$

Here τ_1 and τ_2 (as well as τ_3 and τ_4) do not vanish simultaneously.

A suitable modification furnishes another formulation. Let $q \in \mathbf{P}_n$, $\sigma_1, \sigma_2 \in \mathbf{R}$, $\mu > 0$ and $\gamma \in \mathbf{R}$ with $\gamma \neq 0$. Then, we seek $p \in \mathbf{P}_n$ such that

$$
(7.4.16) \qquad \begin{cases} -p''(\eta_i^{(n)}) + \mu p(\eta_i^{(n)}) = q(\eta_i^{(n)}) & 1 \le i \le n-1, \\ -p''(\eta_0^{(n)}) + \mu p(\eta_0^{(n)}) - \gamma p'(\eta_0^{(n)}) = q(\eta_0^{(n)}) - \gamma\sigma_1, \\ -p''(\eta_n^{(n)}) + \mu p(\eta_n^{(n)}) + \gamma p'(\eta_n^{(n)}) = q(\eta_n^{(n)}) + \gamma\sigma_2. \end{cases}
$$

This is equivalent to

$$(7.4.17)\qquad -p'' + \mu p - \gamma[p'(\eta_0^{(n)}) - \sigma_1]\tilde{l}_0^{(n)} + \gamma[p'(\eta_n^{(n)}) - \sigma_2]\tilde{l}_n^{(n)} = q.$$

This approach was introduced in FUNARO(1986) for Legendre nodes, and in FUNARO (1988) for Chebyshev nodes. For $n = 3$, we present the relative linear system

$$(7.4.18)\qquad \begin{bmatrix} -\tilde{d}_{00}^{(2)} - \gamma\tilde{d}_{00}^{(1)} + \mu & -\tilde{d}_{01}^{(2)} - \gamma\tilde{d}_{01}^{(1)} & -\tilde{d}_{02}^{(2)} - \gamma\tilde{d}_{02}^{(1)} & -\tilde{d}_{03}^{(2)} - \gamma\tilde{d}_{03}^{(1)} \\ -\tilde{d}_{10}^{(2)} & -\tilde{d}_{11}^{(2)} + \mu & -\tilde{d}_{12}^{(2)} & -\tilde{d}_{13}^{(2)} \\ -\tilde{d}_{20}^{(2)} & -\tilde{d}_{21}^{(2)} & -\tilde{d}_{22}^{(2)} + \mu & -\tilde{d}_{23}^{(2)} \\ -\tilde{d}_{30}^{(2)} + \gamma\tilde{d}_{30}^{(1)} & -\tilde{d}_{31}^{(2)} + \gamma\tilde{d}_{31}^{(1)} & -\tilde{d}_{32}^{(2)} + \gamma\tilde{d}_{32}^{(1)} & -\tilde{d}_{33}^{(2)} + \gamma\tilde{d}_{33}^{(1)} + \mu \end{bmatrix}$$

$$\times \begin{bmatrix} p(\eta_0^{(n)}) \\ p(\eta_1^{(n)}) \\ p(\eta_2^{(n)}) \\ p(\eta_3^{(n)}) \end{bmatrix} = \begin{bmatrix} q(\eta_0^{(n)}) - \gamma\sigma_1 \\ q(\eta_1^{(n)}) \\ q(\eta_2^{(n)}) \\ q(\eta_3^{(n)}) + \gamma\sigma_2 \end{bmatrix}.$$

The reason for using these type of boundary conditions will be explained in section 9.4. Alternative techniques are presented in CANUTO(1986).

We now investigate fourth-order problems. A classical example is to find $p \in \mathbf{P}_{n+2}$, $n \geq 2$, such that

$$(7.4.19)\qquad \begin{cases} p^{IV}(\eta_i^{(n)}) = q(\eta_i^{(n)}) \qquad 1 \leq i \leq n-1, \\ p(\eta_0^{(n)}) = \sigma_1, \quad p(\eta_n^{(n)}) = \sigma_2, \\ p'(\eta_0^{(n)}) = \sigma_3, \quad p'(\eta_n^{(n)}) = \sigma_4, \end{cases}$$

where $q \in \mathbf{P}_{n-2}$ and $\sigma_k \in \mathbf{R}$, $1 \leq k \leq 4$. For the purpose of writing the associated linear system, we construct the unique polynomial $s \in \mathbf{P}_3$ that satisfies the same boundary conditions as p:

(7.4.20) $$s(x) := \frac{1}{4}\Big[2\sigma_1 + 2\sigma_2 + \sigma_3 - \sigma_4 + (3\sigma_2 - 3\sigma_1 - \sigma_3 - \sigma_4)x$$
$$+ (\sigma_4 - \sigma_3)x^2 + (\sigma_1 - \sigma_2 + \sigma_3 + \sigma_4)x^3\Big], \qquad \forall x \in \bar{I}.$$

Then, $r := p - s \in \mathbf{P}_{n+2}$, is expressed as follows (see also FUNARO and HEINRICHS (1990)):

(7.4.21) $$r(x) = \sum_{j=1}^{n-1} [p(\eta_j^{(n)}) - s(\eta_j^{(n)})] \frac{1 - x^2}{1 - (\eta_j^{(n)})^2} \tilde{l}_j^{(n)}(x), \qquad \forall x \in \bar{I},$$

where we noted that $r(\pm 1) = r'(\pm 1) = 0$. Deriving this relation four times and evaluating at the nodes, finally leads to the $(n-1) \times (n-1)$ system

(7.4.22) $$\sum_{j=1}^{n-1} \frac{1}{1 - (\eta_j^{(n)})^2} \Big[[1 - (\eta_i^{(n)})^2]\tilde{d}_{ij}^{(4)} - 8\eta_i^{(n)}\tilde{d}_{ij}^{(3)} - 12\tilde{d}_{ij}^{(2)}\Big] r(\eta_j^{(n)}) = q(\eta_i^{(n)}),$$
$$1 \le i \le n-1.$$

Once r is computed, we determine p from the expression (7.4.20).

Let us analyze another example. We wish to find $p \in \mathbf{P}_{n+2}$, $n \ge 2$, such that

(7.4.23) $$\begin{cases} p^{IV}(\eta_i^{(n)}) = q(\eta_i^{(n)}) & 1 \le i \le n-1, \\ p(\eta_0^{(n)}) = \sigma_1, \quad p(\eta_n^{(n)}) = \sigma_2, & \\ p''(\eta_0^{(n)}) = \sigma_3, \quad p''(\eta_n^{(n)}) = \sigma_4, & \end{cases}$$

where $q \in \mathbf{P}_{n-2}$ and $\sigma_k \in \mathbf{R}$, $1 \le k \le 4$. Again, it is sufficient to solve a $(n-1)\times(n-1)$ linear system in the unknowns $r(\eta_j^{(n)})$, $1 \le j \le n-1$, where $r := p - s$ and $s \in \mathbf{P}_3$ is now given by

(7.4.24) $$s(x) := \frac{1}{12}\Big[6\sigma_1 + 6\sigma_2 - 3\sigma_3 - 3\sigma_4 + (6\sigma_2 - 6\sigma_1 + \sigma_3 - \sigma_4)x$$
$$+ (3\sigma_3 + 3\sigma_4)x^2 + (\sigma_4 - \sigma_3)x^3\Big], \qquad \forall x \in \bar{I}.$$

The matrix is given by squaring the equivalent of the matrix in (7.4.11), for a general $n \geq 2$. In fact, (7.4.23) can be decoupled, with the substitution $\hat{p} := p''$, into two consecutive second order systems.

We finally conclude with some examples relative to unbounded domains. Let us consider first the case $I =]0, +\infty[$. Here, problems are formulated in the space of Laguerre functions (see sections 6.7 and 9.5). For $Q \in \mathbf{S}_{n-2}$, $n \geq 2$, $\sigma \in \mathbf{R}$ and $\mu > 0$, we are concerned with finding the solution $P \in \mathbf{S}_{n-1}$ of the problem

$$\text{(7.4.25)} \qquad \begin{cases} -P''(\eta_i^{(n)}) \ + \ \mu P(\eta_i^{(n)}) \ = \ Q(\eta_i^{(n)}), \quad 1 \leq i \leq n-1, \\ P(\eta_0^{(n)}) \ = \ \sigma, \end{cases}$$

where the $\eta_i^{(n)}$'s are the Laguerre Gauss-Radau nodes (see section 3.6). By setting $q(x) := Q(x)e^x$, $p(x) := P(x)e^x$, $\forall x \in [0, +\infty[$, we get an equivalent formulation in the space of polynomials

$$\text{(7.4.26)} \qquad \begin{cases} -p''(\eta_i^{(n)}) \ + \ 2p'(\eta_i^{(n)}) \ + \ (\mu - 1)p(\eta_i^{(n)}) \ = \ q(\eta_i^{(n)}), \quad 1 \leq i \leq n-1, \\ p(\eta_0^{(n)}) \ = \ \sigma. \end{cases}$$

This leads to a $n \times n$ linear system. The entries of the corresponding matrix are obtained in the usual way. Similar considerations hold for other types of boundary conditions. We remark that we only have one boundary point for a second-order equation. Nevertheless, problem (7.4.25) admits a unique solution. Actually, another condition is implicitely assumed by observing that P decays to zero at infinity exponentially fast. More details are given in section 9.5.

For the Hermite case there are no boundary points. Let us discuss an example. The generating problem will be given in section 10.2. Let $Q \in \mathbf{S}_{n-1}$, $n \geq 1$, be a Hermite function and $\mu > 0$. We want to find $P \in \mathbf{S}_{n-1}$ such that

$$\text{(7.4.27)} \qquad -P''(\xi_i^{(n)}) \ - \ 2\xi_i^{(n)} P'(\xi_i^{(n)}) \ + \ \mu P(\xi_i^{(n)}) \ = \ Q(\xi_i^{(n)}), \quad 1 \leq i \leq n,$$

where the $\xi_i^{(n)}$'s are the zeroes of H_n. Boundary conditions are replaced by the relation $\lim_{x\to\pm\infty} P(x) = 0$. By the substitution of $q(x) := Q(x)e^{x^2}$, $p(x) := P(x)e^{x^2}$, $x \in \mathbf{R}$, into (7.4.27) we recover the corresponding problem in $\mathbf{P}_{n-1}$, i.e.

$$(7.4.28) \qquad -p''(\xi_i^{(n)}) + 2\xi_i^{(n)} p'(\xi_i^{(n)}) + \mu\, p(\xi_i^{(n)}) = q(\xi_i^{(n)}), \qquad 1 \le i \le n.$$

This leads to a non singular $n \times n$ linear system in the unknowns $p(\xi_j^{(n)})$, $1 \le j \le n$.

7.5 Derivatives of scaled functions

Following the suggestions given in FUNARO(1990a) and in section 1.6, with the help of the scaling function (1.6.13), we can in part reduce the problems associated with the numerical evaluation of high degree Laguerre or Hermite polynomials.
For example, starting from the matrix in (7.2.15), we define a new matrix by setting

$$(7.5.1) \qquad \hat{d}_{ij}^{(1)} := \tilde{d}_{ij}^{(1)}\, S_n^{(\alpha)}(\eta_i^{(n)}) \left[S_n^{(\alpha)}(\eta_j^{(n)})\right]^{-1}, \quad 0 \le i \le n-1,\ 0 \le j \le n-1.$$

Let p be a polynomial in $\mathbf{P}_{n-1}$, $n \ge 1$. Then we define $\hat{p} := S_n^{(\alpha)} p$. Now, one has

$$(7.5.2) \qquad \sum_{j=0}^{n-1} \hat{d}_{ij}^{(1)}\, \hat{p}(\eta_j^{(n)}) = S_n^{(\alpha)}(\eta_i^{(n)})\, p'(\eta_i^{(n)}), \qquad 0 \le i \le n-1.$$

Thus, the matrix $\hat{D}_n := \{\hat{d}_{ij}^{(1)}\}_{\substack{0\le i\le n-1\\ 0\le j\le n-1}}$ acts on the scaled polynomial $\hat{p}$ and leads to its scaled derivative. The values at the nodes of these functions are in general more appropriate for numerical purposes. Moreover, let us better examine the entries of $\hat{D}_n$. From (1.6.12), (1.6.13) and (7.5.1), we have

$$(7.5.3)\qquad \hat{d}_{ij}^{(1)} = \begin{cases} -\dfrac{n-1}{\alpha+2} & i=j=0, \\[2ex] \dfrac{\alpha+1}{\eta_i^{(n)}}\ \hat{L}_n^{(\alpha)}(\eta_i^{(n)}) & 1\le i\le n-1,\ \ j=0, \\[2ex] \dfrac{-1}{(\alpha+1)\ \eta_j^{(n)}\ \hat{L}_n^{(\alpha)}(\eta_j^{(n)})} & i=0,\ \ 1\le j\le n-1, \\[2ex] \dfrac{\hat{L}_n^{(\alpha)}(\eta_i^{(n)})}{\hat{L}_n^{(\alpha)}(\eta_j^{(n)})}\ \dfrac{1}{\eta_i^{(n)}-\eta_j^{(n)}} & 1\le i\le n-1,\ 1\le j\le n-1,\ i\ne j, \\[2ex] \dfrac{\eta_i^{(n)}-\alpha}{2\ \eta_i^{(n)}} & 1\le i=j\le n-1. \end{cases}$$

All the coefficients are now computed in terms of the scaled Laguerre functions. Second-order derivative operators are obtained by squaring $\hat{D}_n$. The corresponding entries are denoted by $\hat{d}_{ij}^{(2)}$, $0\le i\le n-1$, $0\le j\le n-1$.
It is clear that problem (7.4.26) is equivalent to the following one:

$$(7.5.4)\qquad \begin{cases} \displaystyle\sum_{j=0}^{n-1}\left[-\hat{d}_{ij}^{(2)}+2\hat{d}_{ij}^{(1)}+(\mu-1)\delta_{ij}\right]\hat{p}(\eta_j^{(n)}) \ = \ \hat{q}(\eta_i^{(n)}), \quad 1\le i\le n-1, \\ \hat{p}(\eta_0^{(n)}) \ = \ \sigma\ S_n^{(\alpha)}(0), \end{cases}$$

where $\hat{q} := S_n^{(\alpha)} q$.
Hence, we obtain a linear system in the unknowns $\hat{p}(\eta_j^{(n)})$, $0\le j\le n-1$, with the advantage of having now a matrix which is less affected by rounding errors.
To evaluate the weighted norm of p, it is not necessary to go back to the quantities $p(\eta_j^{(n)})$, $0\le j\le n-1$. We can instead use (3.10.7).

In the Hermite case, the entries in (7.2.5) are modified as follows:

(7.5.5) $$\hat{d}_{ij}^{(1)} := d_{ij}^{(1)}\ S_{n/2}^{(-1/2)}([\xi_i^{(n)}]^2)\ \left[S_{n/2}^{(-1/2)}([\xi_j^{(n)}]^2)\right]^{-1} \quad \text{if } n \text{ is even,}$$

(7.5.6) $$\hat{d}_{ij}^{(1)} := d_{ij}^{(1)}\ \xi_i^{(n)}\ S_{(n-1)/2}^{(1/2)}([\xi_i^{(n)}]^2)\ \left[\xi_j^{(n)}\ S_{(n-1)/2}^{(1/2)}([\xi_j^{(n)}]^2)\right]^{-1} \quad \text{if } n \text{ is odd,}$$

$$1 \le i \le n,\ 1 \le j \le n.$$

By (1.7.11) and (1.7.12), the above relations imply

(7.5.7) $$\hat{d}_{ij}^{(1)} = \begin{cases} \dfrac{\xi_i^{(n)}\ \hat{H}'_n(\xi_i^{(n)})}{\xi_j^{(n)}\ \hat{H}'_n(\xi_j^{(n)})}\ \dfrac{1}{\xi_i^{(n)} - \xi_j^{(n)}} & \text{if } i \neq j, \\ \xi_i^{(n)} & \text{if } i = j. \end{cases}$$

Therefore, we transform (7.4.28) to an equivalent problem where the unknown is the polynomial p multiplied by the scaling function.

7.6 Numerical solution

Various techniques are available for the numerical solution of the systems presented in the previous sections. Basic algorithms are discussed in all the comprehensive books on numerical analysis. We mention for instance FOX(1964), RALSTON(1965), ISAACSON and KELLER(1966), DAHLQUIST and BJÖRCK(1974), HAGEMAN and YOUNG(1981), COMINCIOLI(1990).

To take full advantage of the accurate results obtainable with spectral methods, we always suggest that one operates with programs written in *double precision* (i.e., 64 bits for a real number storage).

In this book, we are only concerned with the discretization of differential equations in one variable. For this reason, the dimension n of the matrices is in general small.

Only in a very few large-scale computations n is bigger than 100 (over 200 in extreme cases). In most of the applications n does not exceed 30. In this case it is convenient to use a ***direct method***, such as Gauss elimination with maximal pivots, rather than an ***iterative method***. Direct factorization (such as LU decomposition) is recommended when the same system has to be solved several times with different data sets. It is well known that the global cost of these procedures is proportional to n^3.

As we shall see in chapter thirteen, for partial differential equations, the derivative matrices in each variable have basically the same structure as those described here. This is because spectral methods are usually applied to boundary value problems, defined on domains obtained by direct product of intervals (rectangles, cubes, and so forth). In this context, to save both computer memory and CPU time, an iterative procedure is often preferable to a direct approach. A survey of the most popular iterative algorithms in spectral methods is given in CANUTO, HUSSAINI, QUARTERONI and ZANG(1988), chapter 5, and BOYD(1989), chapter 12. The development of these techniques is based on the analysis of problems in one variable. Therefore, a detailed study of the simplest situations will be helpful in finding new strategies.

A lot of problems are defined in the physical space (see section 7.4). The major drawback, however, is that the corresponding matrices are full and non symmetric. With the exception of the Chebyshev case, where the FFT algorithm considerably reduces the amount of operations (see sections 4.3 and 7.2), the cost of a matrix-vector multiplication is proportional to n^2. To be competitive, an iterative method has to reach an accurate solution in very few iterations. To achieve this goal, it is imperative to work with ***preconditioned matrices***. How to construct and implement this kind of matrix is discussed in the next chapter. For the moment, let us briefly recall some basic facts.

Let D be a $n \times n$ matrix ($n \geq 1$) and $\bar{q}$ a given vector in $\mathbf{R}^n$. We want to determine the solution $\bar{p} \in \mathbf{R}^n$ of the linear system

$$(7.6.1) \qquad D\bar{p} = \bar{q}, \qquad \text{where } \det(D) \neq 0.$$

A ***one-step*** iterative method to evaluate $\bar{p}$ is obtained by introducing a new $n \times n$ non singular matrix R and defining an initial guess $\bar{p}^{(0)} \in \mathbf{R}^n$. Successively, one constructs the sequence of vectors such that

$$\bar{p}^{(k+1)} := (I - R^{-1}D)\bar{p}^{(k)} + R^{-1}\bar{q}, \quad k \in \mathbf{N}, \tag{7.6.2}$$

where I is the identity matrix.

Denote by $\rho(M) \geq 0$ the *spectral radius* of the matrix $M := I - R^{-1}D$, i.e., the largest eigenvalue modulus among the, possibly complex, eigenvalues of M. Then we have a classical convergence result (see for instance ISAACSON and KELLER(1966), p.63).

Theorem 7.6.1 - *If* M *admits a diagonal form and* $\rho(M) < 1$, *then we have* $\lim_{k\to+\infty} \bar{p}^{(k)} = \bar{p}$, *where* $\bar{p}$ *is the solution to* (7.6.1).

■

The closer $\rho(M)$ is to zero, the faster the convergence will be, since it depends on the geometric progression $[\rho(M)]^k$, $k \in \mathbf{N}$. Actually, when $R = D$, we obtain the exact solution in one step.

Many famous schemes have their root in formula (7.6.2). The simplest is the *Richardson method* (see RICHARDSON(1910)) obtained by setting $R^{-1} = \theta I$, where $\theta \in \mathbf{R}$. Let us indicate by $\lambda_m \in \mathbf{C}$, $1 \leq m \leq n$, the eigenvalues of D. If D admits a diagonal form and the following hypotheses are satisfied:

$$\mathrm{Re}\lambda_m > 0 \quad \text{and} \quad 0 < \theta < \frac{2\,\mathrm{Re}\lambda_m}{|\lambda_m|^2}, \quad 1 \leq m \leq n, \tag{7.6.3}$$

then $\rho(M) < 1$, and the Richardson scheme is convergent.

To proceed with our investigation, we need to know more about the eigenvalues of the matrices introduced in this chapter. This analysis is given in chapter eight.

Despite its simplicity, the Richardson scheme is useful to help understand the basic principles of iterative methods. Consequently, more elaborate iterative algorithms will not be discussed.

8
EIGENVALUE ANALYSIS

The behavior of the eigenvalues of the matrices introduced in chapter seven is interesting for two reasons. The first is related to the application of iterative methods for the solution of the corresponding systems of equations. The second is to examine the approximation of eigenvalue problems for differential operators, which arise in various applications in mathematical physics.

8.1 Eigenvalues of first derivative operators

All the eigenvalues of the derivative matrices presented in sections 7.1 and 7.2, are equal to zero. Actually, if there existed one eigenvalue different from zero, a corresponding polynomial eigenfunction would not vanish after repeated applications of the derivative operator. This is in contradiction to one of the fundamental theorems of calculus.
The situation changes when we introduce auxiliary constraints, such as boundary conditions. Now, since the matrix is required to be non singular, all the corresponding eigenvalues must be different from zero.

Let us begin by analyzing the eigenvalues of the matrix associated with problem (7.4.1), relative to the Jacobi case with $n \geq 1$, $\alpha > -1$ and $\beta > -1$. More precisely, we are concerned with finding $n+1$ complex polynomials $p_{n,m}$, $0 \leq m \leq n$, of degree at most n, and $n+1$ complex numbers $\lambda_{n,m}$, $0 \leq m \leq n$, such that

$$(8.1.1) \qquad \begin{cases} p'_{n,m}(\eta_i^{(n)}) = \lambda_{n,m}\, p_{n,m}(\eta_i^{(n)}), & 1 \leq i \leq n, \\ p_{n,m}(\eta_0^{(n)}) = \lambda_{n,m}\, p_{n,m}(\eta_0^{(n)}), & \end{cases}$$

$$0 \leq m \leq n.$$

Each $p_{n,m}$, $0 \leq m \leq n$ (different from the zero function) is determined up to a constant factor. Let us suppose that $p_{n,m}(\eta_0^{(n)}) \neq 0$, for $m = 0$. Then, it is easy to realize that $\lambda_{n,0} = 1$. On the contrary, the remaining $\lambda_{n,m}$, $1 \leq m \leq n$, are eigenvalues of the $n \times n$ matrix, obtained by assuming $\sigma = 0$ in (7.4.1) (in (7.4.3) we have an example for $n = 3$). Therefore, except for $m = 0$, the eigenfunctions satisfy $p_{n,m}(\eta_0^{(n)}) = 0$.

In figures 8.1.1 to 8.1.4, we give the plot of the eigenvalues $\lambda_{n,m}$, $1 \leq m \leq n$, in the window $[-4, 12] \times [-12, 12]$ of the complex plane. We considered respectively the four cases: $\alpha = \beta = -.7,\ -.5,\ 0,\ .5$. The integer n varies from 4 to 10. The eigenvalues, corresponding to the same value of n, have been joined together by segments.

In the first three cases, all the eigenvalues lie in the half plane of the complex numbers with positive real part. In general, we are induced to conjecture that $\mathrm{Re}\lambda_{n,m} > 0$, $1 \leq m \leq n$, when $\alpha > -1, -1 < \beta \leq 0$. Unfortunately, we do not have the proof of this fact. The statement has been proven in SOLOMONOFF and TURKEL(1989), for the Chebyshev case ($\alpha = \beta = -.5$), and could be adapted to other situations. Anyway, the proof is quite technical and we omit it.

We further characterize the eigenvalues with the following proposition.

Theorem 8.1.1 - *For* $1 \leq m \leq n$, $\lambda_{n,m}$ *satisfies the relation*

$$(8.1.2) \qquad \lambda_{n,m}^n + \sum_{j=1}^{n} \left(\frac{d^j}{dx^j} \tilde{l}_0^{(n)}(-1) \right) \lambda_{n,m}^{n-j} = 0,$$

where the Lagrange polynomial $\tilde{l}_0^{(n)}$ *is defined in* (3.2.8).

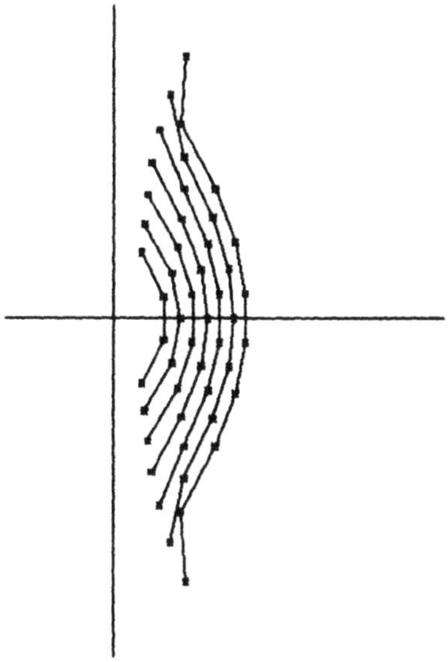

Figure 8.1.1 - Eigenvalues $\{\lambda_{n,m}\}_{1\leq m\leq n}$ for $4 \leq n \leq 10$ and $\alpha = \beta = -.7$.

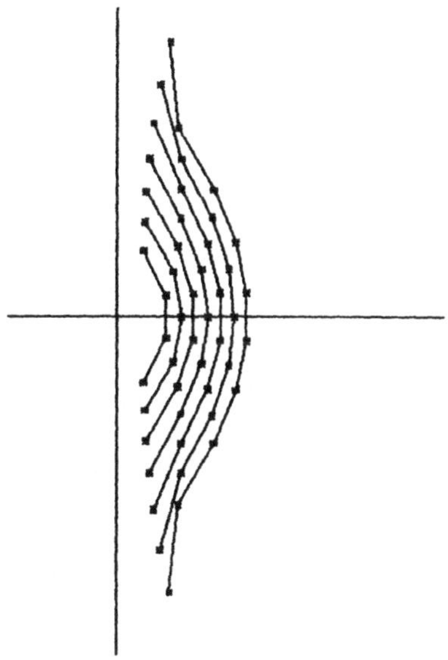

Figure 8.1.2 - Eigenvalues $\{\lambda_{n,m}\}_{1\leq m\leq n}$ for $4 \leq n \leq 10$ and $\alpha = \beta = -.5$.

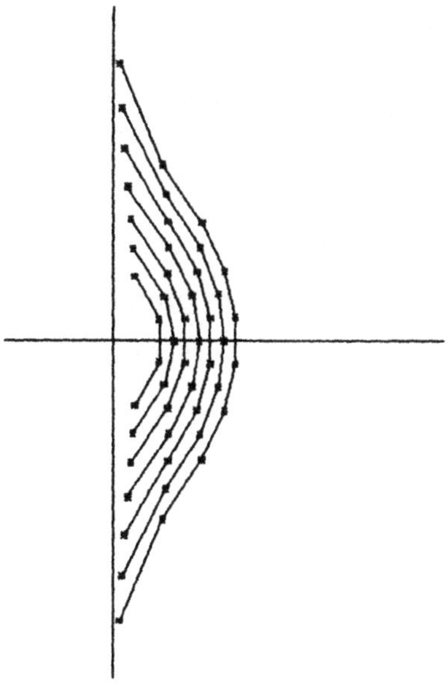

Figure 8.1.3 - Eigenvalues $\{\lambda_{n,m}\}_{1\leq m\leq n}$ for $4 \leq n \leq 10$ and $\alpha = \beta = 0$.

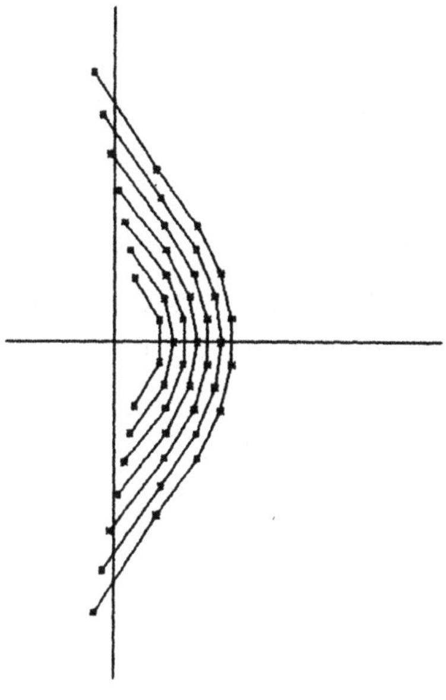

Figure 8.1.4 - Eigenvalues $\{\lambda_{n,m}\}_{1\leq m\leq n}$ for $4 \leq n \leq 10$ and $\alpha = \beta = .5$.

Proof - The eigenfunctions are solutions of the equation

$$(8.1.3)\qquad p'_{n,m} = \lambda_{n,m}p_{n,m} + p'_{n,m}(-1)\,\tilde{l}_0^{(n)}, \quad \text{in } \bar{I}=[-1,1],\ 1\le m\le n.$$

By differentiating the above expression k times, we get by recursion

$$(8.1.4)\qquad \frac{d^{k+1}}{dx^{k+1}}p_{n,m} = \lambda^{k+1}_{n,m}p_{n,m} + p'_{n,m}(-1)\left[\lambda^k_{n,m}\,\tilde{l}_0^{(n)} + \sum_{j=1}^{k}\left(\frac{d^j}{dx^j}\tilde{l}_0^{(n)}\right)\lambda^{k-j}_{n,m}\right],$$

$$1\le m\le n,\ k\ge 1.$$

Recalling that the derivative of order $n+1$ of a polynomial of degree n is zero, and $p_{n,m}(-1)=0,\ 1\le m\le n$, we take $k=n$ in (8.1.4) and evaluate at the point $x=-1$. This yields (8.1.2) (it is evident from (8.1.3) that $p'_{n,m}(-1)\ne 0,\ 1\le m\le n$).

■

In other words, the $\lambda_{n,m}$'s are the roots of the characteristic polynomial of degree n associated with our eigenvalue problem. By directly solving the differential equation in (8.1.3), we obtain the following explicit expression for the eigenfunctions:

$$(8.1.5)\qquad p_{n,m}(x) = p'_{n,m}(-1)\,e^{\lambda_{n,m}x}\int_{-1}^{x} e^{-\lambda_{n,m}t}\,\tilde{l}_0^{(n)}(t)\,dt, \qquad x\in\bar{I},\ 1\le m\le n.$$

Some properties can be deduced from formulas (8.1.2) and (8.1.5), but the most interesting conjectures follow from numerical experiments. We will give other specifications in section 8.3. We must pay attention when we carry out tests on problem (8.1.1). Indeed, in TREFETHEN and TRUMMER(1987), the authors note that desagreeable rounding errors occur when computing the eigenvalues without appropriate machine accuracy. Conclusions from careless computations could sometimes be misleading.

Let us procced with our investigation. The eigenvalue problem associated with (7.4.4) is

$$(8.1.6)\qquad \begin{cases} p'_{n,m}(\eta_i^{(n)}) = \lambda_{n,m}\,p_{n,m}(\eta_i^{(n)}), & 1\le i\le n,\\[2mm] p'_{n,m}(\eta_0^{(n)}) + \gamma p_{n,m}(\eta_0^{(n)}) = \lambda_{n,m}\,p_{n,m}(\eta_0^{(n)}), \end{cases}$$

$$0\le m\le n.$$

We find from numerical tests that the $\lambda_{n,m}$'s have positive real part when $-1 < \beta \leq 0$ and $\gamma > \gamma_n$, where γ_n, $n \geq 1$, is a positive real constant. This result is found in FUNARO and GOTTLIEB(1988) for $\alpha = \beta = -1/2$. Here, we give a simplified version in the Legendre case.

Theorem 8.1.2 - *Let* $\alpha = \beta = 0$. *Then, for any* $n \geq 1$, *there exists a constant* $\gamma_n > 0$, *such that, for* $\gamma > \gamma_n$, *all the eigenvalues of problem* (8.1.6) *satisfy* $\text{Re}\lambda_{n,m} > 0$, $0 \leq m \leq n$.

Proof - We first note that $p_{n,m}(\eta_0^{(n)}) \neq 0$, $0 \leq m \leq n$. Otherwise, we would have the differential equation $p'_{n,m} = \lambda_{n,m} p_{n,m}$ which is satisfied only for $p_{n,m} \equiv \mathbf{0}$. Then, for $0 \leq m \leq n$, we get the relations (the overbar denotes the complex conjugate)

$$(8.1.7) \qquad \frac{d}{dx}(|p_{n,m}|^2)(\eta_i^{(n)}) = (p_{n,m}\bar{p}'_{n,m} + p'_{n,m}\bar{p}_{n,m})(\eta_i^{(n)})$$

$$= (\bar{\lambda}_{n,m} p_{n,m}\bar{p}_{n,m} + \lambda_{n,m} p_{n,m}\bar{p}_{n,m})(\eta_i^{(n)}) = 2[|p_{n,m}|(\eta_i^{(n)})]^2 \text{Re}\lambda_{n,m}, \quad 1 \leq i \leq n,$$

$$(8.1.8) \qquad 2\gamma[|p_{n,m}|(\eta_0^{(n)})]^2 + \frac{d}{dx}(|p_{n,m}|^2)(\eta_0^{(n)}) = 2[|p_{n,m}|(\eta_0^{(n)})]^2 \text{Re}\lambda_{n,m}.$$

We multiply the expressions in (8.1.7), (8.1.8) by the weights $\tilde{w}_i^{(n)}$, $0 \leq i \leq n$, given in (3.5.6). By summing over i, one obtains

$$(8.1.9) \qquad 2\gamma[|p_{n,m}|(\eta_0^{(n)})]^2 \tilde{w}_0^{(n)} + \sum_{i=0}^{n} \frac{d}{dx}(|p_{n,m}|^2)(\eta_i^{(n)})\, \tilde{w}_i^{(n)}$$

$$= 2\text{Re}\lambda_{n,m} \sum_{i=0}^{n} [|p_{n,m}|(\eta_i^{(n)})]^2 \tilde{w}_i^{(n)}, \qquad 0 \leq m \leq n.$$

Since the weights are positive, it is sufficient to show that the term on the left-hand side of (8.1.9) is also positive. By virtue of formula (3.5.1) and theorem 3.5.1 ($w \equiv 1$), considering that $\frac{d}{dx}(|p_{n,m}|^2) \in \mathbf{P}_{2n-1}$, we get

$$(8.1.10)\qquad 2\gamma[p_{n,m}|(\eta_0^{(n)})]^2\tilde{w}_0^{(n)} + \sum_{i=0}^{n}\frac{d}{dx}(|p_{n,m}|^2)(\eta_i^{(n)})\ \tilde{w}_i^{(n)}$$

$$= 2\gamma[|p_{n,m}|(\eta_0^{(n)})]^2\tilde{w}_0^{(n)} + \int_{-1}^{1}\frac{d}{dx}|p_{n,m}|^2\ dx$$

$$= (2\gamma\tilde{w}_0^{(n)} - 1)[|p_{n,m}|(\eta_0^{(n)})]^2 + [|p_{n,m}|(\eta_n^{(n)})]^2, \qquad 0 \le m \le n.$$

By taking $\gamma_n := (2\tilde{w}_0^{(n)})^{-1} = \frac{1}{4}n(n+1)$, we conclude the proof.

■

We note that γ_n must be proportional to n^2, to ensure that the eigenvalues have positive real part. This is also true for the Chebyshev case. In general, there is a real eigenvalue growing linearly with γ. When γ tends to infinity, this eigenvalue diverges and the remaining n eigenvalues approach those of problem (8.1.1) for $1 \le m \le n$. Heuristically, for $\gamma = +\infty$, we are forcing the eigenfunctions to satisfy a vanishing boundary condition at the point $x = -1$.

Setting $q = \lambda p$ in (7.4.5) and arguing as in theorem 8.1.1, we obtain, for any $\gamma \in \mathbf{R}$

$$(8.1.11)\qquad \lambda_{n,m}^{n+1} + \gamma\lambda_{n,m}^{n} + \gamma\sum_{j=1}^{n}\left(\frac{d^j}{dx^j}\tilde{l}_0^{(n)}(-1)\right)\lambda_{n,m}^{n-j} = 0, \qquad 0 \le m \le n.$$

For boundary conditions at $x = 1$, similar statements hold. Now, the counterparts of (8.1.1) and (8.1.6) are respectively the following ones:

$$(8.1.12)\qquad \begin{cases} -p'_{n,m}(\eta_i^{(n)}) = \lambda_{n,m}\ p_{n,m}(\eta_i^{(n)}), & 1 \le i \le n, \\ p_{n,m}(\eta_n^{(n)}) = \lambda_{n,m}\ p_{n,m}(\eta_n^{(n)}), \end{cases}$$

$$(8.1.13)\qquad \begin{cases} -p'_{n,m}(\eta_i^{(n)}) = \lambda_{n,m}\ p_{n,m}(\eta_i^{(n)}), & 1 \le i \le n, \\ -p'_{n,m}(\eta_n^{(n)}) + \gamma p_{n,m}(\eta_n^{(n)}) = \lambda_{n,m}\ p_{n,m}(\eta_n^{(n)}), \end{cases}$$

$$0 \le m \le n.$$

The change in the sign ensures that the eigenvalues verify $\mathrm{Re}\lambda_{n,m} > 0$, $0 \le m \le n$, provided that $-1 < \alpha \le 0$, $\beta > -1$, and $\gamma > \gamma_n > 0$.

Finally, the analysis of the eigenvalue problems associated to (7.4.7) and (7.4.8) leads to positive real parts when $\alpha > -1$, $-1 < \beta \le 0$, $\gamma > \gamma_n > 0$, and when A is a non negative function.

8.2 Eigenvalues of higher-order operators

Let us begin with the discussion of second-order problems in the Jacobi case. The eigenvalue problem corresponding to (7.4.9) consists in finding $n+1$ complex polynomials $p_{n,m}$, $0 \le m \le n$, of degree at most n ($n \ge 2$), and $n+1$ complex numbers $\lambda_{n,m}$, $0 \le m \le n$, such that

$$(8.2.1) \qquad \begin{cases} -p''_{n,m}(\eta_i^{(n)}) &= \lambda_{n,m} p_{n,m}(\eta_i^{(n)}), \qquad 1 \le i \le n-1, \\ p_{n,m}(\eta_0^{(n)}) &= \lambda_{n,m} p_{n,m}(\eta_0^{(n)}), \\ p_{n,m}(\eta_n^{(n)}) &= \lambda_{n,m} p_{n,m}(\eta_n^{(n)}), \end{cases}$$

$$0 \le m \le n.$$

We assume that $p_{n,0}(\eta_0^{(n)}) \ne 0$ and $p_{n,n}(\eta_n^{(n)}) \ne 0$. This implies that $\lambda_{n,0} = \lambda_{n,n} = 1$. The other $n-1$ eigenvalues are obtained from the reduced system (7.4.9) after setting $\sigma_1 = \sigma_2 = 0$ (see for instance (7.4.11), corresponding to the case $n = 3$). The relative eigenfunctions satisfy $p_{n,m}(\eta_0^{(n)}) = p_{n,m}(\eta_n^{(n)}) = 0$, $1 \le m \le n-1$.

To construct the characteristic polynomial, we argue as in theorem 8.1.1. We briefly outline the proof. This time, we start from the relation

$$(8.2.2) \qquad -p''_{n,m} = \lambda_{n,m} p_{n,m} + p''_{n,m}(-1)\, \tilde{l}_0^{(n)} + p''_{n,m}(1)\, \tilde{l}_n^{(n)}, \qquad 1 \le m \le n-1.$$

We differentiate both the terms of equality (8.2.2), until the left-hand side vanishes. Evaluating the resulting expression at the points $x = \pm 1$, leads to a 2×2 linear system in the unknowns $p''_{n,m}(-1)$ and $p''_{n,m}(1)$. The eigenfunction $p_{n,m}$ in (8.2.2) is not uniquely determined, since it can differ by a multiplicative constant. Hence, the determinant of the 2×2 system must vanish. Imposing this final condition yields the characteristic polynomial.

Also for problem (8.2.1), we have eigenvalues with positive real part. We first prove a preliminary result in the ultraspherical case.

Lemma 8.2.1 - *Let* $\nu := \alpha = \beta$ *with* $-1 < \nu \leq 1$. *Then we can find a constant* $C > 0$ *such that, for any* $n \geq 2$ *and* $p \in \mathbf{P}_n$ *satisfying* $p(\pm 1) = 0$, *one has*

$$\int_{-1}^{1} p'(pw)'\ dx \ \geq\ C \int_{-1}^{1} [p']^2 w\ dx. \tag{8.2.3}$$

Proof - We follow the guideline of the proof given in CANUTO and QUARTERONI(1981), in the case $\nu = -1/2$. First we have

$$\int_{-1}^{1} p'(pw)'\ dx \ =\ \int_{-1}^{1} [p']^2 w\ dx \ -\ \frac{1}{2}\int_{-1}^{1} p^2 w''\ dx, \tag{8.2.4}$$

where we integrated by parts, using the condition $p(\pm 1) = 0$. If $0 \leq \nu \leq 1$, the proof is complete with $C = 1$, since $w'' \leq 0$. On the other hand, let $-1 < \nu < 0$ and define $\tau := \frac{\nu - 1}{2\nu} > 0$. Then, we have

$$\int_{-1}^{1} p'(pw)'\ dx \ =\ \int_{-1}^{1} [p']^2 w\ dx \ +\ 2\tau \int_{-1}^{1} pp'w'\ dx \tag{8.2.5}$$

$$+\ \tau^2 \int_{-1}^{1} p^2 (w')^2 w^{-1}\ dx \ +\ (\tfrac{1}{2} - \tau) \int_{-1}^{1} [p^2]' w'\ dx \ -\ \tau^2 \int_{-1}^{1} p^2 (w')^2 w^{-1}\ dx$$

$$=\ \int_{-1}^{1} (p'w \ +\ \tau p w')^2 w^{-1}\ dx \ +\ \int_{-1}^{1} p^2 \left[(\tau - \tfrac{1}{2})w'' \ -\ \tau^2 (w')^2 w^{-1}\right] dx.$$

Considering that

$$(\tau - \tfrac{1}{2})w'' \ -\ \tau^2 (w')^2 w^{-1} \ \geq\ -\ \frac{1+\nu}{4\nu}\ w'', \tag{8.2.6}$$

formula (8.2.5) yields

$$\int_{-1}^{1} p'(pw)'\ dx \ \geq\ -\ \frac{1+\nu}{4\nu} \int_{-1}^{1} p^2 w''\ dx. \tag{8.2.7}$$

Combining with (8.2.4), we finally get

$$(8.2.8) \qquad \int_{-1}^{1} p'(pw)'\ dx \ \geq \ \int_{-1}^{1} [p']^2 w\ dx \ + \ \frac{2\nu}{1+\nu}\int_{-1}^{1} p'(pw)'\ dx.$$

Inequality (8.2.3) follows with $C = \frac{1+\nu}{1-\nu}$.

■

Another proof of lemma 8.2.1 is given in BERNARDI and MADAY(1989).
When the polynomials are complex, the proof of lemma 8.2.1 is identical to that given above. In this way, we can find a constant $C > 0$, such that

$$(8.2.9) \qquad \mathrm{Re}\left(\int_{-1}^{1} p'(\bar{p}w)'\ dx\right) \ \geq \ C\int_{-1}^{1} |p'|^2 w\ dx.$$

We can now state the next proposition.

Theorem 8.2.2 - *Let* $\nu := \alpha = \beta$ *with* $-1 < \nu \leq 1$. *Then, for any* $n \geq 2$, *the eigenvalues of problem* (8.2.1) *satisfy* $\mathrm{Re}\lambda_{n,m} > 0,\ 1 \leq m \leq n-1$.

Proof - We remark that formula (3.5.1) and theorem 3.5.1 also apply to complex polynomials. It suffices to argue with the real and imaginary parts separately.
Recalling that $p_{n,m}(\pm 1) = 0$, we use integration by parts to obtain

$$(8.2.10) \qquad \int_{-1}^{1} p'_{n,m}(\bar{p}_{n,m}w)'\ dx \ = \ -\int_{-1}^{1} p''_{n,m}\bar{p}_{n,m}w\ dx$$

$$= -\sum_{i=0}^{n} p''_{n,m}(\eta_i^{(n)})\bar{p}_{n,m}(\eta_i^{(n)})\tilde{w}_i^{(n)} \ = \ \lambda_{n,m}\sum_{i=0}^{n}[|p_{n,m}|(\eta_i^{(n)})]^2\tilde{w}_i^{(n)}, \qquad 1 \leq m \leq n-1.$$

We first note that $p'_{n,m} \not\equiv 0$ for all m. Otherwise, the vanishing conditions at the boundaries would imply $p_{n,m} \equiv 0$ for some m. Recalling (8.2.9), we easily conclude the proof, since the real part of the left-hand side in (8.2.10) is strictly positive.

■

A straightforward and meaningful corollary is obtained by examining relation (8.2.10).

Theorem 8.2.3 - *Let* $\alpha = \beta = 0$. *Then, for any* $n \geq 2$, *the eigenvalues of problem* (8.2.1) *are real and strictly positive.*

Proof - Given that $w \equiv 1$, it is sufficient to recall the equality

$$(8.2.11) \qquad \int_{-1}^{1} p'_{n,m}(\bar{p}_{n,m}w)'\,dx = \int_{-1}^{1} |p'_{n,m}|^2\,dx > 0, \qquad 1 \leq m \leq n-1.$$

■

For different values of the parameters α and β, problem (8.2.1) still furnishes real eigenvalues, though the proof of this statement is more involved than for the Legendre case. By studying the roots of the characteristic polynomial, GOTTLIEB and LUSTMAN(1983) give the proof in the Chebyshev case. To this end, they analyze the eigenvalues of two auxiliary first-order problems. Other cases are still an open question. For $-1 < \alpha < 1$ and $-1 < \beta < 1$, the numerical experiments generate real eigenvalues, and some complex eigenvalues with relatively small imaginary part. In general, the analysis of the eigenvalue problem related to (7.4.12) does not lead to real eigenvalues, due to the presence of the first-order derivative (see previous section). However, when A and B are sufficiently close to zero, the second order operator dominates, and the eigenvalues are a small perturbation of those in (8.2.1). Thus, their real part is positive.

We examine the eigenvalue problems associated to (7.4.13) and (7.4.16). Namely, for $0 \leq m \leq n$, we have

$$(8.2.12) \qquad \begin{cases} -p''_{n,m}(\eta_i^{(n)}) + \mu p_{n,m}(\eta_i^{(n)}) = \lambda_{n,m} p_{n,m}(\eta_i^{(n)}), & 1 \leq i \leq n-1, \\ -p'_{n,m}(\eta_0^{(n)}) = \lambda_{n,m} p_{n,m}(\eta_0^{(n)}), & \\ p'_{n,m}(\eta_n^{(n)}) = \lambda_{n,m} p_{n,m}(\eta_n^{(n)}), & \end{cases}$$

$$(8.2.13) \qquad \begin{cases} -p''_{n,m}(\eta_i^{(n)}) + \mu p_{n,m}(\eta_i^{(n)}) = \lambda_{n,m} p_{n,m}(\eta_i^{(n)}), \qquad 1 \leq i \leq n-1, \\ -p''_{n,m}(\eta_0^{(n)}) + \mu p_{n,m}(\eta_0^{(n)}) - \gamma p'_{n,m}(\eta_0^{(n)}) = \lambda_{n,m} p_{n,m}(\eta_0^{(n)}), \\ -p''_{n,m}(\eta_n^{(n)}) + \mu p_{n,m}(\eta_n^{(n)}) + \gamma p'_{n,m}(\eta_n^{(n)}) = \lambda_{n,m} p_{n,m}(\eta_n^{(n)}), \end{cases}$$

where $\mu > 0$ and $\gamma \in \mathbf{R}$ with $\gamma \neq 0$.

We note that constant polynomials are eigenfunctions of problem (8.2.13), corresponding to the eigenvalue μ. Very little theory has been developed. We only present the following result.

Theorem 8.2.4 - *Let* $\alpha = \beta = 0$. *Then, for any* $n \geq 2$, *if* $\gamma = \frac{1}{2}n(n+1)$, *the eigenvalues of problem* (8.2.13) *are real and strictly positive.*

Proof - Noting that $\gamma\tilde{w}_0^{(n)} = \gamma\tilde{w}_n^{(n)} = 1$ (see (3.5.6)), we get

$$(8.2.14) \qquad \int_{-1}^{1} |p'_{n,m}|^2 dx \ + \ \mu \sum_{i=0}^{n} [|p_{n,m}|(\eta_i^{(n)})]^2 \tilde{w}_i^{(n)}$$

$$= [p'_{n,m}\bar{p}_{n,m}](1) \ - \ [p'_{n,m}\bar{p}_{n,m}](-1) \ - \ \int_{-1}^{1} p''_{n,m}\bar{p}_{n,m} \ dx \ + \ \mu \sum_{i=0}^{n} [|p_{n,m}|(\eta_i^{(n)})]^2 \tilde{w}_i^{(n)}$$

$$= \gamma\tilde{w}_n^{(n)}[p'_{n,m}\bar{p}_{n,m}](1) - \gamma\tilde{w}_0^{(n)}[p'_{n,m}\bar{p}_{n,m}](-1) + \sum_{i=0}^{n} [(-p''_{n,m} + \mu p_{n,m})\bar{p}_{n,m}](\eta_i^{(n)})\tilde{w}_i^{(n)}$$

$$= \lambda_{n,m} \sum_{i=0}^{n} [|p_{n,m}|(\eta_i^{(n)})]^2 \tilde{w}_i^{(n)}, \qquad 0 \leq m \leq n.$$

The proof follows easily.

■

Real or complex eigenvalues, with a relatively small imaginary part, are obtained for many other values of the parameters α and β, both for problems (8.2.12) and (8.2.13). In the latter case, γ is required to be positive and larger than a constant γ_n, which is in general proportional to n^2. To our knowledge, currently, there are no theoretical results to support these conjectures.

It is worthwhile to mention that an incorrect specification of the sign of the equation, when imposing boundary conditions (see for instance (7.4.13)), can result in a matrix with at least one negative eigenvalue. For example, replacing the second equation in (8.2.12) by $p'_{n,m}(\eta_0^{(n)}) = \lambda_{n,m}p_{n,m}(\eta_0^{(n)})$, we still obtain real eigenvalues, but one of them is negative.

Let us investigate the case of fourth-order operators. Here, $\lambda_{n,m}$, $1 \leq m \leq n-1$, and $r_{n,m}$, $1 \leq m \leq n-1$, are respectively the eigenvalues and the eigenfunctions associated to problem (7.4.22). Therefore, we can write $r_{n,m}^{IV}(\eta_i^{(n)}) = \lambda_{n,m} r_{n,m}(\eta_i^{(n)})$, $1 \leq i \leq n-1$, and $r_{n,m}(\pm 1) = r'_{n,m}(\pm 1) = 0$.
Before proceeding with our analysis, we state the following proposition.

Lemma 8.2.5 - *Let* $\nu := \alpha = \beta$ *with* $-1 < \nu \leq 1$. *Then, we can find a constant* $C > 0$, *such that, for any* $n \geq 2$ *and* $p \in \mathbf{P}_{n+2}$ *satisfying* $p(\pm 1) = p'(\pm 1) = 0$, *one has*

$$\int_{-1}^{1} p''(pw)''dx \ \geq \ C \int_{-1}^{1} [p'']^2 w \ dx. \tag{8.2.15}$$

■

The proof of the above result can be found in BERNARDI and MADAY(1990). It is quite technical, but very similar to that of theorem 8.2.1. The same proof, adapted to the case of complex polynomials, yields the inequality

$$\mathrm{Re}\left(\int_{-1}^{1} p''(\bar{p}w)''dx\right) \ \geq \ C \int_{-1}^{1} |p''|^2 w \ dx. \tag{8.2.16}$$

We are now ready to prove the following proposition.

Theorem 8.2.6 - *Let* $\nu := \alpha = \beta$ *with* $-1 < \nu \leq 1$. *Then, for any* $n \geq 2$, *the eigenvalues relative to the system* (7.4.22) *satisfy* $\mathrm{Re}\lambda_{n,m} > 0$, $1 \leq m \leq n-1$. *Moreover, when* $\nu = 0$, *we have* $\lambda_{n,m} \in \mathbf{R}$, $1 \leq m \leq n-1$.

Proof - One has

$$\sum_{i=0}^{n} r_{n,m}^{IV}(\eta_i^{(n)})\bar{r}_{n,m}(\eta_i^{(n)})\tilde{w}_i^{(n)} \ = \ \lambda_{n,m} \sum_{i=0}^{n} [|r_{n,m}|(\eta_i^{(n)})]^2 \tilde{w}_i^{(n)}. \tag{8.2.17}$$

We must show that the left-hand side of (8.2.17) has positive real part. Now, for $1 \leq m \leq n-1$, we can find a complex number $\zeta_{n,m} \in \mathbf{C}$ such that

$$\frac{1}{1-x^2}\, r_{n,m}(x) \ = \ \zeta_{n,m} u_n(x) \ + \ \{\textit{lower degree terms}\}, \tag{8.2.18}$$

(8.2.19) $$(1-x^2)\, r_{n,m}^{IV}(x) \;=\; n(n^2-1)(n+2)\zeta_{n,m}u_n(x) \;+\; \{\textit{lower degree terms}\},$$

where $u_n = P_n^{(\nu,\nu)}$. These relations can be checked by equating the coefficients of the monomials x^n. With the same arguments used in section 3.8 (we recall for instance (3.8.12) and (3.8.15)), it is easy to show that

(8.2.20) $$\sum_{i=0}^{n} r_{n,m}^{IV}(\eta_i^{(n)})\bar{r}_{n,m}(\eta_i^{(n)})\tilde{w}_i^{(n)}$$

$$= \int_{-1}^{1} r_{n,m}^{IV}\bar{r}_{n,m}\; w\; dx \;+\; (\|u_n\|_{w,n}^2 - \|u_n\|_w^2)\; n(n^2-1)(n+2)|\zeta_{n,m}|^2.$$

Having $\|u_n\|_{w,n} \geq \|u_n\|_w$, $\forall n \geq 1$ (see (2.2.10) and (3.8.3)), integration by parts and formula (8.2.16) yield

(8.2.21) $$\mathrm{Re}\left(\sum_{i=0}^{n} r_{n,m}^{IV}(\eta_i^{(n)})\bar{r}_{n,m}(\eta_i^{(n)})\tilde{w}_i^{(n)}\right) \;\geq\; \mathrm{Re}\left(\int_{-1}^{1} r_{n,m}^{IV}\bar{r}_{n,m}w\; dx\right)$$

$$= \mathrm{Re}\left(\int_{-1}^{1} r_{n,m}''(\bar{r}_{n,m}w)''dx\right) \;\geq\; C\int_{-1}^{1} |r_{n,m}''|^2 w\; dx \;>0, \qquad 1 \leq m \leq n-1.$$

We note that, since $r_{n,m}(\pm 1) = r_{n,m}'(\pm 1) = 0$, the last integral in (8.2.21) does not vanish (otherwise we would have $r_{n,m} \equiv 0$). Hence, by (8.2.17) we deduce $\mathrm{Re}\lambda_{n,m} > 0$, $1 \leq m \leq n-1$.

When $\nu = 0$ (thus $w \equiv 1$), the right-hand side in (8.2.20) turns out to be real, after integrating by parts twice.

■

Real and positive eigenvalues are also observed for other values of the parameters α and β, and also for different type of boundary conditions such as those in (7.4.23). Hints for the determination of the characteristic polynomial are given in FUNARO and HEINRICHS(1990).

Similar arguments hold for the case of unbounded domains. Let us see some example. Here, we denote by $\lambda_{n,m}$ and $p_{n,m}$, $0 \leq m \leq n-1$, respectively the eigenvalues and the eigenfunctions related to problem (7.4.26). In particular, the $p_{n,m}$'s are complex polynomials of degree at most $n-1$. Again, we have $\lambda_{n,0} = 1$ and $p_{n,m}(0) = 0$, for $1 \leq m \leq n-1$.

We recall that the space of Laguerre functions $\mathbf{S}_n$ has been introduced in section 6.7. As in the previous cases, we need a preparatory result.

Lemma 8.2.7 - *Let* $-1 < \alpha < 1$ *and* $v(x) := x^\alpha e^x$, $x \in]0, +\infty[$. *Then, we can find a constant* $C > 0$ *such that, for any* $n \geq 1$ *and* $P \in \mathbf{S}_{n-1}$, *satisfying* $P(0) = 0$, *one has*

$$(8.2.22) \qquad \int_0^{+\infty} P'(Pv)'dx \; + \; \mu \int_0^{+\infty} P^2 v \; dx \; \geq C \int_0^{+\infty} [(P')^2 + P^2] \; v \; dx,$$

where $\mu > \frac{1}{2} \max\{1, \frac{1}{1-\alpha}\}$.

■

For the proof we refer to KAVIAN(1990). Combining this lemma with (7.2.25) and the experience we have gained in the previous examples, we obtain the following result.

Theorem 8.2.8 - *Let* $-1 < \alpha < 1$ *and* $\mu > \frac{1}{2} \max\{1, \frac{1}{1-\alpha}\}$. *Then, for any* $n \geq 1$, *the eigenvalues relative to the system* (7.4.26) *satisfy* $\mathrm{Re}\lambda_{n,m} > 0$, $0 \leq m \leq n-1$.

■

Actually, in the Laguerre case, the eigenvalues of the operator considered are real. Nevertheless, there is no proof of this.

Concerning problem (7.5.4), where the unknowns are the point values of a scaled function, the eigenvalues are identical to those of problem (7.4.26). Actually, definition (7.5.1) corresponds to a linear change of variables which does not alter the spectrum of the derivative matrix.

For the matrix in (7.2.15), with the vanishing boundary condition at the point $x = 0$, the eigenvalues are complex, but their real part is positive provided that $\alpha \leq 0$.

In particular, when $\alpha = 0$, the real part of these eigenvalues is constantly equal to $\frac{1}{2}$. The proof of this fact is quite simple and is left as exercise.

Finally, we consider problem (7.4.28) in which the eigenvalues and eigenfunctions are respectively denoted by $\lambda_{n,m}$ and $p_{n,m}$, $1 \leq m \leq n$. This time, the theory is simple. Comparing (7.4.28) with the differential equation (1.7.1), we get $\lambda_{n,m} = 2(m-1) + \mu$ and $p_{n,m} = H_{m-1}$, $1 \leq m \leq n$. Therefore, the eigenvalues are real and positive when $\mu > 0$.

8.3 Condition number

The *condition number* is a useful quantity to estimate the speed of convergence of an iterative method for the solution of linear systems.

It is well-known that a real $n \times n$ matrix D corresponds to a linear operator from $\mathbf{R}^n$ to $\mathbf{R}^n$. Denoting by $\mathcal{L}(\mathbf{R}^n, \mathbf{R}^n)$ the space of linear mappings from $\mathbf{R}^n$ in $\mathbf{R}^n$, we can define the norm of D (see for instance ISAACSON and KELLER(1966), p.37) as follows:

$$(8.3.1) \qquad \|D\|_{\mathcal{L}(\mathbf{R}^n,\mathbf{R}^n)} := \sup_{\substack{\bar{x}\in\mathbf{R}^n \\ \bar{x}\neq 0}} \frac{\|D\bar{x}\|}{\|\bar{x}\|} = \sup_{\substack{\bar{x}\in\mathbf{R}^n \\ \|\bar{x}\|=1}} \|D\bar{x}\|,$$

where by $\|\cdot\|$ we indicate the canonical norm in $\mathbf{R}^n$. The expression in (8.3.1) actually satisfies all the properties required in section 2.1. In practice, the norm of D measures the maximal deformation, when applying the matrix to the vectors of the unitary sphere of $\mathbf{R}^n$.

We remark that, for any $n \times n$ matrix D and any $\bar{x} \in \mathbf{R}^n$, one has the inequality

$$(8.3.2) \qquad \|D\bar{x}\| \leq \|D\|_{\mathcal{L}(\mathbf{R}^n,\mathbf{R}^n)} \|\bar{x}\|.$$

At this point, we can define the condition number of a nonsingular matrix D. This is the real positive number given by

$$(8.3.3) \qquad \kappa(D) := \|D\|_{\mathcal{L}(\mathbf{R}^n,\mathbf{R}^n)} \|D^{-1}\|_{\mathcal{L}(\mathbf{R}^n,\mathbf{R}^n)},$$

from which one can easily deduce $\kappa(D) \geq 1$.
This quantity gives an idea of the distribution of the eigenvalues of D in the complex plane. When $\kappa(D)$ is large, we generally expect scattered eigenvalues with considerable variations in their magnitude. On the other hand, when $\kappa(D)$ is close to 1, the moduli of the eigenvalues are gathered together in a small interval. We can be more precise for symmetric matrices. In this situation, we have the relation

$$\kappa(D) = \pi(D) := \frac{\max_{1\leq m\leq n} |\lambda_m|}{\min_{1\leq m\leq n} |\lambda_m|}, \tag{8.3.4}$$

where the λ_m's are the eigenvalues of D.
When D is not symmetric we have the inequality: $1 \leq \pi(D) \leq \kappa(D)$.
Nearly all the matrices analyzed in the sequel are not symmetric. Nevertheless, for these matrices, the quantities $\pi(D)$ and $\kappa(D)$ display more or less the same behavior. So that, allowing ourselves an abuse of terminology, the two concepts will often be interchanged.

To better understand the utility of the condition number, we return to the iterative method described in section 7.6. The speed of convergence of the scheme (7.6.2) depends on the spectral radius $\rho(M)$, where $M := I - R^{-1}D$. In the Richardson method we have

$$\rho(M) = \max_{1\leq m\leq n} |1 - \theta\lambda_m|, \tag{8.3.5}$$

where the λ_m's are the eigenvalues of D, satisfying the conditions in (7.6.3).
If the eigenvalues are clustered around a particular real value, then we can find θ such that $\rho(M)$ is near zero. This will result in a fast convergence. On the contrary, if the eigenvalues are scattered, the parameter θ cannot control all of them, and $\rho(M)$ will be dangerously close to 1. In short, we have to beware of matrices with a large condition number. These are called *ill-conditioned* matrices.
Unfortunately, all the matrices considered in section 7.4 are ill-conditioned. As an example, we examine the eigenvalue problem (8.1.1). Since the $\lambda_{n,m}$'s are distinct, the corresponding matrix admits a diagonal form. From figures 8.1.1 to 8.1.4, it is clear that

the magnitude of the eigenvalues grows with n. It is especially for the eigenvalues with the largest modulus that the ratio between the real part and the imaginary part is small. According to (7.6.3), this yields a very restrictive condition on θ. For instance, in the Chebyshev case, the parameter θ is required to be proportional to $1/n^2$ to achieve convergence when applying the Richardson method. This makes the spectral radius in (8.3.5) very close to 1. Consequently, the Richardson scheme is very ineffective. The situation is worse for the Legendre case.

Concerning second-order or fourth-order problems, no improvements are observed. Although the eigenvalues are in general real and positive, they are quite scattered. For instance, referring to problem (8.2.1), it is not difficult to obtain the estimate

$$c_1 \leq |\lambda_{n,m}| \leq c_2\, n^4, \qquad 1 \leq m \leq n-1, \tag{8.3.6}$$

where c_1 and c_2 are positive constants.
Moreover, assuming that the $\lambda_{n,m}$'s are real and in increasing order, we find that $\lambda_{n,1}$ converges to a real number for $n \to +\infty$, while $\lambda_{n,n-1}$ grows like n^4. Sharper estimates are given in WEIDEMAN and TREFETHEN(1988) in the Chebyshev case, and in VANDEVEN (1990) in the Legendre case. For the proof of (8.3.6), when $-1 < \alpha < 1$, $-1 < \beta < 1$, we refer the reader to section 8.6.
Thus, the condition number of the matrix corresponding to the system (7.4.9) is expected to be very large. The minimal spectral radius obtained from (8.3.5) is $\rho = (\lambda_{n,n-1} - \lambda_{n,1})/(\lambda_{n,n-1} + \lambda_{n,1})$, relative to the choice $\theta := 2/(\lambda_{n,n-1} + \lambda_{n,1})$ (see ISAACSON and KELLER(1966), p.84). Therefore, ρ is very close to 1, and the convergence of the scheme is unbelievably slow. The situation becomes extremely bad for fourth-order problems. In HEINRICHS(1989), the author suggests a way to replace problem (7.4.9) with an equivalent one, whose matrix has a reduced condition number. Such an improvement is still insufficient when applied within the framework of iterative techniques.

It is time now to introduce the idea of ***preconditioning***. We note that the scheme (7.6.2) can be interpreted as a byproduct of the Richardson method with $\theta = 1$, applied to the system

$$(8.3.7) \qquad (R^{-1}D)\bar{p} = \bar{q}^*, \quad \text{where} \quad \bar{q}^* := R^{-1}\bar{q}.$$

Of course (8.3.7) is equivalent to (7.6.1). The basic difference is that, by an appropriate choice of the matrix R, we can modify $\kappa(R^{-1}D)$ to get a well-behaved system. Thus, we first compute the vector $\bar{q}^*$, then we solve (8.3.7) with an iterative approach. In this context, R is named *preconditioning matrix* (or *preconditioner*). Such a matrix must fulfill two requirements. First, R must be easily invertible, using an inexpensive algorithm. Second, the condition number of $R^{-1}D$ has to be as close as possible to 1. The choice $R = I$ (which leads to the Richardson method) satisfies the first restriction but not the latter. The choice $R = D$ gives $\kappa(R^{-1}D) = 1$, but the computation of R^{-1} brings us back to the initial problem. The desired R is the right balance between these two extreme cases.

The next several sections describe how to construct preconditioners for the matrices previously analyzed.

8.4 Preconditioners for second-order operators

We begin by studying problem (7.4.9). The case of first-order derivatives is more delicate and is considered later.

Let D denote the $(n+1)\times(n+1)$ matrix corresponding to the system (7.4.9) (in (7.4.10) we have an example for $n = 3$). We recall that D coincides with the second derivative matrix $-\tilde{D}_n^2$ (see section 7.2), except for two rows, which have been suitably modified to take care of the boundary conditions.

As remarked in section 8.3, the condition number of D grows at least as n^4. We are concerned with finding an appropriate preconditioner R for D. To this end, for any $n \geq 2$, we define $h_j^{(n)} := \eta_j^{(n)} - \eta_{j-1}^{(n)}$, $1 \leq j \leq n$, and $\hat{h}_j^{(n)} := \frac{1}{2}(\eta_{j+1}^{(n)} - \eta_{j-1}^{(n)})$, $1 \leq j \leq n-1$.

Next, following ORSZAG(1980), we consider the matrix $R := \{r_{ij}\}_{\substack{0\le i\le n\\ 0\le j\le n}}$, with entries

$$(8.4.1) \qquad r_{ij} := \begin{cases} 1 & \text{if } i=j=0 \text{ or } i=j=n, \\[2ex] \dfrac{-1}{h_i^{(n)}\hat{h}_i^{(n)}} & \text{if } 1\le i=j+1\le n-1, \\[2ex] \dfrac{2}{h_{i+1}^{(n)}h_i^{(n)}} & \text{if } 1\le i=j\le n-1, \\[2ex] \dfrac{-1}{h_{i+1}^{(n)}\hat{h}_i^{(n)}} & \text{if } 1\le i=j-1\le n-1, \\[2ex] 0 & \text{elsewhere.} \end{cases}$$

The matrix R represents the second-order centered finite-differences operator defined on the grid points $\eta_j^{(n)}$, $0\le j\le n$. If f is a given regular function, the matrix R applied to the vector $\{f(\eta_j^{(n)})\}_{0\le j\le n}$ furnishes an approximation of $-f''(\eta_i^{(n)})$, $1\le i\le n-1$. We remark that this kind of discretization is of local type (see section 7.2), hence it does not possess the good convergence behavior of spectral methods, even if the nodes of a Gauss formula are involved. However, R is tridiagonal, thus, the computation of R^{-1} can be performed in a fast way. Therefore, R satisfies the first requirement to be a good preconditioning matrix. Now, we must study the spectrum of $R^{-1}D$. We denote by $\Lambda_{n,m}$, $0\le m\le n$, the eigenvalues obtained after preconditioning. Two of them, say $\Lambda_{n,0}$ and $\Lambda_{n,n}$, are equal to 1. Numerical experiments indicate the existence of a constant $c>1$ (depending on α and β) such that, for any $n\ge 2$, one has

$$(8.4.2) \qquad 1 \le |\Lambda_{n,m}| < c, \qquad 1\le m\le n-1.$$

Comparing this with (8.3.6), the improvement is impressive. Furthermore, the $\Lambda_{n,m}$'s are real and positive and c is in general less than 2.5 . We can be more precise in the Chebyshev case. In fact, in HALDENWANG, LABROSSE, ABBOUDI and DEVILLE(1984), an exact expression is given for the preconditioned eigenvalues. Namely, we have

$$(8.4.3) \qquad \Lambda_{n,m} = m(m+1) \frac{\sin^2 \frac{\pi}{2n} \cos \frac{\pi}{2n}}{\sin \frac{m\pi}{2n} \sin \frac{(m+1)\pi}{2n}}, \qquad 1 \leq m \leq n-1.$$

Thus, we obtain $1 \leq \Lambda_{n,m} < \frac{\pi^2}{4}$, $0 \leq m \leq n$.

We apply the Richardson method to the system in (8.3.7) with $\theta := 2/(\pi^2/4 + 1)$. This choice minimizes the spectral radius of $M = I - \theta R^{-1} D$, which takes the value $\rho = (\pi^2/4 - 1)/(\pi^2/4 + 1) \approx 0.42$. Starting from the initial guess $\bar{p}^{(0)} = R^{-1}\bar{q}$ (see (7.6.2)), one achieves the exact solution to the system (to machine accuracy in double precision) in about twenty iterations. Though we do not have access to the explicit expression of the preconditioned eigenvalues for different values of α and β, the behavior is more or less the same.

We note that, in practice, the matrix $R^{-1}D$ is not actually required. Each iteration consists of two steps. In the first step, we evaluate the matrix-vector multiplication $\bar{p} \to D\bar{p}$. This can be carried out with the help of the FFT (see section 4.3) in the Chebyshev case. In the second step, we compute $D\bar{p} \to R^{-1}D\bar{p}$ by solving a tridiagonal linear system. In this way, we avoid storing the whole matrix R^{-1}. Moreover, the cost of this computation is proportional to n (see ISAACSON and KELLER(1966), p.55).

Symmetric preconditioners, based on finite element discretizations, have been proposed in DEVILLE and MUND(1985) and in CANUTO and QUARTERONI(1985). They allow us to apply a variation of the *conjugate gradient* iterative method for solving (8.3.7). A standard trick is to accelerate the convergence by updating the parameter θ at each iteration. This can be done in several ways. For brevity, we do not investigate this here. A survey of the most used preconditioned iterative techniques in spectral methods is given in CANUTO, HUSSAINI, QUARTERONI and ZANG(1988), p.148. Many practical suggestions are provided in BOYD(1989).

We note that R is ill-conditioned. This introduces small rounding errors when evaluating R^{-1}. Hence, after the preconditioned Richardson method has reached a steady solution, we suggest that it be refined by performing a few iterations of the unpreconditioned Richardson method to remove the rounding errors.

The matrix in (7.4.11) is obtained after eliminating the first and the last unknowns from system (7.4.10). A $(n-1)\times(n-1)$ preconditioner for this matrix is obtained by deleting the first and the last columns and rows from the matrix in (8.4.1). The preconditioned eigenvalues $\Lambda_{n,m}$, $1 \leq m \leq n-1$, coincide with those considered above. The same preconditioner R in (8.4.1) can be used for the matrix of system (7.4.12). The resulting preconditioned spectrum does not seem to be much affected by lower-order terms, provided the coefficients A and B are not too large. As an alternative, one can use a preconditioner based on a centered finite-difference discretization of the whole differential operator $-\frac{d^2}{dx^2} + A\frac{d}{dx} + B$ (see ORSZAG(1980)).

For other types of boundary conditions we could run into some difficulty. Let D denote the matrix of system (7.4.13). In this situation, tridiagonal finite-difference preconditioners seem less effective. All the preconditioned eigenvalues satisfy relation (8.4.2), except for one of them, which is real, positive, but tends to zero as $1/n^2$. This badly affects the condition number. A more appropriate $(n+1)\times(n+1)$ preconditioner $R := \{r_{ij}\}_{\substack{0\leq i\leq n\\ 0\leq j\leq n}}$ is obtained as follows:

(8.4.4)
$$r_{ij} := \begin{cases} -d_{ij}^{(1)} & \text{if } i=0,\ 0\leq j\leq n, \\[2mm] \dfrac{-1}{h_i^{(n)}\hat{h}_i^{(n)}} & \text{if } 1\leq i=j+1\leq n-1, \\[2mm] \dfrac{2}{h_{i+1}^{(n)}h_i^{(n)}} + \mu & \text{if } 1\leq i=j\leq n-1, \\[2mm] \dfrac{-1}{h_{i+1}^{(n)}\hat{h}_i^{(n)}} & \text{if } 1\leq i=j-1\leq n-1, \\[2mm] d_{ij}^{(1)} & \text{if } i=n,\ 0\leq j\leq n, \\[2mm] 0 & \text{elsewhere.} \end{cases}$$

All the resulting preconditioned eigenvalues are now well-behaved. Unfortunately, the matrix R is sparse but not banded. However, it can be decomposed by a product of three well-structured matrices, so that the global cost of the computation $\bar{p} \to R^{-1}\bar{p}$ is proportional to n (see BERTOLUZZA and FUNARO(1991)). Slight modifications to (8.4.4) lead to a suitable preconditioner for the matrix of system (7.4.16).

Finite-difference preconditioning matrices for fourth-order problems have been investigated in FUNARO and HEINRICHS(1990), both for the boundary-value problems (7.4.19) and (7.4.23). The analysis has been carried out in the Chebyshev case, but similar conclusions also apply to other situations. A little care has to be payed when dealing with the finite-difference schemes at the boundary points. Before preconditioning, the ratio between the largest and the smallest eigenvalues is proportional to n^8. The preconditioned eigenvalues satisfy relation (8.4.2). In particular, in the Chebyshev case, we can take $c = \frac{\pi^4}{16} \approx 6.08$.

For matrices related to problems defined in unbounded domains, we can use the same arguments. However, since it is not advisable to work with high degree Laguerre or Hermite polynomials (see sections 1.6 and 1.7), the matrices are not very large in practical applications. Therefore, we suggest direct methods to solve these systems.

8.5 Preconditioners for first-order operators

Denote by D the $(n+1)\times(n+1)$ matrix corresponding to the system (7.4.1). Here, the idea of using finite-differences to construct a preconditioner for D does not offer the same features emphasized in section 8.4. For instance, let $R := \{r_{ij}\}_{\substack{0\le i\le n\\ 0\le j\le n}}$ be defined as follows:

$$(8.5.1) \qquad r_{ij} := \begin{cases} 1 & \text{if } i=j=0, \\[2mm] \dfrac{-1}{h_i^{(n)}} & \text{if } 1\le i=j+1\le n, \\[2mm] \dfrac{1}{h_i^{(n)}} & \text{if } 1\le i=j\le n, \\[2mm] 0 & \text{elsewhere.} \end{cases}$$

Now, we have a bidiagonal preconditioner. Though the spread of eigenvalues of the preconditioned matrix is reduced, we are still far from the results of the previous section. The eigenvalues of $R^{-1}D$ are complex with positive real part, but they approach the imaginary axis with increasing n.

A more effective preconditioner was proposed in FUNARO(1987). First introduce the linear operator $T : \mathbf{R}^n \to \mathbf{R}^n$. For any polynomial $p \in \mathbf{P}_{n-1}$, the operator T maps the vector $\{p(\xi_j^{(n)})\}_{1\leq j\leq n}$ into the vector $\{p(\eta_i^{(n)})\}_{1\leq i\leq n}$. The way to perform this transformation is described by formula (4.4.1). The coefficients $l_j^{(n)}(\eta_i^{(n)})$, $1 \leq i \leq n$, $1 \leq j \leq n$, are the entries of the $n \times n$ matrix relative to the mapping T. Then, we define $Z := \{z_{ij}\}_{\substack{0\leq i\leq n\\ 0\leq j\leq n}}$, to be the $(n+1) \times (n+1)$ matrix

$$z_{ij} := \begin{cases} 1 & \text{if } i = j = 0, \\ l_j^{(n)}(\eta_i^{(n)}) & \text{if } 1 \leq i \leq n, \ 1 \leq j \leq n, \\ 0 & \text{elsewhere.} \end{cases} \tag{8.5.2}$$

We claim that ZR is a good preconditioning matrix for D. We remark that Z is a full matrix. On the other hand, we can find explicitly the entries of Z^{-1}. Indeed, one easy checks the relation

$$p(\xi_i^{(n)}) = \sum_{j=1}^{n} p(\eta_j^{(n)}) \frac{1+\eta_j^{(n)}}{1+\xi_i^{(n)}} \tilde{l}_j^{(n)}(\xi_i^{(n)}), \qquad 1 \leq i \leq n, \tag{8.5.3}$$

which is true for any $p \in \mathbf{P}_{n-1}$.

Clearly, (8.5.3) is the inverse of the transformation T. Noting that $(ZR)^{-1} = R^{-1}Z^{-1}$, we can evaluate $\bar{p} \to (ZR)^{-1}\bar{p}$ with a cost proportional to n^2. The computing time is further reduced, in the Chebyshev case, by using the FFT (see section 4.4). Let us examine the eigenvalues $\Lambda_{n,m}$, $0 \leq m \leq n$, of $(ZR)^{-1}D$. We have a precise answer in the Chebyshev case. Actually, one has

$$\Lambda_{n,0} = 1, \qquad \Lambda_{n,m} = m \frac{\sin \frac{\pi}{2n}}{\sin \frac{m\pi}{2n}}, \qquad 1 \leq m \leq n. \tag{8.5.4}$$

Therefore, we get $1 \leq \Lambda_{n,m} < \frac{\pi}{2}$, $0 \leq m \leq n$.

The improvement is explained by noting that the finite-differences operator R is a good approximation of the operator $Z^{-1}D$, which maps the values $p(\eta_j^{(n)})$, $0 \leq j \leq n$, into the values $p'(\xi_i^{(n)}) \approx (p(\eta_i^{(n)}) - p(\eta_{i-1}^{(n)}))/h_i^{(n)}$, $1 \leq i \leq n$. Thus, $R^{-1}(Z^{-1}D)$ is very close to the identity operator. For different values of α and β, the results are similar.

With minor modifications, the same preconditioner can be used to deal with the matrix of system (7.4.4). For more details, we refer the reader to FUNARO and GOTTLIEB(1988).

Though the results obtained with the preconditioning matrix ZR are impressive, for more general problems, such as (7.4.7), the use of a full preconditioner is not recommended. For example, consider $A \equiv 1$ in (7.4.7). The matrix $ZR + I$ turns out to be an appropriate preconditioner for $D + I$, i.e. $\kappa((ZR + I)^{-1}(D + I))$ is close to 1. Nevertheless, we cannot cheaply invert $ZR + I$. An alternative is to approximate the operator Z by a banded matrix $\hat{Z}$. In this way $\hat{Z}R + I$ will be also banded. This idea has been developed in FUNARO and ROTHMAN(1989). In that paper, the operator T is substituted by a mapping $\hat{T}$, which extrapolates the point values $p(\eta_i^{(n)})$, $1 \leq i \leq n$, with the help of second-degree polynomials. The resulting preconditioner is four-diagonal and the eigenvalues of the matrix $(\hat{Z}R + I)^{-1}(D + I)$ are complex, but clustered in a neighborhood of the real unity. The preconditioned condition number seems to be suitable for numerical applications.

8.6 Convergence of eigenvalues

We are now ready to consider the first application of spectral methods to solve several problems in physics. For example, let us analyze the following eigenvalue problem:

$$(8.6.1) \qquad \begin{cases} -\phi'' = \lambda\, \phi & \text{in } I =]-1,1[, \\ \phi(-1) = \phi(1) = 0. \end{cases}$$

It is well-known that (8.6.1) has a countable set of positive eigenvalues $\lambda_m = \frac{\pi^2}{4}m^2$, $m \geq 1$, corresponding to the normalized eigenfunctions: $\phi_m(x) = \sin\frac{\pi}{2}mx,\ x \in \bar{I}$, m even, $\phi_m(x) = \cos\frac{\pi}{2}mx,\ x \in \bar{I}$, m odd. Besides, for any $n \geq 2$, we have the eigenvalues $\lambda_{n,m}$, $1 \leq m \leq n-1$, of problem (8.2.1). These will be assumed to be real, distinct and in increasing order.

We are concerned with the asymptotic behavior of the $\lambda_{n,m}$'s when n tends to infinity. Actually, we expect the following convergence result:

$$\text{(8.6.2)} \qquad \forall m \geq 1, \qquad \lim_{n\to+\infty} \lambda_{n,m} = \lambda_m.$$

This would suggest a technique for approximating the eigenvalues of differential operators, which is a crucial problem in many applications of mathematical physics (see for instance COURANT and HILBERT(1953), Vol.1). This procedure is investigated for the solution of a large number of problems in CALOGERO and FRANCO (1985). In that paper, discretizations of the derivative operators (denoted by *Lagrangian differentiations*) are obtained starting from an arbitrary set of nodes, not necessarily related to Gauss type formulas. Heuristic estimates of the rate of convergence of the limit in (8.6.2) are presented in CALOGERO(1983) for equispaced nodes and DURAND(1985) for Gauss type nodes. According to these papers the convergence is exponential, and, as pointed out by the authors in their numerical experiments, the results are definitely better than those obtained by other techniques, such as the finite-differences method.

Remarks about the behavior of the eigenvalues in the Chebyshev case can be found in WEIDEMAN and TREFETHEN(1988). The authors make the following observation. For any fixed $n \geq 2$, about two thirds of the eigenvalues in (8.2.1) are very close to the corresponding eigenvalues in (8.6.1). The remaining part of the spectrum deviates sharply. This behavior justifies the estimate (8.3.6), since $\lambda_{n,n-1}$ turns out to be proportional to n^4, while λ_n just grows like n^2. This fact is explained by the poor approximation properties of the eigenfunctions with high frequency oscillations, where the number of nodes is not sufficient to recover a satisfactory resolution (see section 6.8). This does not imply that (8.6.2) is false. In fact, for fixed m, the eigenvalue $\lambda_{n,m}$ begins to converge for sufficiently large n.

In figure 8.6.1, for the Legendre case, in the interval $[0, 600]$, we show the exact eigenvalues λ_m, $1 \leq m \leq 15$ (first line) and the computed eigenvalues $\lambda_{n,m}$, $1 \leq m \leq n-1$ (next lines), when n varies from 6 to 11.

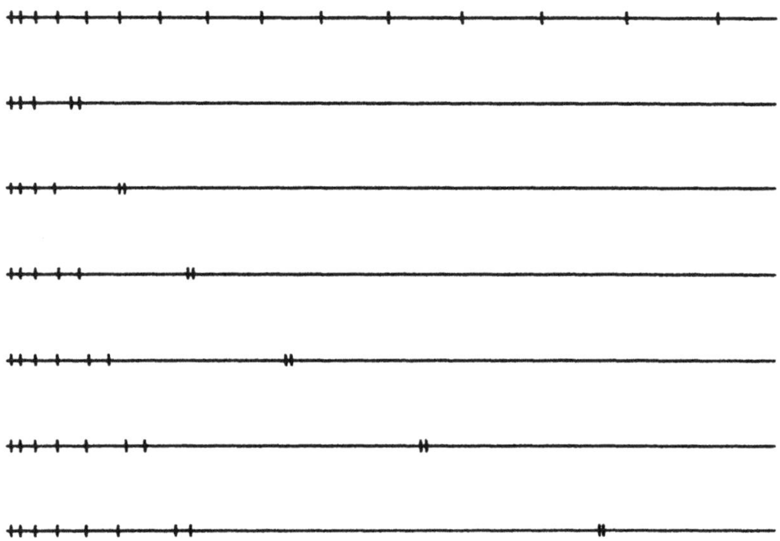

Figure 8.6.1 - Behavior of the eigenvalues of problem (8.2.1) in the Legendre case when $6 \leq n \leq 11$.

In order to check (8.3.6), we first need the following result. We recall that the polynomial space $\mathbf{P}_n^0$, $n \geq 2$, is defined in (6.4.1).

Lemma 8.6.1 - *Let* $-1 < \alpha < 1$ *and* $-1 < \beta < 1$. *Then we can find two constants* $C_1 > 0$, $C_2 > 0$, *such that, for any* $n \geq 2$ *and* $p \in \mathbf{P}_n^0$, *one has*

$$\left\| \frac{p}{1-x^2} \right\|_w \leq C_1 \, \|p'\|_w, \tag{8.6.3}$$

(8.6.4) $$\left(\int_{-1}^{1} p'(pw)'\,dx\right)^{\frac{1}{2}} \leq C_2\,\|p'\|_w.$$

Proof - Inequality (8.6.3) is a byproduct of (5.7.5). We provide however some details. With the help of the Schwarz inequality, one obtains

(8.6.5) $$\left\|\frac{p}{1-x^2}\right\|_w^2 = \int_{-1}^{1}\left(\int_{-1}^{x} p'(s)ds\right)^2 \frac{w(x)}{(1-x^2)^2}\,dx$$

$$\leq \int_{-1}^{1}\left(\int_{-1}^{x}[p'(s)]^2 w(s)ds \cdot \int_{-1}^{x}\frac{1}{w(s)}\,ds\right)\frac{w(x)}{(1-x^2)^2}\,dx$$

$$\leq \|p'\|_w^2 \int_{-1}^{1}\left(\int_{-1}^{x}\frac{1}{w(s)}\,ds\right)\frac{w(x)}{(1-x^2)^2}\,dx.$$

The last integral in (8.6.5) is finite, by virtue of the hypotheses on α and β. Concerning (8.6.4), the Schwarz inequality yields

(8.6.6) $$\int_{-1}^{1} p'(pw)'\,dx = \|p'\|_w^2 + \int_{-1}^{1}\left(p\,\frac{w'}{\sqrt{w}}\right)(p'\sqrt{w})\,dx$$

$$\leq \|p'\|_w^2 + \left(\int_{-1}^{1} p^2\,(w')^2 w^{-1}\,dx\right)^{\frac{1}{2}} \|p'\|_w.$$

By noting that $(w')^2 w^{-1} \leq 4(|\alpha| + |\beta|)^2(1-x^2)^{-2}w$, we can use (8.6.3) to conclude.

∎

Inequality (8.3.6) is now straightforward. Starting from relation (8.2.10), we use lemma 8.2.1, inequality (8.6.3) and theorem 3.8.2, to prove that $|\lambda_{n,m}| \geq c_1$. On the contrary, from (8.6.4), (6.3.6) and theorem 3.8.2, we obtain $|\lambda_{n,m}| \leq c_2 n^4$.

As far as we know, a full analysis on the convergence of both the eigenvalues and the relative eigenfunctions is not yet available. A general approach, based on the properties of *compact operators*, is studied in VAINIKKO(1964), VAINIKKO(1967) and OSBORN(1975), as well as in many other papers. A self-contained exposition with a comprehensive list of references is provided in CHATELIN(1983), where hints for the treatment of problems similar to the one considered here are given in chapter 4B.

We can say something more about the Legendre case. Combining the proof of theorem 2.2.1 with that of theorem 8.2.2, one checks that, for $\alpha = \beta = 0$, the eigenfunctions $p_{n,m}$, $1 \leq m \leq n-1$, as well as their derivatives, are orthogonal with respect to the inner product $(\cdot,\cdot)_{w,n}$ (see section 3.8). This is true if we assume that the eigenvalues $\lambda_{n,m}$, $1 \leq m \leq n-1$, are distinct.
Let $\mathbf{X}_m$, $1 \leq m \leq n-1$, denote a subspace of $\mathbf{P}_n^0$ of dimension m. Then, by virtue of the orthogonality of the eigenfunctions, the following characterization, known as *Min-Max principle*, holds:

$$\lambda_{n,m} = \min_{\mathbf{X}_m \subset \mathbf{P}_n^0} \left\{ \max_{\substack{p \in \mathbf{X}_m \\ p \not\equiv 0}} \frac{\|p'\|_w^2}{\|p\|_{w,n}^2} \right\}, \qquad 1 \leq m \leq n-1. \tag{8.6.7}$$

Applications of formula (8.6.7), in the field of finite-element approximations, are presented for instance in STRANG and FIX(1973), chapter six.
Other situations may be considered. The problem

$$\begin{cases} -\phi'' + \mu\phi = \lambda\,\phi & \text{in } I =]-1,1[, \quad \mu > 0, \\ \phi'(-1) = \phi'(1) = 0, \end{cases} \tag{8.6.8}$$

has a countable set of eigenvalues $\lambda_m := \frac{\pi^2}{4}m^2 + \mu$, $m \in \mathbf{N}$. These are effectively approximated by the eigenvalues of problem (8.2.13). We only remark that two of the $\lambda_{n,m}$'s are proportional to the parameter γ and are out of interest.

Different arguments apply to problem (8.2.12). Here, the $\lambda_{n,m}$'s do not approach the eigenvalues of problem (8.6.8). If we replace the second equation in (8.2.12) by $p'_{n,m}(\eta_0^{(n)}) = \lambda_{n,m} p_{n,m}(\eta_0^{(n)})$, we get a negative eigenvalue, say $\lambda_{n,0}$. Another eigenvalue, say $\lambda_{n,1}$, is such that $\lim_{n\to+\infty}(\lambda_{n,0} + \lambda_{n,1}) = -1$. The remaining eigenvalues approximate those of problem (8.6.8), with the exception of $\lambda_0 = \mu$.

These considerations demonstrate the importance of the treatment of boundary conditions in spectral methods. A non correct specification may give raise to unexpected results. Though this is true in all numerical methods, the situation seems more crucial for spectral methods, due to their global nature. On the other hand, if the boundary conditions are well treated, results are really competitive. An example of a second-order eigenvalue problem is examined in section 12.4.

Let us finally consider the eigenvalues associated to the system (7.4.22). These converge to the eigenvalues of the problem

$$(8.6.9)\qquad \begin{cases} \phi^{IV} \;=\; \lambda\,\phi & \text{in } I =]-1,1[, \\ \phi(\pm 1) \;=\; \phi'(\pm 1) \;=\; 0. \end{cases}$$

Here, λ assumes a countable number of positive values corresponding to the roots of the equation $\cosh 2\sqrt[4]{\lambda}\,\cos 2\sqrt[4]{\lambda} = 1$ (see COURANT and HILBERT(1953), Vol.1, p.296).

No asymptotic estimates can be recovered for the eigenvalues of problem (8.1.1). In fact, the equation $\phi' = \lambda\phi$ in $]-1,1[$, $\phi(-1) = \sigma$, has only the solution $\phi \equiv 0$ when $\sigma = 0$.

8.7 Multigrid method

We already noticed that the speed of convergence of the Richardson method strongly depends on the location of the eigenvalues λ_m, $1 \le m \le n$, of the matrix D in (7.6.1). Let $\bar{p}_m \in \mathbf{C}^n$, $1 \le m \le n$, denote the eigenvectors of D, suitably normalized. In the computation $\bar{p} \to M\bar{p}$, different choices of the parameter θ will result in a different treatment of the components (*modes*) of the vector $\bar{p}$, along the directions selected by $\bar{p}_m$, $1 \le m \le n$. For instance, let us assume that the λ_m's are real, distinct, positive and in increasing order. By examining relation (8.3.5), we deduce that the maximum damping of the components corresponding to a large integer m (*high modes*), is obtained for small values of θ. Conversely, *low modes* decay fast for larger values of the parameter θ, provided the assumptions in (7.6.3) are satisfied. Therefore, for ill-conditioned matrices we cannot expect a strong reduction of all the modes.

This can be partly overcome by using the *multigrid method*. We illustrate this procedure with an example. We denote by $D^{(n)}$, $n \ge 2$, the $(n+1)\times(n+1)$ matrix of the system in (7.4.9). This is obtained starting from the set of grid-points $\eta_j^{(n)}$, $0 \le j \le n$ (*fine-grid*). Another set of points in $[-1,1]$ (*coarse-grid*) is given by the nodes $\eta_j^{(k)}$, $0 \le j \le k$, where $2 \le k \le n$. A polynomial $p \in \mathbf{P}_n$, determined by its values on the fine-grid, is evaluated at the coarse-grid by using formula (3.2.7). On the other hand, we can go from the coarse-grid to the fine-grid by extrapolation, i.e.

$$p(\eta_i^{(n)}) = \sum_{j=0}^{k} p(\eta_j^{(k)})\, \tilde{l}_j^{(k)}(\eta_i^{(n)}), \qquad 0 \le i \le n. \tag{8.7.1}$$

In short, the philosophy of the multigrid method is the following. Since the eigenvalues of $D^{(n)}$ are not available (the cost of their computation is very high), an alternative to varying the parameter θ is to accelerate the convergence by using different grids. A good treatment of low modes is obtained by applying the Richardson method to the matrix $D^{(k)}$ (k small) defined on the coarse-grid. Then, we can switch to the fine-grid by (8.7.1), and work with the matrix $D^{(n)}$ to damp the high modes with a relatively small θ. In general, the algorithm applies to different sets of grid-points, selected in turn, according to some prescribed rule (the most classical one is the *V-cycle*). There are many ways to advance the method. These depend on the problem and the competence of the user. A comparative analysis is given in HEINRICHS(1988). Other results are provided in MUÑOZ(1990). Early applications within the framework of spectral methods, for more specialized problems, are examined, for instance, in ZANG, WONG and HUSSAINI(1982), STREETT, ZANG and HUSSAINI(1985), BRANDT, FULTON and TAYLOR(1985), ZANG and HUSSAINI(1986). For a general overview of the method, we refer to HACKBUSCH(1985).

Using the multigrid method for the preconditioned systems in place of a plain preconditioned iterative scheme, does not improve very much the rate of convergence for problems in one dimension. More interesting are the applications in two or three dimensions. Performances also depend on the velocity and the cost to transfer the data from one grid to the next. Particular efficiency is obtained in the Chebyshev case. When n is even and $k = n/2$, recalling relation (3.8.16), all the points of the coarse-grid are also points of the fine-grid. Furthermore, the inverse formula (8.7.1) is implemented via FFT.

9
ORDINARY DIFFERENTIAL EQUATIONS

The large amount of material accumulated in the previous chapters is now directed towards the study of the convergence of approximate solutions to differential equations. Beginning with the simplest linear equations, we then extend our analysis to other, more complex, problems. Throughout this chapter, however, we remain focused on different techniques, rather than on solving complicated equations.

9.1 General considerations

Assume that $I \subset \mathbf{R}$ is a bounded open interval. Without loss of generality we set $I :=]-1,1[$. In fact, any other interval is mapped into $]-1,1[$ by a linear change of variables and a translation. Note that the space of polynomials $\mathbf{P}_n$, $n \in \mathbf{N}$, is invariant under these transformations.
Let $f : \bar{I} \to \mathbf{R}$ be a given continuous function. A very simple differential equation is

$$(9.1.1) \qquad \begin{cases} U' = f & \text{in }]-1,1], \\ U(-1) = \sigma, \end{cases}$$

where $\sigma \in \mathbf{R}$, and $U : \bar{I} \to \mathbf{R}$ is the unknown. Of course, we know the solution explicitly, i.e.

$$U(x) = \sigma + \int_{-1}^{x} f(s)\, ds, \qquad \forall x \in \bar{I}. \tag{9.1.2}$$

We are concerned with the convergence analysis of polynomial approximations of U. We give an overview of different techniques in the coming sections.
A generalization of (9.1.1) is

$$\begin{cases} U' + AU = f & \text{in }]-1,1], \\ U(-1) = \sigma, \end{cases} \tag{9.1.3}$$

where $A : \bar{I} \to \mathbf{R}$ is a given continuous function.

Other elementary problems can be studied. We are mainly interested in ***boundary-value problems***. Two typical examples of second-order equations are

$$\begin{cases} -U'' = f & \text{in } I, \\ U(-1) = \sigma_1, \quad U(1) = \sigma_2, \end{cases} \tag{9.1.4}$$

$$\begin{cases} -U'' + \mu U = f & \text{in } I, \\ U'(-1) = \sigma_1, \quad U'(1) = \sigma_2, \end{cases} \tag{9.1.5}$$

where $\sigma_1, \sigma_2 \in \mathbf{R}$ and $\mu > 0$. In particular, (9.1.4) and (9.1.5) are known as *Dirichlet problem* and *Neumann problem* respectively.
Within the framework of fourth-order problems, we consider for instance the equation

$$\begin{cases} U^{IV} = f & \text{in } I, \\ U(-1) = \sigma_1, \quad U(1) = \sigma_2, \\ U'(-1) = \sigma_3, \quad U'(1) = \sigma_4, \end{cases} \tag{9.1.6}$$

with $\sigma_i \in \mathbf{R}$, $1 \le i \le 4$.

Theoretical considerations of the problems proposed above are presented in many books on ordinary differential equations. We refer for instance to GOLOMB and SHANKS (1965), BRAUER and NOHEL(1967), SIMMONS(1972), etc.. Once we gain enough experience with the numerical treatment of these elementary equations, we will turn to more realistic problems. Several non trivial examples are discussed in chapter twelve.

9.2 Approximation of linear equations

A technique to approximate the solution of problem (9.1.1) is the so called ***tau method.*** Early applications were considered in LANCZOS(1956). Theoretical developments are given for instance in GOTTLIEB and ORSZAG(1977), for Chebyshev and Legendre expansions. The unknown function U is approximated by a sequence of polynomials $p_n \in \mathbf{P}_n$, $n \geq 1$, which are expressed in the frequency space relative to some Jacobi weight function w. For any $n \geq 1$, these are required to be solutions of the following problem:

$$\text{(9.2.1)} \qquad \begin{cases} p_n' = \Pi_{w,n-1} f & \text{in }]-1,1], \\ p_n(-1) = \sigma. \end{cases}$$

The operator $\Pi_{w,n}$, $n \in \mathbf{N}$, is the same operator introduced in section 2.4. In this scheme the boundary condition is the same as in (9.1.1), but the function f is replaced by a truncation of the series (6.2.11). In practice, we *project* the differential equation in the space $\mathbf{P}_{n-1}$. Following the suggestions provided in section 7.3, where now $q := \Pi_{w,n-1} f$, we can reduce (9.2.1) to a linear system. The vector on the right-hand side contains the first n Fourier coefficients of the function f and the datum σ. The unknown vector consists of the Fourier coefficients of the polynomial p_n. Note that these do not coincide with the first $n+1$ Fourier coefficients of the function U.
The question is to check whether $\lim_{n \to +\infty} p_n = U$, and how to interpret this limit. By (9.1.2), we easily get

$$(9.2.2) \qquad U(x) - p_n(x) = \int_{-1}^{x} (f - \Pi_{w,n-1}f)(s)\, ds, \qquad \forall x \in \bar{I}.$$

At this point, an estimate of the rate of convergence to zero of the error $U - p_n$, $n \geq 1$, in some suitable norm, is straightforward after recalling the results of section 6.2. For example, when $-1 < \alpha < 1$ and $-1 < \beta < 1$, it is easy to obtain uniform convergence with the help of the Schwarz inequality and of theorem 6.2.4. Indeed, for $k \in \mathbf{N}$, one has

$$(9.2.3) \qquad \sup_{x \in [-1,1]} |U(x) - p_n(x)| \leq C \left(\frac{1}{n}\right)^k \|f\|_{H_w^k(I)}, \qquad \forall n > k.$$

The space $H_w^k(I)$, $k \in \mathbf{N}$, is defined in (5.7.1). As already observed in chapter six, the more regular the function f, the faster the decay of the error is.

Another approach is to construct the approximation scheme in the physical space. This leads to the *collocation* (or *pseudospectral*) method. Now, the approximating polynomials $p_n \in \mathbf{P}_n$, $n \geq 1$, satisfy

$$(9.2.4) \qquad \begin{cases} p_n'(\eta_i^{(n)}) = f(\eta_i^{(n)}) & 1 \leq i \leq n, \\ p_n(\eta_0^{(n)}) = \sigma. \end{cases}$$

Here, we impose the boundary condition at $x = -1$ and we *collocate* the differential equation at the nodes $\eta_i^{(n)}$, $1 \leq i \leq n$. We are left with a linear system in the unknowns $p_n(\eta_j^{(n)})$, $0 \leq j \leq n$. The corresponding matrix is obtained following the guideline of section 7.4. We remark that the computed values $p_n(\eta_j^{(n)})$, $0 \leq j \leq n$, do not coincide in general with the values $U(\eta_j^{(n)})$, $0 \leq j \leq n$.

We can study the convergence of the collocation method by arguing as follows. We define $r(x) := 1 + x$, $x \in \bar{I}$. Then, we note that (9.2.4) is equivalent to

$$(9.2.5) \qquad \begin{cases} p_n' = [\tilde{I}_{w,n}(fr)]r^{-1} & \text{in }]-1,1], \\ p_n(-1) = \sigma, \end{cases}$$

where $\tilde{I}_{w,n}$, $n \geq 1$, is the interpolation operator at the points $\eta_j^{(n)}$, $0 \leq j \leq n$ (see section 3.3). It turns out that $[\tilde{I}_{w,n}(fr)]r^{-1}$ is the interpolant of f at the nodes $\eta_j^{(n)}$, $1 \leq j \leq n$. This shows the equivalence of (9.2.5) with (9.2.4). The analysis of the error $U - p_n$, $n \geq 1$, is then recovered from the estimates of the error $[fr - \tilde{I}_{w,n}(fr)]r^{-1}$. These results have been developed in section 6.6.

The two approximation methods described here differ in the treatment of the function f. Projection and interpolation operators have been used respectively. A small difference is due to the aliasing error (see section 4.2), but the approximating solutions behave similarly. The first approach is more effective in terms of computational cost.

Tau and collocation methods are respectively used to approximate the solution of problem (9.1.3) according to the following schemes:

$$\text{(9.2.6)} \qquad \begin{cases} p_n' + \Pi_{w,n-1}(Ap_n) = \Pi_{w,n-1}f & \text{in }]-1,1], \\ p_n(-1) = \sigma , \end{cases}$$

$$\text{(9.2.7)} \qquad \begin{cases} p_n'(\eta_i^{(n)}) + A(\eta_i^{(n)})p_n(\eta_i^{(n)}) = f(\eta_i^{(n)}) & 1 \leq i \leq n, \\ p_n(\eta_0^{(n)}) = \sigma , \end{cases}$$

$$p_n \in \mathbf{P}_n, \quad n \geq 1.$$

The corresponding linear systems are given by (7.3.3) and (7.4.7). Here, the second approach is preferable, having the coefficients of the matrix relative to the tau method a complicate expression, for a non-constant function A.

The analysis of convergence is now more involved and general results are not available. We examine for instance the collocation method in the Legendre case. We recall that the norm in the space $L_w^2(I)$ is defined in (5.2.4).

Theorem 9.2.1 - *Let* $\alpha = \beta = 0$ *and* $A : \bar{I} \to \mathbf{R}$, *be a continuous positive function satisfying* $A(x) \geq \epsilon$, $\forall x \in \bar{I}$, *for a given* $\epsilon > 0$. *Then*

$$\lim_{n \to +\infty} \|p_n - U\|_{L^2_w(I)} = 0, \tag{9.2.8}$$

where p_n, $n \geq 1$, *is the solution of* (9.2.7) *and* U *is the solution of* (9.1.3).

Proof - We first note that, by the triangle inequality (2.1.5), one has

$$\|p_n - U\|_{L^2_w(I)} \leq \|p_n - \tilde{I}_{w,n}U\|_{L^2_w(I)} + \|\tilde{I}_{w,n}U - U\|_{L^2_w(I)}, \qquad n \geq 1. \tag{9.2.9}$$

The last term of the right-hand side tends to zero in view of theorem 6.6.1. After defining $s_n := p_n - \tilde{I}_{w,n}U \in \mathbf{P}_n$, $n \geq 1$, we require an estimate for the error $\|s_n\|_{L^2_w(I)}$, $n \geq 1$. Equations (9.1.3) and (9.2.7) yield

$$\begin{cases} s_n'(\eta_i^{(n)}) + A(\eta_i^{(n)})s_n(\eta_i^{(n)}) = (U - \tilde{I}_{w,n}U)'(\eta_i^{(n)}) & 1 \leq i \leq n, \\ s_n(\eta_0^{(n)}) = 0. \end{cases} \tag{9.2.10}$$

Using the quadrature formula (3.5.1) with $w \equiv 1$, we obtain

$$\int_I s_n' s_n \, dx + \sum_{i=0}^{n} A(\eta_i^{(n)}) s_n^2(\eta_i^{(n)}) \, \tilde{w}_i^{(n)} \tag{9.2.11}$$

$$= \sum_{i=0}^{n} (U - \tilde{I}_{w,n}U)'(\eta_i^{(n)}) \, s_n(\eta_i^{(n)}) \, \tilde{w}_i^{(n)}, \qquad n \geq 1.$$

Next, we observe that $\int_I s_n' s_n dx = \frac{1}{2}s_n^2(1) \geq 0$, $n \geq 1$. Therefore, from the assumptions on A and the Schwarz inequality, one gets

$$\epsilon \, \|s_n\|_{w,n} \leq \|\tilde{I}_{w,n}U' - (\tilde{I}_{w,n}U)'\|_{w,n}, \qquad n \geq 1, \tag{9.2.12}$$

where we note that $\tilde{I}_{w,n}U'$ coincides with U' at the nodes. The norm $\|\cdot\|_{w,n}$, $n \geq 1$, is given in (3.8.2). By virtue of theorem 3.8.2, this norm is equivalent to the $L^2_w(I)$ norm. Besides, when $U \in C^1(\bar{I})$, the right-hand side of (9.2.12) tends to zero. Actually, by (3.8.6), one has

$$(9.2.13)\qquad \|\tilde{I}_{w,n}U' - (\tilde{I}_{w,n}U)'\|_{w,n} \leq \gamma_1^{-1} \|\tilde{I}_{w,n}U' - (\tilde{I}_{w,n}U)'\|_{L^2_w(I)}$$

$$\leq \gamma_1^{-1} \left\{\|\tilde{I}_{w,n}U' - U'\|_{L^2_w(I)} + \|(\tilde{I}_{w,n}U - U)'\|_{L^2_w(I)}\right\}, \qquad n \geq 1.$$

Finally, we use inequalities like (6.6.1) and (6.6.8) to estimate the last term. This shows that $\|s_n\|_{L^2_w(I)}$ tends to zero for $n \to +\infty$. Thus, we get (9.2.8).

■

As seen from the proof given above, the rate of convergence in (9.2.8) is the typical one of spectral approximations. In particular, for analytic solutions U, the decay of the error is exponential. The extension of theorem 9.2.1 to other Jacobi weights is not straightforward. Convergence estimates for the Chebyshev case can be derived from the results in SOLOMONOFF and TURKEL(1989). The use of *collocation nodes* relative to Gauss type formulas is helpful in the theoretical analysis, since it allows us to substitute summations with integrals as in (9.2.11). Unfortunately, when the weight function w is singular at the boundary, the integrals cannot be manipulated further. This is the major drawback in the analysis of approximations involving Chebyshev polynomials. For example, the quantity $\int_I \phi'\phi w dx$, $\phi \in \mathbf{P}_n$, $\phi(-1) = 0$ is not in general positive when $w(x) = 1/\sqrt{1-x^2}$, $x \in I$ (see GOTTLIEB and ORSZAG(1977), p.89).

First, we show the results of numerical experiment when U is a very smooth function. We choose $\sigma = 0$ and $A(x) := 1+x^2$, $x \in \bar{I}$. The datum f in (9.1.3) is such that $U(x) := \sin(1+x)$, $x \in \bar{I}$. In table 9.2.1, we give the error $E_n := \|p_n - \tilde{I}_{w,n}U\|_{L^2_w(I)}$ for various n, when p_n is obtained with the scheme (9.2.7) relative to the Legendre nodes. It is evident that the convergence is extremely fast. Note that, from (9.2.9), the error $\|p_n - U\|_{L^2_w(I)}$ is bounded by E_n plus a term which does not depend on the polynomial p_n. Therefore, the examination of E_n gives sufficient information about the efficiency of the approximation technique. In addition, according to formula (3.8.10), the computation of E_n is obtained from the values of p_n at the nodes. These values are the actual solutions of (9.2.7), after inverting the linear system corresponding to (7.4.7).

n	E_n	n	E_n
2	$.8249 \times 10^{-1}$	8	$.3399 \times 10^{-7}$
3	$.1814 \times 10^{-1}$	9	$.2784 \times 10^{-8}$
4	$.1232 \times 10^{-2}$	10	$.8594 \times 10^{-10}$
5	$.1697 \times 10^{-3}$	11	$.5817 \times 10^{-11}$
6	$.8560 \times 10^{-5}$	12	$.1511 \times 10^{-12}$
7	$.8874 \times 10^{-6}$	13	$.8770 \times 10^{-14}$

Table 9.2.1 - Errors relative to the Legendre collocation approximation of problem (9.1.3).

We consider another example. In figure 9.2.1, we plot the approximating polynomials p_n, $n = 4,\ 10,\ 16$, obtained by the collocation method (9.2.7) at the Legendre nodes.

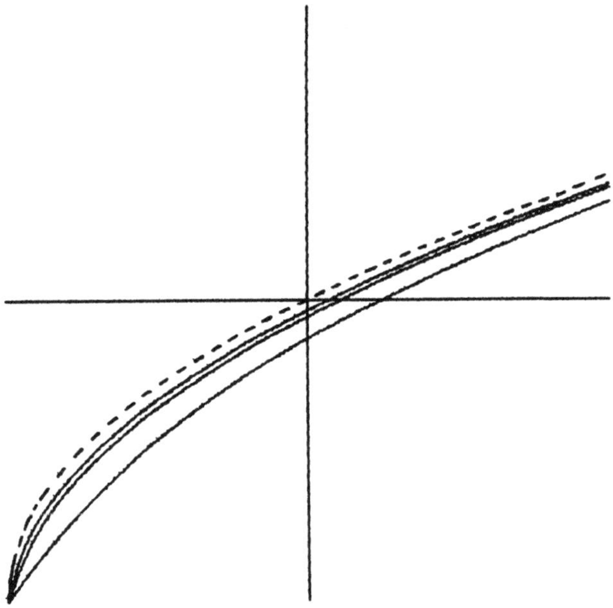

Figure 9.2.1 - Legendre collocation approximations for $n = 4,\ 10,\ 16$.

The exact solution (dashed line in figure 9.2.1) is $U(x) := \sqrt{1+x} - 1$, $x \in \bar{I}$, corresponding to the data: $\sigma = -1$, $A \equiv \frac{1}{2}$ and $f(x) := \frac{2+x}{2\sqrt{1+x}} - \frac{1}{2}$, $x \in]-1,1]$. We observe that $f \notin C^0(\bar{I})$ and that U has a singularity in the derivative at $x = -1$. Nevertheless, convergence is still achieved. The same experiment, using the collocation method at the Chebyshev nodes, displays a similar behavior.

The numerical schemes introduced here can be also generalized to other type of equations. For example, tau and collocation methods applied to problem (9.1.4), give respectively

$$\text{(9.2.14)} \qquad \begin{cases} -p_n'' = \Pi_{w,n-2} f & \text{in } I, \\ p_n(-1) = \sigma_1, \quad p_n(1) = \sigma_2, \end{cases}$$

$$\text{(9.2.15)} \qquad \begin{cases} -p_n''(\eta_i^{(n)}) = f(\eta_i^{(n)}) & 1 \le i \le n-1, \\ p_n(\eta_0^{(n)}) = \sigma_1, \quad p_n(\eta_n^{(n)}) = \sigma_2, \end{cases}$$

$$p_n \in \mathbf{P}_n, \quad n \ge 2.$$

The systems corresponding to (9.2.14) an (9.2.15) have been investigated in sections 7.3 and 7.4, respectively. Similar considerations hold for the remaining differential equations proposed in section 9.1. In all cases, we expect convergence of the sequence of polynomials p_n to the exact solution U, as n tends to infinity. To provide convergence estimates, it is useful to restate the original problems in an appropriate way. We study how to do this in the next sections.

9.3 The weak formulation

An effective approach for the study of solutions of differential equations, and how to approximate them, is the *variational method*. Here, the problem is conveniently formulated in the so-called *weak form*. For a given equation, there are several ways to proceed.

Classical formulations are illustrated for instance in LIONS and MAGENES(1972). Alternative formulations, specifically addressed to applications in the field of spectral methods, are introduced in CANUTO and QUARTERONI(1981), MADAY and QUARTERONI(1981), CANUTO and QUARTERONI(1982b). Following these papers, we present several useful results. Let us begin by examining the Dirichlet problem (9.1.4). Without loss of generality, we can assume that $\sigma_1 = \sigma_2 = 0$ (*homogeneous* boundary conditions). Actually, solutions corresponding to non-homogeneous boundary conditions are obtained by adding to U the function $r(x) := \frac{1}{2}(1-x)\sigma_1 + \frac{1}{2}(1+x)\sigma_2$, $x \in \bar{I}$.

Let $\mathbf{X}$ be a suitable functional space, for now the space $C_0^\infty(I)$ (see section 5.4). Recall that $\phi \in \mathbf{X}$ satisfies $\phi(\pm 1) = 0$. Thus, one obtains the identity

$$(9.3.1) \qquad \int_I U'(\phi w)' \, dx = \int_I f\phi w \, dx, \qquad \forall \phi \in \mathbf{X}.$$

The above relation is obtained by multiplying both sides of equation (9.1.4) by ϕw and integrating in I. The final form follows after integration by parts using the condition $\phi(\pm 1) = 0$. In this context, ϕ is called a *test function*. It is an easy exercise to check that, due to the freedom to choose the test functions, we start from (9.3.1) and deduce the original equation (9.1.4). Thus, one concludes that, when U is solution to (9.3.1) with $U(\pm 1) = 0$, then it is also solution to (9.1.4) with $\sigma_1 = \sigma_2 = 0$.
The advantage of replacing (9.1.4) with (9.3.1) is that the new formulation can be suitably generalized by weakening the hypotheses on the data and the solution. To this end, we note that, according to (9.3.1), U is not required to have a second derivative. This regularity assumption is substituted by appropriate integrability conditions on U' and ϕ'. In order to proceed with our investigation, we need to use the functional spaces introduced in section 5.7. Henceforth, w will be the Jacobi weight function in the ultraspherical case, where $\nu := \alpha = \beta$ satisfies $-1 < \nu < 1$. We shall seek the solution U in the space $H^1_{0,w}(I)$ (see (5.7.6)), which is also a natural choice for the space of test functions. Therefore, we set $\mathbf{X} \equiv H^1_{0,w}(I)$. Finally, we take $f \in L^2_w(I)$ (see (5.2.1)). As we will see, these assumptions are sufficient to guarantee the existence of the integrals in (9.3.1), which are now considered in the Lebesgue sense.
The next step is to insure that, in spite of such generalizations, problem (9.3.1) is still

well-posed. This means that it admits a unique *weak solution* $U \in \mathbf{X}$. We remark that two solutions of (9.3.1) which differ on a set of measure zero are in the same class of equivalence in $\mathbf{X}$.

We first introduce some notation. Consider the *bilinear form* $B_w : \mathbf{X} \times \mathbf{X} \to \mathbf{R}$ defined as follows:

$$B_w(\psi, \phi) := \int_I \psi'(\phi w)' dx, \qquad \forall \psi, \phi \in \mathbf{X}. \tag{9.3.2}$$

The mapping B_w is linear in each argument (this justifies the term *bilinear*). We also define the linear operator $F_w : \mathbf{X} \to \mathbf{R}$ according to

$$F_w(\phi) := \int_I f \phi w \, dx, \qquad \forall \phi \in \mathbf{X}. \tag{9.3.3}$$

With the above definitions we restate (9.3.1). We are concerned with finding $U \in \mathbf{X}$ such that

$$B_w(U, \phi) = F_w(\phi), \qquad \forall \phi \in \mathbf{X}. \tag{9.3.4}$$

The answer to our question relies on the following general result.

Theorem 9.3.1 *(Lax & Milgram) - Let* $\mathbf{X}$ *be a Hilbert space. Let* $B : \mathbf{X} \times \mathbf{X} \to \mathbf{R}$ *be a bilinear form and* $F : \mathbf{X} \to \mathbf{R}$ *be a linear operator. Let us assume that there exist three positive constants* C_1, C_2, C_3 *such that*

$$|B(\psi, \phi)| \leq C_1 \, \|\psi\|_{\mathbf{X}} \, \|\phi\|_{\mathbf{X}}, \qquad \forall \psi, \phi \in \mathbf{X}, \tag{9.3.5}$$

$$B(\psi, \psi) \geq C_2 \, \|\psi\|_{\mathbf{X}}^2, \qquad \forall \psi \in \mathbf{X}, \tag{9.3.6}$$

$$|F(\phi)| \leq C_3 \, \|\phi\|_{\mathbf{X}}, \qquad \forall \phi \in \mathbf{X}. \tag{9.3.7}$$

Then, there exists a unique solution $U \in \mathbf{X}$ *of problem*

$$B(U,\phi) = F(\phi), \qquad \forall \phi \in \mathbf{X}. \tag{9.3.8}$$

Moreover, we can find a positive constant $C_4 > 0$ *such that*

$$\|U\|_{\mathbf{X}} \leq C_4 \sup_{\substack{\phi \in \mathbf{X} \\ \phi \not\equiv 0}} \frac{|F(\phi)|}{\|\phi\|_{\mathbf{X}}}. \tag{9.3.9}$$

■

The Lax-Milgram theorem is a powerful tool for proving existence and uniqueness for a large class of linear differential problems. Though the proof is not difficult, we omit it for simplicity. Applications and extensions have been considered in many publications. We suggest the following authors: LAX and MILGRAM(1954), BREZIS(1983), BREZZI and GILARDI(1987). Inequality (9.3.6) plays a fundamental role in the proof of theorem 9.3.1. It is in general known as a *coerciveness condition* and indicates that (9.3.8) is an *elliptic problem.*

The final step is to check wether the Lax-Milgram theorem can be applied to study the solution of (9.3.4). It is quite easy to show that $\mathbf{X} \equiv H^1_{0,w}(I) \subset C^0(\bar{I})$ is a Hilbert space (see section 5.3). In addition, inequalities (9.3.5), (9.3.6) and (9.3.7) are directly obtained from the following result (we recall that the norm in $H^1_{0,w}(I)$ is given by (5.7.7)).

Theorem 9.3.2 - *Let* $\nu := \alpha = \beta$ *with* $-1 < \nu < 1$. *Then, we can find three positive constants* C_1, C_2, C_3 *such that*

$$\left|\int_I \psi'(\phi w)' dx\right| \leq C_1 \left(\int_I [\psi']^2 w \, dx\right)^{\frac{1}{2}} \left(\int_I [\phi']^2 w \, dx\right)^{\frac{1}{2}}, \qquad \forall \psi, \phi \in H^1_{0,w}(I), \tag{9.3.10}$$

$$\int_I \psi'(\psi w)' dx \geq C_2 \int_I [\psi']^2 w \, dx, \qquad \forall \psi \in H^1_{0,w}(I), \tag{9.3.11}$$

$$\left|\int_I f\phi w \, dx\right| \leq C_3 \|f\|_{L^2_w(I)} \left(\int_I [\phi']^2 w \, dx\right)^{\frac{1}{2}}, \qquad \forall \phi \in H^1_{0,w}(I). \tag{9.3.12}$$

Proof - We start by considering $\psi, \phi \in \mathbf{P}_n^0 \subset H_{0,w}^1(I)$, for some $n \geq 2$. Then, the first inequality is obtained by applying the Schwarz inequality (2.1.7) to the integral $\int_I (\psi' \sqrt{w})[(\phi w)'/\sqrt{w}]dx$. Recalling that $(w')^2/w \leq w/(1-x^2)^2$, we conclude using the results of lemma 8.6.1. The second inequality is proven in lemma 8.2.1. The third one comes from the Schwarz inequality and (5.7.5). The proof of (9.3.10), (9.3.11), (9.3.12) in $H_{0,w}^1(I)$ requires a little care. In short, one approximates $\phi \in H_{0,w}^1(I)$ by a sequence of polynomials $\phi_n \in \mathbf{P}_n^0$, $n \in \mathbf{N}$, converging to ϕ in such a way that $\lim_{n\to+\infty} \|\phi - \phi_n\|_{H_{0,w}^1(I)} = 0$. This also implies: $\lim_{n\to+\infty} \|\phi_n\|_{H_{0,w}^1(I)} = \|\phi\|_{H_{0,w}^1(I)}$. The proposition follows by a limit process which is standard in Lebesgue integration theory (see KOLMOGOROV and FOMIN(1961)). The details are omitted.

■

Thus, (9.3.4) is a well-posed problem. In the special case when $f \in C^0(\bar{I}) \subset L_w^2(I)$, the solution U of (9.3.4) is the classical *strong* solution satisfying (9.1.4).

Generalizations can be examined with little modification. For example, let us consider the boundary-value problem

$$\text{(9.3.13)} \qquad \begin{cases} -U'' + A_1 U' + A_2 U = f & \text{in } I, \\ U(\pm 1) = 0, \end{cases}$$

where $A_1 : \bar{I} \to \mathbf{R}$ and $A_2 : \bar{I} \to \mathbf{R}$ are given bounded integrable functions. Here, we can replace (9.3.2) by the bilinear form

$$\text{(9.3.14)} \qquad B_w(\psi, \phi) = \int_I \psi'(\phi w)' dx + \int_I [A_1 \psi' \phi + A_2 \psi \phi] w \, dx, \qquad \forall \psi, \phi \in \mathbf{X}.$$

One checks that, if the inequality $A_2 w - \frac{1}{2}(A_1 w)' \geq 0$ holds in I, then the hypotheses of theorem 9.3.1 are satisfied. Therefore, the variational formulation of (9.3.13) admits a unique weak solution.

We now propose a weak formulation of the Neumann problem (9.1.5). Let us first discuss the Legendre case. We set $\mathbf{X} \equiv H^1_w(I)$, where $w \equiv 1$. We introduce $B_w : \mathbf{X} \times \mathbf{X} \to \mathbf{R}$ and $F_w : \mathbf{X} \to \mathbf{R}$ such that

(9.3.15) $$B_w(\psi,\phi) := \int_I \psi'\phi' \, dx + \mu \int_I \psi\phi \, dx, \qquad \forall \psi,\phi \in \mathbf{X},$$

(9.3.16) $$F_w(\psi) := \int_I f\phi \, dx + \sigma_2\phi(1) - \sigma_1\phi(-1), \qquad \forall \phi \in \mathbf{X}.$$

According to these new definitions, we consider the problem of finding $U \in \mathbf{X}$ which satisfies (9.3.4).

Theorem 9.3.3 - *Let* $w \equiv 1$ *and* $\mathbf{X} \equiv H^1_w(I)$. *Let* B_w *and* F_w *be respectively defined by* (9.3.15) *and* (9.3.16). *Then, problem* (9.3.4) *has a unique weak solution* $U \in \mathbf{X}$. *Moreover, if* $U \in C^2(\bar{I})$, *then* U *is solution of* (9.1.5).

Proof - It is an easy matter to verify the hypotheses of theorem 9.3.1. In particular, to prove (9.3.7) we note that, by (5.7.4), one has $|\phi(\pm 1)| \leq \|\phi\|_{C^0(I)} \leq K_2 K_1^{-1} \|\phi\|_{\mathbf{X}}$. Then, we have existence and uniqueness of a weak solution $U \in \mathbf{X}$. When U is more regular, we can write

(9.3.17) $$U'(1)\phi(1) - U'(-1)\phi(-1) + \int_I [-U'' + \mu U]\phi \, dx$$

$$= \int_I f\phi \, dx + \sigma_2\phi(1) - \sigma_1\phi(-1), \qquad \forall \phi \in \mathbf{X},$$

which leads to $-U'' + \mu U = f$ in I, by testing on the functions $\phi \in H^1_{0,w}(I) \subset \mathbf{X}$. We can now remove the integrals in (9.3.17) and recover the boundary conditions $U'(-1) = \sigma_1$, $U'(1) = \sigma_2$.

■

We can also study variational formulations of the Neumann problem based on different Jacobi weight functions. We set $\mathbf{X} \equiv H^1_w(I)$, where w is the ultraspherical weight function with $\nu := \alpha = \beta$ satisfying $-1 < \nu < 1$. Then, $B_w : \mathbf{X} \times \mathbf{X} \to \mathbf{R}$ and $F_w : \mathbf{X} \to \mathbf{R}$ are modified to get

(9.3.18)
$$B_w(\psi,\phi) := \int_I (\psi-\tilde{\psi})'[(\phi-\tilde{\phi})w]'dx + \int_I \tilde{\psi}'\tilde{\phi}'\ dx$$
$$+\ \mu\int_I \psi(\phi-\tilde{\phi})w\ dx + \mu\int_I \psi\tilde{\phi}\ dx, \qquad \forall\psi,\phi\in\mathbf{X},$$

(9.3.19)
$$F_w(\phi) := \int_I f(\phi-\tilde{\phi})w\ dx + \int_I f\tilde{\phi}\ dx + \sigma_2\phi(1) - \sigma_1\phi(-1), \qquad \forall\phi\in\mathbf{X}.$$

In the above expressions, for a given function $\chi\in\mathbf{X}$, we have defined $\tilde{\chi}\in\mathbf{X}$ to be the function $\tilde{\chi}(x) := \frac{1}{2}(1-x)\chi(-1)+\frac{1}{2}(1+x)\chi(1)$, $x\in\bar{I}$. We note that $\tilde{\chi}\in\mathbf{P}_1$ and $\chi-\tilde{\chi}\in H^1_{0,w}(I)$. For $\nu=0$, (9.3.18) and (9.3.19) give respectively (9.3.15) and (9.3.16). As usual, we are interested in finding the solution $U\in\mathbf{X}$ of (9.3.4).
Without entering into details of the proof, we claim that it is possible to find a constant $\mu^*>0$ such that, for any $\mu\in]0,\mu^*]$, the bilinear form B_w in (9.3.18) fulfills the requirements (9.3.5) and (9.3.6). This result is given in FUNARO(1988) for the Chebyshev case and can be easily generalized to other Jacobi weights. This implies existence and uniqueness of a weak solution U. When $f\in C^0(\bar{I})$, we can relate U to problem (9.1.5) by arguing as follows. Considering that $\tilde{\phi}'\in\mathbf{P}_0$ is a constant function, we write

(9.3.20)
$$\int_I \tilde{U}'\tilde{\phi}'\ dx = (\tilde{U}(1)-\tilde{U}(-1))\tilde{\phi}' = (U(1)-U(-1))\tilde{\phi}'$$
$$= \int_I U'\tilde{\phi}'\ dx = U'(1)\tilde{\phi}(1) - U'(-1)\tilde{\phi}(-1) - \int_I U''\tilde{\phi}\ dx.$$

Therefore, since $\tilde{U}''\equiv\mathbf{0}$ and $\tilde{\phi}(\pm1)=\phi(\pm1)$, one has

(9.3.21)
$$B_w(U,\phi) = \int_I(-U''+\mu U)(\phi-\tilde{\phi})w\ dx + \int_I(-U''+\mu U)\tilde{\phi}\ dx$$
$$+\ U'(1)\phi(1) - U'(-1)\phi(-1) = F_w(\phi), \qquad \forall\phi\in\mathbf{X}.$$

One first uses test functions ϕ belonging to $H^1_{0,w}(I)$ (therefore $\tilde{\phi}\equiv\mathbf{0}$). This implies the differential equation $-U''+\mu U=f$ in I. Finally, we eliminate the integrals and obtain the boundary conditions.

As pointed out in section 9.2, the use of ultraspherical weight functions with $\nu \neq 0$, can lead to some trouble in the treatment of the integrals. This is why the bilinear form B_w has been decomposed into two components. In the one containing the weight function w, integration by parts is allowed since $\phi - \tilde{\phi}$ vanishes at the points $x = \pm 1$. The other component does not contain the weight function and takes into account the boundary constraints.

Variational formulations also exist for fourth-order equations. For example, we examine problem (9.1.6). We assume $\sigma_i = 0,\ 1 \leq i \leq 4$. Otherwise, we replace U by $U - s$, where $s \in \mathbf{P}_3$ is given in (7.4.20). Then, for $f \in L^2_w(I)$, we define

(9.3.22) $$B_w(\psi, \phi) := \int_I \psi''(\phi w)'' dx, \qquad \forall \psi, \phi \in \mathbf{X},$$

(9.3.23) $$F_w(\phi) := \int_I f \phi w \, dx, \qquad \forall \phi \in \mathbf{X}.$$

A natural choice for the functional space is

(9.3.24) $$\mathbf{X} \equiv H^2_{0,w}(I) := \{\phi |\ \phi \in H^2_w(I),\ \ \phi(\pm 1) = \phi'(\pm 1) = 0\}.$$

It is left for the reader to check that B_w and F_w satisfy the hypotheses of Lax-Milgram theorem, when w is the ultraspherical weight function with $-1 < \nu < 1$. In particular, coerciveness is a consequence of lemma 8.2.5. This property provides a unique weak solution $U \in \mathbf{X}$ of the problem $B_w(U, \phi) = F_w(\phi),\ \forall \phi \in \mathbf{X}$. In the standard way, one verifies that U is solution to (9.1.6) whenever $f \in C^0(\bar{I})$. More properties and results are discussed in BERNARDI and MADAY(1990) and FUNARO and HEINRICHS(1990).

In a more advanced setting, we can construct variational formulations of the problems considered above, where two different spaces are used for the solution and the test functions. In this case, one defines an appropriate bilinear form $B : \mathbf{X} \times \mathbf{Y} \to \mathbf{R}$, and a linear operator $F : \mathbf{Y} \to \mathbf{R}$, in order that $U \in \mathbf{X}$ is solution of the weak problem $B(U, \phi) = F(\phi),\ \forall \phi \in \mathbf{Y}$. A generalization of theorem 9.3.1 is obtained by modifying condition (9.3.6) as follows:

$$\text{(9.3.25)} \qquad \begin{cases} \forall\psi \in \mathbf{X}, \;\; \exists\phi \in \mathbf{Y}: \;\; |B(\psi,\phi)| \geq C_2 \, \|\psi\|_{\mathbf{X}}\|\phi\|_{\mathbf{Y}}, \\ \forall\phi \in \mathbf{Y} \;\; \phi \not\equiv 0, \;\; \exists\psi \in \mathbf{X}: \;\; B(\psi,\phi) \neq 0. \end{cases}$$

Extensions of this kind are studied, for instance, in NEČAS(1962). This approach was used in CANUTO and QUARTERONI(1984) to provide an alternative variational formulation to problem (9.1.5).

9.4 Approximation of problems in the weak form

Let us see how to use the results of the previous section to carry out the convergence analysis for the approximation schemes (9.2.14) and (9.2.15). If B_w and F_w are respectively given by (9.3.2) and (9.3.3) ($\mathbf{X} \equiv H^1_{0,w}(I)$), we already noticed that (9.3.4) has a unique weak solution and generalizes problem (9.1.4) with $\sigma_1 = \sigma_2 = 0$. We also write (9.2.14) in a variational form.

For any $n \geq 2$, we consider the finite dimensional space $\mathbf{P}^0_n \subset \mathbf{X}$ (see (6.4.11)). We remark that $\mathbf{P}^0_n$ is a Hilbert space with the same inner product defined in $\mathbf{X}$. Then, we introduce the linear operator $F_{w,n} : \mathbf{P}^0_n \to \mathbf{R}$ as follows:

$$\text{(9.4.1)} \qquad F_{w,n}(\phi) \;:=\; \int_I (\Pi_{w,n-2}f)\phi w \; dx, \qquad \forall\phi \in \mathbf{P}^0_n,$$

where $f \in L^2_w(I)$.

For the sake of simplicity, let w be the ultraspherical weight function with $\nu := \alpha = \beta$ satisfying $-1 < \nu < 1$. With the help of theorem 9.3.2, we can apply theorem 9.3.1 to obtain a unique solution $p_n \in \mathbf{P}^0_n$ of the problem

$$\text{(9.4.2)} \qquad B_w(p_n,\phi) \;=\; F_{w,n}(\phi), \qquad \forall\phi \in \mathbf{P}^0_n,$$

where B_w has been introduced in (9.3.2). We note that the space in which we seek the solution is rather small now. On the other hand, we require (9.4.2) to be satisfied only for test functions belonging to a subspace of $\mathbf{X}$.

It is very easy to check that p_n is solution to (9.2.14) where $\sigma_1 = \sigma_2 = 0$. In fact, setting $a(x) := (1-x^2)w(x)$, $x \in I$, and recalling that $p_n'' + \Pi_{w,n-2} f \in \mathbf{P}_{n-2}$, one has

$$-\int_I p_n'' \phi w \, dx = F_{w,n}(\phi), \qquad \forall \phi \in \mathbf{P}_n^0$$

$$\Updownarrow$$

$$-\int_I p_n'' \chi a \, dx = \int_I (\Pi_{w,n-2} f) \chi a \, dx, \qquad \forall \chi \in \mathbf{P}_{n-2}$$

$$\Updownarrow$$

$$-\Pi_{a,n-2}(p_n'') = \Pi_{a,n-2}(\Pi_{w,n-2} f) \qquad \text{in } I$$

$$\Updownarrow$$

$$-p_n'' = \Pi_{w,n-2} f \qquad \text{in } I.$$

There are several reasons to prefer the new formulation (9.4.2). First of all, inequality (9.3.9) yields

$$\|p_n'\|_w \le C_4 \sup_{\substack{\phi \in \mathbf{P}_n^0 \\ \phi \not\equiv 0}} \frac{F_{w,n}(\phi)}{\|\phi'\|_w} \tag{9.4.3}$$

$$\le C_4 \, \|\Pi_{w,n-2} f\|_w \sup_{\substack{\phi \in \mathbf{P}_n^0 \\ \phi \not\equiv 0}} \frac{\|\phi\|_w}{\|\phi'\|_w} \le C_5 \, \|f\|_{L_w^2(I)}, \qquad \forall n \ge 2.$$

To get (9.4.3), we used the Schwarz inequality, relation (6.2.7) and inequality (5.7.4). According to the usual terminology, (9.4.3) is known as a ***stability condition***. Namely, a certain norm of the solution p_n is bounded independently of n. In addition, p_n converges to the solution U of (9.3.4) when n tends to infinity. The proof is simple, being byproduct of a very general result (see STRANG(1972)).

Theorem 9.4.1 - *Let* $\mathbf{X}$ *be a Hilbert space. Let* $B : \mathbf{X} \times \mathbf{X} \to \mathbf{R}$ *be a bilinear form and* $F : \mathbf{X} \to \mathbf{R}$ *be a linear operator, both satisfying the hypotheses of theorem 9.3.1. Let us denote by* $U \in \mathbf{X}$ *the unique solution of the problem* $B(U, \phi) = F(\phi)$, $\forall \phi \in \mathbf{X}$. *For any* $n \in \mathbf{N}$, *let* $\mathbf{X}_n \subset \mathbf{X}$ *be a finite dimensional subspace of dimension* n *and* $F_n : \mathbf{X}_n \to \mathbf{R}$ *be a linear operator. Let us denote by* $p_n \in \mathbf{X}_n$ *the solution of the problem* $B(p_n, \phi) = F_n(\phi)$, $\forall \phi \in \mathbf{X}_n$.
Then, we can find a constant $C > 0$ *such that, for any* $n \in \mathbf{N}$, *we have*

$$(9.4.4) \qquad \|U - p_n\|_{\mathbf{X}} \leq C \left\{ \inf_{\chi \in \mathbf{X}_n} \|U - \chi\|_{\mathbf{X}} + \sup_{\substack{\phi \in \mathbf{X}_n \\ \phi \not\equiv 0}} \frac{|F(\phi) - F_n(\phi)|}{\|\phi\|_{\mathbf{X}}} \right\}.$$

Proof - For any $n \in \mathbf{N}$, we define $\chi_n \in \mathbf{X}_n$ such that

$$(9.4.5) \qquad B(\chi_n, \phi) = B(U, \phi), \qquad \forall \phi \in \mathbf{X}_n.$$

Existence and uniqueness of χ_n are obtained by applying theorem 9.3.1, after noting that $B(U, \cdot) : \mathbf{X} \to \mathbf{R}$ is a linear operator for any fixed $U \in \mathbf{X}$. We use the triangle inequality to write

$$(9.4.6) \qquad \|U - p_n\|_{\mathbf{X}} \leq \|U - \chi_n\|_{\mathbf{X}} + \|\chi_n - p_n\|_{\mathbf{X}}, \qquad n \in \mathbf{N}.$$

Due to (9.3.6) and (9.4.5), for any $\chi \in \mathbf{X}_n$, we get

$$(9.4.7) \qquad \|U - \chi_n\|_{\mathbf{X}}^2 \leq C_2^{-1} B(U - \chi_n, U - \chi_n)$$

$$= C_2^{-1} B(U - \chi_n, U - \chi) \leq C_1 C_2^{-1} \|U - \chi_n\|_{\mathbf{X}} \|U - \chi\|_{\mathbf{X}},$$

which gives an estimate of the error $U - \chi_n$, $n \in \mathbf{N}$.
On the other hand, (9.3.6) yields

$$(9.4.8) \qquad \|\chi_n - p_n\|_{\mathbf{X}} \leq \frac{B(\chi_n - p_n, \chi_n - p_n)}{C_2 \|\chi_n - p_n\|_{\mathbf{X}}} \leq$$

$$\leq C_2^{-1} \sup_{\substack{\phi \in \mathbf{X}_n \\ \phi \not\equiv 0}} \frac{B(\chi_n - p_n, \phi)}{\|\phi\|_{\mathbf{X}}} = C_2^{-1} \sup_{\substack{\phi \in \mathbf{X}_n \\ \phi \not\equiv 0}} \frac{|F(\phi) - F_n(\phi)|}{\|\phi\|_{\mathbf{X}}}, \qquad n \in \mathbf{N}.$$

Inequality (9.4.4) follows by combining (9.4.6), (9.4.7) and (9.4.8).

■

Returning to problem (9.4.2), we apply the previous theorem to get the estimate

$$(9.4.9) \quad \|U - p_n\|_{H^1_{0,w}(I)} \leq C\left\{ \inf_{\chi \in \mathbf{P}^0_n} \|U - \chi\|_{H^1_{0,w}(I)} + \|f - \Pi_{w,n-2} f\|_{L^2_w(I)} \right\}, \quad \forall n \geq 2.$$

We further bound this error using the results of chapter six. We can consider inequalities (6.4.15) and (6.4.6) after taking $\chi := \Pi^1_{0,w,n} U$ in (9.4.9). We remark that $f = -U''$ in I. Thus, we finally get

$$(9.4.10) \qquad \|U - p_n\|_{H^1_{0,w}(I)} \leq C \left(\frac{1}{n}\right)^k \|f\|_{H^k_w(I)}, \qquad \forall n > k \geq 0.$$

This shows that p_n converges to U with a rate depending on the regularity of f. Due to inequality (5.7.4), this convergence is uniform.

We remark that the function χ_n, $n \geq 2$, defined in the proof of theorem 9.4.1, here coincides with the polynomial $\hat{\Pi}^1_{0,w,n} U \in \mathbf{P}^0_n$, introduced in (6.4.13). The existence of such a polynomial follows from theorem 9.3.1, and estimates of the error $U - \hat{\Pi}^1_{0,w,n} U$ can be recovered from (9.4.7).

The same kind of analysis applies to the collocation method (9.2.15) in the case $\sigma_1 = \sigma_2 = 0$. It is sufficient to recall the quadrature formula (3.5.1) and modify (9.4.1) according to

$$(9.4.11) \qquad F_{w,n}(\phi) := \sum_{j=0}^{n} f(\eta_j^{(n)}) \phi(\eta_j^{(n)}) \, \tilde{w}_j^{(n)}, \qquad \forall \phi \in \mathbf{P}^0_n,$$

where $f \in C^0(\bar{I})$. We recall the Schwarz inequality and theorem 3.8.2 to check (9.3.7). Actually, one has $|F_{w,n}(\phi)| \leq C_3 \|f\|_{C^0(\bar{I})} \|\phi\|_{H^1_{0,w}(I)}$, $\forall \phi \in \mathbf{P}^0_n$, $\forall n \geq 2$.

Choosing as test functions the Lagrange polynomials $\phi := \tilde{l}_i^{(n)}$, $1 \leq i \leq n-1$, we obtain the relations $-p_n''(\eta_i^{(n)}) = f(\eta_i^{(n)})$, $1 \leq i \leq n-1$ (see (9.2.15) and (7.4.9)). To show that p_n converges to U when $n \to +\infty$, we use theorem 9.4.1. Therefore, we must bound the error $|F_w(\phi) - F_{w,n}(\phi)|$, $\phi \in \mathbf{P}_n^0$. Recalling relation (3.8.15) one gets

$$(9.4.12) \qquad \left|F_w(\phi) - F_{w,n}(\phi)\right| = \left|(f - \tilde{I}_{w,n}f, \phi)_w + [(\tilde{I}_{w,n}f, \phi)_w - (\tilde{I}_{w,n}f, \phi)_{w,n}]\right|$$

$$\leq \|f - \tilde{I}_{w,n}f\|_w \|\phi\|_w + \left| \frac{\|u_n\|_{w,n}^2 - \|u_n\|_w^2}{\|u_n\|_{w,n}^2} \frac{(\tilde{I}_{w,n}f, u_n)_{w,n}}{\|u_n\|_{w,n}} \frac{(\phi, u_n)_{w,n}}{\|u_n\|_{w,n}} \right|$$

$$\leq \left(\|f - \tilde{I}_{w,n}f\|_w + C \frac{|(\tilde{I}_{w,n}f, u_n)_{w,n}|}{\|u_n\|_{w,n}} \right) \|\phi\|_w, \qquad \forall \phi \in \mathbf{P}_n^0,$$

where $u_n := P_n^{(\alpha,\beta)}$, $n \in \mathbf{N}$, and $C > 0$ does not grow with n (see section 3.8). By formula (3.5.1) and orthogonality, one has $(\Pi_{w,n-1}\psi, u_n)_{w,n} = (\Pi_{w,n-1}\psi, u_n)_w = 0$, $\forall \psi \in \mathbf{P}_n$. This implies that

$$(9.4.13) \qquad \frac{|(\tilde{I}_{w,n}f, u_n)_{w,n}|}{\|u_n\|_{w,n}} \leq \|\tilde{I}_{w,n}f - \Pi_{w,n-1}(\tilde{I}_{w,n}f)\|_{w,n}$$

$$\leq \gamma_1^{-1} \left[\|\tilde{I}_{w,n}f - f\|_w + \|f - \Pi_{w,n-1}f\|_w + \|\Pi_{w,n-1}(f - \tilde{I}_{w,n}f)\|_w \right], \qquad n \geq 2.$$

For the last inequality we used (3.8.6) and (2.1.5). By combining (9.4.12) and (9.4.13), final estimates are obtained using the results of sections 6.2 and 6.6. The results of a numerical experiment are documented in table 11.2.1.

The study of the approximation of elliptic problems, on the basis of their variational formulations, is a standard technique in finite element or finite-differences methods. Celebrated examples are considered for instance in AUBIN (1972), STRANG and FIX(1973), CIARLET(1978), FLETCHER(1984), and in many other books. The simplest approximation technique is in general represented by the so called *Galerkin method.* Within the framework of polynomial approximations, this procedure consists in finding $p_n \in \mathbf{P}_n^0$ such that

$$(9.4.14) \qquad B_w(p_n, \phi) = F_w(\phi), \qquad \forall \phi \in \mathbf{P}_n^0,$$

where B_w and F_w are respectively given in (9.3.2) and (9.3.3). The convergence analysis is straightforward, since now the linear operator on the right-hand side does not depend on n. In addition, (9.4.14) is equivalent to the problem

$$(9.4.15) \qquad \begin{cases} -p_n'' = \Pi_{a,n-2} f & \text{in } I, \\ p_n(\pm 1) = 0, \end{cases}$$

where $a(x) := (1-x^2)w(x)$, $x \in I$. The solution in this particular case has the form $p_n = \hat{\Pi}^1_{0,w,n} U$, $\forall n \geq 2$ (see (6.4.13)). However, the linear system associated to (9.4.15) is not easy to recover when solving the problem in the frequency space relative to the weight function w. Usually, this system is obtained by *testing* (9.4.14) on the set of polynomials: $\phi(x) := P_k^{(\alpha,\beta)}(x) - \frac{1}{2}(1-x)P_k^{(\alpha,\beta)}(-1) - \frac{1}{2}(1+x)P_k^{(\alpha,\beta)}(1)$, $x \in \bar{I}$, $2 \leq k \leq n$. By the orthogonality of Jacobi polynomials, we obtain the corresponding $(n-1) \times (n-1)$ matrix which does not explicitly contain the rows relative to the boundary conditions as in the tau method (clarifying remarks about this point are given in BOYD(1989), chapter six).

Let us consider now the approximation of other differential equations, such as (9.1.5). For $\mathbf{X} \equiv H_w^1(I)$, a variational formulation is given by (9.3.4), where B_w and F_w are defined in (9.3.18) and (9.3.19) respectively. Polynomial approximations of the solution U can be determined in several ways. We leave the analysis of tau and Galerkin methods to the reader. We examine the collocation method.
We require $p_n \in \mathbf{P}_n$, $n \geq 2$, to be solution of

$$(9.4.16) \qquad B_{w,n}(p_n, \phi) = F_{w,n}(\phi), \qquad \forall \phi \in \mathbf{P}_n.$$

Here, $B_{w,n} : \mathbf{P}_n \times \mathbf{P}_n \to \mathbf{R}$ and $F_{w,n} : \mathbf{P}_n \to \mathbf{R}$ are respectively defined by

$$(9.4.17)\qquad B_{w,n}(\psi,\phi) := \int_I (\psi-\tilde{\psi})'[(\phi-\tilde{\phi})w]'dx + \int_I \tilde{\psi}'\tilde{\phi}'\,dx$$

$$+\ \mu\int_I (I_{a,n-1}\psi)(\phi-\tilde{\phi})w\,dx + \mu\int_I (I_{a,n-1}\psi)\tilde{\phi}\,dx, \qquad \forall\psi,\phi\in\mathbf{P}_n,$$

$$(9.4.18)\qquad F_{w,n}(\phi) := \int_I (I_{a,n-1}f)(\phi-\tilde{\phi})w\,dx + \int_I (I_{a,n-1}f)\tilde{\phi}\,dx$$

$$+\sigma_2\phi(1)-\sigma_1\phi(-1), \qquad \forall\phi\in\mathbf{P}_n,$$

where $a(x) := (1-x^2)w(x)$, $x\in I$, and the interpolation operator $I_{a,n-1}$, $n\geq 2$, is defined in section 3.3.

Our aim is to show that (9.4.16) leads to a collocation type approximation. First, assume $\phi\in\mathbf{P}_n^0$ so that $\tilde{\phi}\equiv\mathbf{0}$. With the help of integration by parts and quadrature formula (3.5.1), we obtain

$$(9.4.19)\qquad -\sum_{j=0}^{n} p_n''(\eta_j^{(n)})\phi(\eta_j^{(n)})\tilde{w}_j^{(n)} + \mu\sum_{j=0}^{n}(I_{a,n-1}p_n)(\eta_j^{(n)})\phi(\eta_j^{(n)})\tilde{w}_j^{(n)}$$

$$= \sum_{j=0}^{n}(I_{a,n-1}f)(\eta_j^{(n)})\phi(\eta_j^{(n)})\tilde{w}_j^{(n)}, \qquad \forall\phi\in\mathbf{P}_n^0.$$

Choosing $\phi := \tilde{l}_i^{(n)}$, $1\leq i\leq n-1$, (9.4.19) implies

$$(9.4.20)\quad -p_n''(x) + \mu(I_{a,n-1}p_n)(x) = (I_{a,n-1}f)(x) \quad \text{at } x=\eta_i^{(n)},\ 1\leq i\leq n-1.$$

Note that, for any $g\in C^0(\bar{I})$, one has $(I_{a,n-1}g)(\eta_i^{(n)}) = (\tilde{I}_{w,n}g)(\eta_i^{(n)}) = g(\eta_i^{(n)})$, $1\leq i\leq n-1$. This gives $-p_n''(\eta_i^{(n)})+\mu p_n(\eta_i^{(n)}) = f(\eta_i^{(n)})$, $1\leq i\leq n-1$. Moreover, for a general $\phi\in\mathbf{P}_n$, we have a relation like (9.3.21). Both $-p_n''+\mu I_{a,n-1}p_n$ and $I_{a,n-1}f$ are polynomials of degree $n-2$. Since they coincide at $n-1$ points, (9.4.20) holds for any $x\in\bar{I}$. This enables us to remove the integrals and recover the boundary conditions $-p_n'(-1) = -\sigma_1$ and $p_n'(1) = \sigma_2$. To obtain the corresponding linear system we follow the suggestions of section 7.4 (see (7.4.13)).

To develop the convergence analysis, an extension of theorem 9.4.1 is needed, since the bilinear form now depends on n. For the sake of simplicity, we do not discuss this generalization. The reader finds indications to proceed in this investigation in STRANG(1972) and CIARLET(1978).

A different treatment of the boundary conditions can be taken into consideration. First, we examine the Legendre case ($w \equiv 1$). In (9.4.16) define $B_{w,n} : \mathbf{P}_n \times \mathbf{P}_n \to \mathbf{R}$ and $F_{w,n} : \mathbf{P}_n \to \mathbf{R}$ such that

$$(9.4.21) \qquad B_{w,n}(\psi,\phi) := \int_I \psi'\phi' \, dx + \mu \sum_{j=0}^{n} \psi(\eta_j^{(n)})\phi(\eta_j^{(n)})\tilde{w}_j^{(n)}, \qquad \forall \psi,\phi \in \mathbf{P}_n,$$

$$(9.4.22) \qquad F_{w,n}(\phi) := \sum_{j=0}^{n} f(\eta_j^{(n)})\phi(\eta_j^{(n)})\tilde{w}_j^{(n)} + \sigma_2\phi(1) - \sigma_1\phi(-1), \qquad \forall \phi \in \mathbf{P}_n.$$

We replaced the integrals in (9.3.15) and (9.3.16) by the summations. The modification does not alter the scheme at the nodes inside the interval I. In fact, using the test functions $\phi := \tilde{l}_i^{(n)}$, $1 \leq i \leq n-1$, we still get the set of equations $-p_n''(\eta_i^{(n)}) + \mu p_n(\eta_i^{(n)}) = f(\eta_i^{(n)})$, $1 \leq i \leq n-1$. At the boundary points one gets

$$(9.4.23) \qquad \begin{cases} -p_n''(-1) + \mu p_n(-1) - \gamma p_n'(-1) = f(-1) - \gamma\sigma_1, \\ \\ -p_n''(1) + \mu p_n(1) + \gamma p_n'(1) = f(1) + \gamma\sigma_2, \end{cases}$$

where $\gamma := [\tilde{w}_0^{(n)}]^{-1} = [\tilde{w}_n^{(n)}]^{-1} = \frac{1}{2}n(n+1)$. Thus, we obtained the system (7.4.16) with $q := \tilde{I}_{w,n}f$. Note that in (9.4.20) the equation is also collocated at the points $x = \pm 1$, and the boundary conditions are used as corrective terms.

We can make an interesting remark. In the previous examples, the variational formulations have been presented as an intermediate step to obtain convergence results by virtue of general abstract theorems. Here, we followed an inverse path. Starting from an appropriate weak formulation, we recovered a new approximation scheme with different conditions at the boundary. These relations may not have a direct physical interpretation. Nevertheless, convergence is still achieved. As a matter of fact, one shows that $\lim_{n\to+\infty} p_n'(-1) = \sigma_1$, $\lim_{n\to+\infty} p_n'(1) = \sigma_2$ (see FUNARO(1988)).

There are several reasons for preferring conditions (9.4.23) to exact boundary conditions. For instance, as noticed in section 8.6, the eigenvalues of the matrix corresponding to system (7.4.16) are more appropriate for computation than the eigenvalues of problem (8.6.8). In addition, for the approximation of the solution of (9.1.5), the new approach gives better numerical results. We compare the two ways of imposing the boundary conditions in table 9.4.1. Take $\mu := 1$ and f, σ_1, σ_2 such that the exact solution of (9.1.5) is $U(x) := \cos(e^x)$, $x \in \bar{I}$. Then, compute the error $E_n := \|p_n - \tilde{I}_{w,n}U\|_w$ for various n. In the first column, p_n is obtained by the collocation method with the conditions $p_n'(-1) = \sigma_1$, $p_n'(1) = \sigma_2$. In the second column, p_n is obtained by the collocation method with the conditions (9.4.23).

n	*Column 1*	*Column 2*
6	$.1090 \times 10^{-1}$	$.2009 \times 10^{-3}$
8	$.4130 \times 10^{-2}$	$.1840 \times 10^{-4}$
10	$.2723 \times 10^{-3}$	$.5941 \times 10^{-6}$
12	$.6896 \times 10^{-5}$	$.1495 \times 10^{-7}$
14	$.1404 \times 10^{-5}$	$.1471 \times 10^{-8}$
16	$.7122 \times 10^{-7}$	$.4405 \times 10^{-10}$

Table 9.4.1 - Errors corresponding to the Legendre collocation approximation of problem (9.1.5).

For other Jacobi ultraspherical weights, we generalize (9.4.21) and (9.4.22) by defining

(9.4.24) $$B_{w,n}(\psi,\phi) := \int_I (\psi - \tilde{\psi})'[(\phi - \tilde{\phi})w]'dx + \int_I \tilde{\psi}'\tilde{\phi}'\,dx$$

$$+ \mu \sum_{j=0}^{n} \psi(\eta_j^{(n)})\,(\phi - \tilde{\phi})(\eta_j^{(n)})\,\tilde{w}_j^{(n)} + \mu \sum_{j=0}^{n} \psi(\eta_j^{(n)})\,\tilde{\phi}(\eta_j^{(n)})\,\chi_j^{(n)}, \qquad \forall \psi, \phi \in \mathbf{P}_n,$$

$$(9.4.25) \qquad F_{w,n}(\phi) := \sum_{j=0}^{n} f(\eta_j^{(n)})\,(\phi - \tilde{\phi})(\eta_j^{(n)})\,\tilde{w}_j^{(n)}$$

$$+ \mu \sum_{j=0}^{n} f(\eta_j^{(n)})\,\tilde{\phi}(\eta_j^{(n)})\,\chi_j^{(n)} + \sigma_2\phi(1) - \sigma_1\phi(-1), \qquad \forall\phi \in \mathbf{P}_n.$$

The weights $\chi_j^{(n)}$, $0 \leq j \leq n$, are defined in section 3.7. When $\nu = 0$, we obtain (9.4.21) and (9.4.22) by noting that $\chi_j^{(n)} = \tilde{w}_j^{(n)}$, $0 \leq j \leq n$.
Basically, $B_{w,n}$ and $F_{w,n}$ are derived from (9.3.18) and (9.3.19) respectively, by considering quadrature formulas in place of integrals. The errors introduced decay spectrally, giving the convergence of the method. We remark that, although Clenshaw-Curtis formula does not have the accuracy of Gaussian formulas, it is only used for polynomials of low degree (we have $\tilde{\phi} \in \mathbf{P}_1$ and $\psi\tilde{\phi} \in \mathbf{P}_{n+1}$).
The corresponding variational problem is equivalent to system (7.4.16) where we set $q := \tilde{I}_{w,n}f$ and $\gamma := [\chi_0^{(n)}]^{-1} = [\chi_n^{(n)}]^{-1}$. In general, if γ grows at least as n^2, one gets convergence of p_n to the solution of (9.1.5) when n tends to infinity (see FUNARO(1988)).

The idea of collocating the equation at the boundary points has been also proposed in FUNARO and GOTTLIEB(1988) for first-order equations, such as (9.1.1). This leads to a collocation scheme like that in (7.4.4), where γ is proportional to n^2. A proof of convergence is easily given for the Legendre case following the guideline of theorem 9.2.1.

The techniques introduced and analyzed in this section are also used for the approximation of the solution of fourth-order equations. The collocation method for computing the solution of problem (9.1.6) is

$$(9.4.26) \qquad \begin{cases} p_n^{IV}(\eta_i^{(n)}) = f(\eta_i^{(n)}), & 1 \leq i \leq n-1, \\ p_n(-1) = \sigma_1, \quad p_n(1) = \sigma_2, \\ p_n'(-1) = \sigma_3, \quad p_n'(1) = \sigma_4, \end{cases}$$

where p_n is a polynomial in $\mathbf{P}_{n+2}$, $n \geq 2$.

The way to write the linear system associated with (9.4.26) is described in section 7.4. A proof of convergence for ultraspherical weights ($\nu := \alpha = \beta,\ -1 < \nu < 1$) is provided in BERNARDI and MADAY(1990). In the same paper, another collocation scheme is proposed and analyzed. In the latter approach the nodes $\eta_i^{(n)},\ 1 \leq i \leq n-1$, in (9.4.26) are replaced by the nodes $\xi_i^{(n-1)},\ 1 \leq i \leq n-1$, which are the zeroes of $P_{n-1}^{(\nu+2,\nu+2)},\ n \geq 2,\ -1 < \nu < 1$. In both cases, error estimates in the space $H^2_{0,w}(I)$ (see (9.3.24)) are given. The theory is developed with the help of quadrature formulas having a double multiplicity at the boundary points (see also GAUTSCHI and LI(1991)). Other boundary conditions are analyzed in FUNARO and HEINRICHS(1990).

9.5 Approximation in unbounded domains

In this section, we discuss examples where the domain $I \subseteq \mathbf{R}$ is not bounded. Let us assume $I :=]0, +\infty[$. A typical second-order problem consists in finding $U : I \to \mathbf{R}$ such that

(9.5.1)
$$\begin{cases} -U'' + \mu U = g & \text{in } I, \\ U(0) = \sigma, \end{cases}$$

where $\sigma \in \mathbf{R}$, $\mu > 0$, and $g : I \to \mathbf{R}$ are given.

In order to determine a unique solution of (9.5.1), we need an extra condition. This is obtained by prescribing the behavior of the unknown U at infinity. For instance, we can require the convergence of U to a given value. Without loss of generality we can impose the condition $\lim_{x\to+\infty} U(x) = 0$. To develop the analysis, we make further assumptions on the rate of decay of U. This is obtained by assuming that U belongs to some functional space $\mathbf{X}$. Thus, the best way to argue is to use a suitable variational formulation of (9.5.1). We remark that it is not restrictive to assume that $\sigma = 0$.

Let $v(x) := x^\alpha e^x,\ x \in I,\ -1 < \alpha < 1$, be a weight function. Let us define the Hilbert space:

$$(9.5.2) \qquad \mathbf{X} \equiv H^1_{0,v}(I) := \{\phi \mid \phi \in H^1_v(I), \ \ \phi(0) = 0\}.$$

According to section 5.7, the norm in $\mathbf{X}$ is given by $\|\phi\|_{\mathbf{X}} := \left(\int_I \phi^2 v dx + \int_I [\phi']^2 v dx\right)^{\frac{1}{2}}$, $\forall \phi \in \mathbf{X}$. Then, for $g \in L^2_v(I)$, we are concerned with finding $U \in \mathbf{X}$ such that

$$(9.5.3) \qquad \int_0^{+\infty} U'(\phi v)' \, dx + \mu \int_0^{+\infty} U\phi v \, dx = \int_0^{+\infty} g\phi v \, dx, \qquad \forall \phi \in \mathbf{X}.$$

With the same arguments of section 9.3, using the Lax-Milgram theorem, we get existence and uniqueness of a weak solution of (9.5.3), provided the parameter μ is larger than $\frac{1}{2}\max\{1, \frac{1}{1-\alpha}\}$. The crucial part of the proof is to show coerciveness (see (9.3.6)). This follows by virtue of lemma 8.2.7.

One checks that $U \in \mathbf{X}$ implies that $\lim_{x \to +\infty} U(x) = 0$, and the decay is exponential. Therefore, due to theorem 6.1.4, we have the requisites to approximate the solution of problem (9.5.3) by Laguerre functions (see section 6.7). To this end, we introduce the subspace $\mathbf{S}^0_n := \{\phi \mid \ \phi \in \mathbf{S}_n, \ \ \phi(0) = 0\} \subset \mathbf{X}$.

For example, when $g \in C^0(\bar{I})$, we discretize (9.5.3) as follows. For any $n \geq 2$, we seek $P_n \in \mathbf{S}^0_{n-1}$ such that

$$(9.5.4) \quad \int_0^{+\infty} P_n'(\phi v)' \, dx + \mu \int_0^{+\infty} P_n \phi v \, dx = \int_0^{+\infty} (\tilde{I}^*_{v,n} g)\phi v \, dx, \qquad \forall \phi \in \mathbf{S}^0_{n-1}.$$

The interpolation operator $\tilde{I}^*_{v,n} : C^0(\bar{I}) \to \mathbf{S}_{n-1}, \ \ n \geq 2$, is defined by the relation $\tilde{I}^*_{v,n} g := [\tilde{I}_{w,n}(ge^x)]e^{-x}, \ \ \forall g \in C^0(\bar{I})$ (see also section 6.7), where $\tilde{I}_{w,n}, \ n \geq 1$, is the Laguerre Gauss-Radau interpolation operator (see section 3.3). We now apply theorem 9.4.1. Estimating the error $\|g - \tilde{I}^*_{v,n} g\|_{L^2_v(I)}$, one deduces that P_n converges to U, as n tends to infinity, in the norm of the space $\mathbf{X}$. The convergence is also uniform because of the inequality

$$(9.5.5) \qquad \sup_{x \in I} |\phi(x) e^{x/2}| \leq C \, \|\phi\|_{\mathbf{X}}, \qquad \forall \phi \in \mathbf{X}.$$

The next step is to determine the solution to (9.5.4). We consider the substitutions: $p_n(x) := P_n(x)e^x, \ x \in \bar{I}, \ n \geq 2$, and $f(x) := g(x)e^x, \ x \in \bar{I}$. Therefore, integration by parts allows us to write (note that the boundary terms are vanishing):

$$(9.5.6)\qquad \int_0^{+\infty}(-p_n''+2p_n'-p_n)\phi w\,dx=\int_0^{+\infty}(\tilde{I}_{w,n}f)\phi w\,dx,\qquad \forall\phi\in\mathbf{P}_{n-1}^0.$$

In (9.5.6), $w(x):=x^\alpha e^{-x}$, $x\in I$, $-1<\alpha<1$, is the Laguerre weight function. At this point, we use the quadrature formula (3.6.1). Taking as test functions the Lagrange polynomials $\tilde{l}_i^{(n)}$, $1\le i\le n-1$, we obtain the collocation scheme

$$(9.5.7)\qquad \begin{cases} -p_n''(\eta_i^{(n)})+2p_n'(\eta_i^{(n)})+(\mu-1)p_n(\eta_i^{(n)})=f(\eta_i^{(n)}), & 1\le i\le n-1,\\ p_n(\eta_0^{(n)})=0, \end{cases}$$

which is the same one presented in (7.4.26) with $\sigma=0$, $q:=I_{a,n-1}f$, and $a(x):=xw(x)$, $x\in I$. For the numerical implementation we suggest taking into account the results of sections 3.10 and 7.5.

Other differential equations are approximated in a similar way. Examples are found in MADAY, PERNAUD-THOMAS and VANDEVEN(1985), and MAVRIPLIS(1989). For the domain $I\equiv\mathbf{R}$, Hermite functions are a suitable basis for computations when the solution decays like e^{-x^2} at infinity. Results in this direction are given in FUNARO and KAVIAN(1988).

The literature offers several other numerical techniques for the treatment of problems in unbounded domains by spectral methods. For instance, one can map the domain I in the interval $]-1,1[$ and approximate the new problem by Jacobi polynomial expansions. The main difficulty is to find coordinate transformations preserving the smoothness of the problem to avoid singularities and retain spectral accuracy. Early results were given in GROSCH and ORSZAG(1977). Further indications and comparisons are provided in BOYD(1982) and BOYD(1989), chapter 13. Using the mapping approach, the solution of the original problem is not required to decay exponentially at infinity as in the case of Laguerre or Hermite approximations.

The idea of truncating the domain is often used in applications. Here, it is not easy in general to detect the appropriate size of the computational domain. In addition, one introduces *artificial boundaries*, where a non correct specification of the behavior of the approximating polynomials may cause loss in accuracy. We describe an example in section 12.2.

9.6 Other techniques

Variations of the principal methods exposed in the previous sections have been suggested by different authors, with the aim of preserving spectral accuracy and facilitate numerical implementation.

For example, we already noted that discretizations by the collocation method of linear differential operators with non-constant coefficients, are much easier to implement than those using tau method. The discrepancy is more remarkable for nonlinear equations (see sections 3.3 and 9.8). However, we can still work in the frequency space by combining different techniques. For instance, we modify problem (9.2.6) by setting

$$(9.6.1) \qquad \begin{cases} p'_n \ + \ \Pi_{w,n-1}[\tilde{I}_{w,n}(Ap_n)] \ = \ \Pi_{w,n-1}f \quad \text{in }]-1,1], \\ p_n(-1) \ = \ \sigma. \end{cases}$$

We can solve the linear system corresponding to (9.6.1) by an iterative method. In this way, whenever we need to evaluate the product Ap_n , we can go to the physical space by Fourier transform (see section 4.1) and perform this operation therein. We return to the frequency space by the inverse transform. Calculations are faster in the Chebyshev case, where we can use the FFT (see section 4.3). We remark that this algorithm does not coincide with a genuine tau method. The computations, however, exhibit a spectral accuracy.

Other methods have been developed starting from variational techniques. We consider for instance problem (9.1.4), which has a weak formulation given by (9.3.2)-(9.3.3)-(9.3.4) for $\sigma_1 = \sigma_2 = 0$. The Galerkin approximation is presented in (9.4.14) which is equivalent to (9.4.15). Assuming that $w \equiv 1$, from the orthogonality of Legendre polynomials, we easily write the corresponding linear system, where the unknowns are the Fourier coefficients of p_n, $n \geq 2$. Using the same weight function, a different choice of the polynomial basis leads to the so called *spectral element method* introduced in PATERA(1984) (see also section 11.2). In this method, p_n is expanded in terms of Chebyshev polynomials, i.e. $p_n = \sum_{k=0}^{n} c_k T_k$. Therefore, substituting in (9.4.14), one obtains

$$(9.6.2) \qquad \sum_{k=0}^{n} c_k \int_{-1}^{1} T_k' \, \phi' \, dx = \int_{-1}^{1} f\phi \, dx, \qquad \forall \phi \in \mathbf{P}_n^0.$$

The coefficients c_k, $0 \le k \le n$, are determined from the boundary conditions (these imply $\sum_{k=0}^{n} c_k = \sum_{k=0}^{n}(-1)^k c_k = 0$) and testing (9.6.2) on the set of polynomials $\phi(x) := T_j(x) - \frac{1}{2}(-1)^j(1-x) - \frac{1}{2}(1+x)$, $x \in \bar{I}$, $2 \le j \le n$. The matrix corresponding to the linear system is full and its entries are obtained by computing the integrals $I_{kj} := \int_{-1}^{1} T_k' T_j' dx$. We can provide an explicit expression of these quantities by recalling (1.5.7) and arguing as in section 2.6. In this approach, we can combine the benefits of working with Chebyshev polynomials, with a variational formulation which does not involve singular weight functions. The collocation scheme is instead obtained by taking $p_n = \sum_{j=1}^{n-1} p_n(\eta_j^{(n)}) \tilde{l}_j^{(n)}$ in (9.4.14), and testing on $\phi := \tilde{l}_j^{(n)}$, $1 \le j \le n-1$.

9.7 Boundary layers

Small perturbations of certain differential operators can force the solution to degenerate in a singular behavior near the boundary points, which is often recognized as a *boundary layer*. The terminology was first introduced in PRANDTL(1905). General motivations and theoretical results are given in ECKHAUS(1979), CHANG and HOWES(1984), LAGERSTROM(1988). A typical example is given by the equation

$$(9.7.1) \qquad \begin{cases} -\epsilon \, U_\epsilon'' + U_\epsilon = 0 & \text{in } I :=]-1,1[, \\ U_\epsilon(-1) = 0, \quad U_\epsilon(1) = 1, \end{cases}$$

where $\epsilon > 0$ is a given parameter. The solution of (9.7.1) is

$$(9.7.2) \qquad U_\epsilon(x) = \frac{e^{(1+x)/\sqrt{\epsilon}} - e^{-(x+1)/\sqrt{\epsilon}}}{e^{2/\sqrt{\epsilon}} - e^{-2/\sqrt{\epsilon}}}, \qquad x \in \bar{I}.$$

When ϵ tends to zero, U_ϵ converges to the discontinuous function: $U_0 \equiv 0$ in $[-1,1[$, $U_0(1) = 1$. For very small values of ϵ, various physical phenomena are related to the function U_ϵ, which displays sharp derivatives in a neighborhood of the point $x = 1$. Due to this behavior, approximations of U_ϵ by low degree global polynomials, are affected by oscillations. Nevertheless, we are not in the situation that characterizes the Gibbs phenomenon (see sections 6.2 and 6.8, as well GOTTLIEB and ORSZAG(1977), p.40). Actually, since the solution is analytic for $\epsilon > 0$, uniform convergence with an exponential rate is obtained with the usual techniques. Tau, Galerkin and collocation methods have been studied in CANUTO(1988). In particular, the tau method is equivalent to finding $p_{\epsilon,n} \in \mathbf{P}_n$, $n \geq 2$, $\epsilon > 0$, satisfying problem (7.3.5), where: $A \equiv 0$, $B \equiv \frac{1}{\epsilon}$, $q \equiv 0$, $\sigma_1 = 0$, $\sigma_2 = 1$. Thus, with the notations of section 7.1, the corresponding linear system is

$$\text{(9.7.3)} \qquad \begin{cases} -\epsilon\, c_k^{(2)} + c_k = 0 & 0 \leq k \leq n-2, \\ \sum_{k=0}^n c_k u_k(-1) = 0, & \sum_{k=0}^n c_k u_k(1) = 1. \end{cases}$$

The coefficients $c_k^{(2)}$, $0 \leq k \leq n$, are given for $\nu := \alpha = \beta$ in KARAGEORGIS and PHILLIPS(1989) (see also (7.1.9) and (7.1.10)). These are all positive for $\nu > -1$. Therefore, it is easy to deduce that $c_k > 0$, $0 \leq k \leq n$. This property is used in CANUTO(1988) to estimate the maximum norm of $p_{\epsilon,n}$. For instance, from (1.3.9), we have

$$\text{(9.7.4)} \qquad |p_{\epsilon,n}(x)| = \left|\sum_{k=0}^n c_k u_k(x)\right| \leq \sum_{k=0}^n c_k |u_k(x)|$$

$$\leq \sum_{k=0}^n c_k \binom{n+\nu}{n} = \sum_{k=0}^n c_k u_k(1) = p_{\epsilon,n}(1) = 1, \qquad \forall x \in \bar{I},\ \forall \epsilon > 0,\ \forall n \geq 2.$$

In addition, using relation (1.4.8), in the Legendre case one gets

$$\text{(9.7.5)} \qquad |p_{\epsilon,n}(x)| \leq \sum_{k=0}^n c_k |P_k(x)| \leq \frac{1}{2}(1+x^2) \sum_{k=0}^n c_k P_k(1) = \frac{1}{2}(1+x^2),$$

$$\forall x \in \bar{I},\ \forall \epsilon > 0,\ \forall n \geq 2.$$

By refining the above inequalities, we can further investigate the behavior of $p_{\epsilon,n}$ near the boundary layer.

In figures 9.7.1 and 9.7.2, we plot the polynomial $p_{\epsilon,n}$ for $n = 16$ and $n = 22$, corresponding to the approximation by the tau method of problem (9.7.1) in the Legendre case ($\nu = 0$) with $\epsilon = 10^{-4}$.

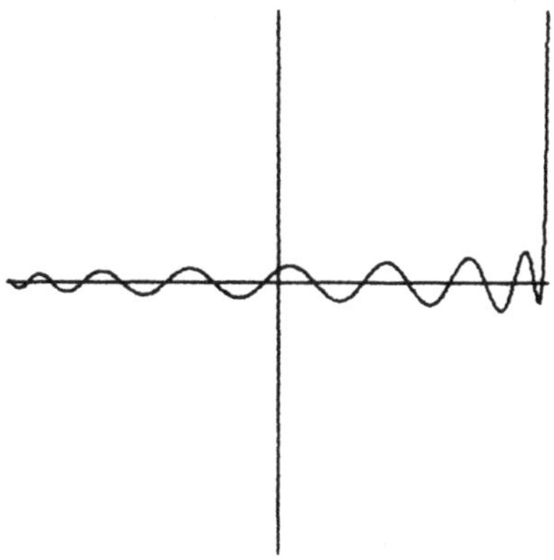

Figure 9.7.1 - Tau-Legendre approximation of problem (9.7.1) for $n = 16$ and $\epsilon = 10^{-4}$.

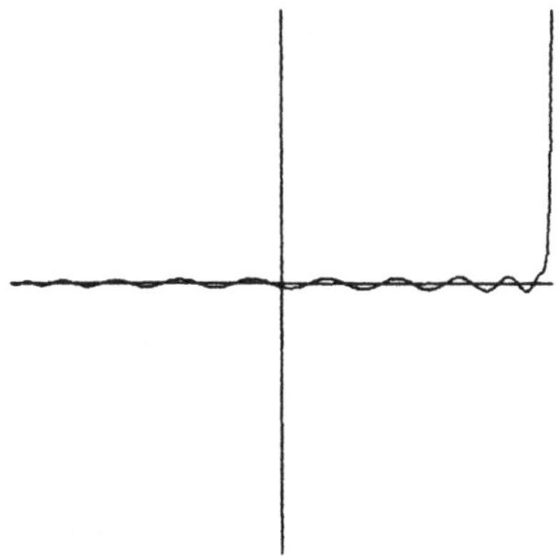

Figure 9.7.2 - Tau-Legendre approximation of problem (9.7.1) for $n = 22$ and $\epsilon = 10^{-4}$.

For higher values of n, we have enough resolution to control the violent deviation of U_ϵ near the point $x = 1$. Then, the oscillations disappear. This happens approximately when n is larger than $C/\sqrt{\epsilon}$, where the constant $C > 0$ does not depend on ϵ.

Similar conclusions follow for the collocation method. Again, we consider the ultraspherical case, where the polynomial $p_{\epsilon,n}$, $n \geq 2$, $\epsilon > 0$, satisfies:

$$(9.7.6)\qquad \begin{cases} -\epsilon\, p''_{\epsilon,n} \;+\; p_{\epsilon,n} \;=\; \left[\dfrac{(n+2\nu)\,c_{n-1}}{(n+\nu)(2n+2\nu-1)} + \dfrac{c_n}{n}x\right] u'_n \qquad \text{in } I, \\[2ex] p_{\epsilon,n}(-1) = 0, \qquad p_{\epsilon,n}(1) = 1, \end{cases}$$

where c_k, $0 \le k \le n$, are the Fourier coefficients of $p_{\epsilon,n}$ and $u_n := P_n^{(\nu,\nu)}$, $\nu > -1$. In fact, the polynomial u'_n vanishes at the points $\eta_i^{(n)}$, $1 \le i \le n-1$, giving the collocation scheme. The corrective term on the right-hand side of the first equation in (9.7.6) is obtained by equating the coefficients of the monomials x^n and x^{n-1} to the respective coefficients of the left-hand side. The Chebyshev case is treated in a similar way.

We can solve (9.7.6) in the frequency space by recalling (2.3.8) and (7.1.3). Due to the positivity of the quantities involved in these computations, we still recover $c_k > 0$, $0 \le k \le n$. Therefore, inequalities (9.7.4) and (9.7.5) also hold when $p_{\epsilon,n}$ is the approximation associated with the collocation method.

The same analysis can be developed for the problem

$$(9.7.7)\qquad \begin{cases} -\epsilon\, U''_\epsilon \;+\; U'_\epsilon \;=\; 0 \qquad \text{in } I, \quad \epsilon > 0, \\[1ex] U_\epsilon(-1) = 0, \quad U_\epsilon(1) = 1. \end{cases}$$

For this example, the results given in CANUTO(1988) in the Chebyshev case point out a different behavior of the approximating polynomials, depending on the parity of their degree.

9.8 Nonlinear equations

In a large number of practical applications, physics events are modelled by the solution of *nonlinear* differential equations. Of course, spectral methods can be employed in these situations as well, although a lot of robust algorithms are already available in the framework of finite-difference computations. The success of a certain approximation technique is strictly dependent on the type of nonlinearity, so that it is difficult to provide general recipes. Some examples are discussed in chapter twelve. As far as the theoretical analysis is concerned, references are very few, both for the difficulty and the modernity of the subject. We give some hints for the implementation. Consider the problem

$$(9.8.1) \qquad \begin{cases} U'(x) = F(x, U(x)) & \forall x \in]-1,1], \\ U(-1) = \sigma, \end{cases}$$

where $F : [-1,1] \times \mathbf{R} \to \mathbf{R}$ is a given function and $\sigma \in \mathbf{R}$. We recover the linear equation (9.1.3) by setting $F(x,y) := f(x) - A(x)y$.
Appropriate conditions on F insure existence, uniqueness and regularity of the solution U (see for instance GOLOMB and SHANKS (1965), BRAUER and NOHEL(1967), SIMMONS(1972), etc.). According to the remarks of section 3.3, approximations in the physical space are preferable to approximations in the frequency space, although one can note a little deterioration in the numerical results, due to the effects of the aliasing errors (see section 4.2 and ORSZAG(1972)). It is common opinion to attribute this phenomenon to the spacing of the collocation grid, when it is not adequate to resolve high frequency oscillations. *De-aliasing* procedures have been proposed by various authors. The reader is addressed to BOYD(1989) for a collection of references.

The collocation method applied to problem (9.8.1) consists in finding $p_n \in \mathbf{P}_n$, $n \geq 1$, such that

$$(9.8.2) \qquad \begin{cases} p_n'(\eta_i^{(n)}) = F(\eta_i^{(n)}, p_n(\eta_i^{(n)})) & 1 \leq i \leq n, \\ p_n(\eta_0^{(n)}) = \sigma. \end{cases}$$

In the Legendre case, a convergence result is obtained by reviewing the proof of theorem 9.2.1. This time, we assume that there exists a constant $\epsilon > 0$ such that

(9.8.3) $-[F(x,y_1)-F(x,y_2)](y_1-y_2) \ \geq \ \epsilon\ (y_1-y_2)^2, \qquad \forall x\in[-1,1],\ \forall y_1,y_2\in\mathbf{R}.$

For a regular F this is equivalent to requiring $-\frac{\partial F}{\partial y}\geq\epsilon$ in $[-1,1]\times\mathbf{R}$.
It is clear that (9.8.2) is not equivalent to a linear system and the unknown vector $\bar{v}:=\{p_n(\eta_i^{(n)})\}_{0\leq i\leq n}$ has to be computed by an iterative approach. We can adopt the Richardson method (see section 7.6) by defining the sequence $\bar{v}^{(k)}\equiv(v_0^{(k)},\cdots,v_n^{(k)})$, $k\in\mathbf{N}$, such that

$$\bar{v}^{(k+1)} \ := \ (I-\theta D)\bar{v}^{(k)} \ + \ \theta\bar{w}^{(k)}, \qquad k\in\mathbf{N}, \tag{9.8.4}$$

where $\bar{v}^{(0)}$ is the initial guess. In (9.8.4), D is the matrix corresponding to the system (7.4.1) and $\bar{w}^{(k)}\equiv\{\sigma,F(\eta_1^{(n)},v_1^{(k)}),\cdots,F(\eta_n^{(n)},v_n^{(k)})\}$, $k\in\mathbf{N}$. If the iterates converge for some $\theta>0$, then one gets $\lim_{k\to+\infty}v_0^{(k)}=\sigma$, $\lim_{k\to+\infty}v_i^{(k)}=p_n(\eta_i^{(n)})$, $1\leq i\leq n$. A faster convergence is realized by using preconditioners (see sections 8.3 and 8.5).

Second-order nonlinear problems can be also considered. An example is given by the equation

$$\begin{cases} -U''(x) \ + \ F(x,U(x),U'(x)) \ = \ 0 \qquad x\in]-1,1[, \\ U(-1) \ = \ \sigma_1, \ \ U(1) \ = \ \sigma_2, \end{cases} \tag{9.8.5}$$

where $F:[-1,1]\times\mathbf{R}^2\to\mathbf{R}$, $\sigma_1\in\mathbf{R}$, $\sigma_2\in\mathbf{R}$, are given.
Spectral type approximations of (9.8.5) are studied in MADAY and QUARTERONI(1982) in the case $F(x,y,z):=\frac{1}{\epsilon}yz-f(x)$, $\sigma_1=\sigma_2=0$, where $\epsilon>0$ and the function $f:[-1,1]\to\mathbf{R}$ is given. A general trick to analyze the solution of problem (9.8.5) and its approximation is to write the equation in the form $U=LF(x,U,U')$, where L is the linear operator which associates to any function f the solution of problem (9.1.4) with $\sigma_1=\sigma_2=0$. Then, the investigation proceeds by using *fixed-point* theorems and the *compactness* of the operator L. More details can be found in BREZZI, RAPPAZ and RAVIART(1980). We further examine the above example in section 10.4.

9.9 Systems of differential equations

Approximation by spectral methods of systems of differential equations in two or more unknowns can be also taken into consideration. We illustrate a simple example which will be also useful in section 12.3. Other systems are examined in section 10.5.

We are concerned with finding the solutions $V : [-1,1] \to \mathbf{R}$ and $W : [-1,1] \to \mathbf{R}$ of the differential problem

$$(9.9.1) \qquad \begin{cases} -V'' + \mu W = f & \text{in }]-1,1[, \\ -W'' - \mu V = g & \text{in }]-1,1[, \\ V(\pm 1) = W(\pm 1) = 0, \end{cases}$$

where f and g are given continuous functions and $\mu \in \mathbf{R}$.

We approximate (9.9.1) by the collocation method. Thus, for $n \geq 1$, we have to find two polynomials $p_n, q_n \in \mathbf{P}_n$ such that

$$(9.9.2) \qquad \begin{cases} -p_n''(\eta_i^{(n)}) + \mu q_n(\eta_i^{(n)}) = f(\eta_i^{(n)}) & 1 \leq i \leq n-1, \\ -q_n''(\eta_i^{(n)}) - \mu p_n(\eta_i^{(n)}) = g(\eta_i^{(n)}) & 1 \leq i \leq n-1, \\ p_n(\eta_0^{(n)}) = q_n(\eta_0^{(n)}) = 0, \\ p_n(\eta_n^{(n)}) = q_n(\eta_n^{(n)}) = 0. \end{cases}$$

When n tends to infinity, p_n converges to V and q_n converges to W. For the proof we can use the techniques developed in section 9.4. The convergence is uniform in $[-1,1]$.

As usual, the determination of the approximating polynomials is equivalent to solving a linear system which can be written by defining the vectors

$$\bar{p}_n \equiv (p_n(\eta_1^{(n)}), \cdots, p_n(\eta_{n-1}^{(n)})), \qquad \bar{q}_n \equiv (q_n(\eta_1^{(n)}), \cdots, q_n(\eta_{n-1}^{(n)})),$$

$$\bar{f}_n \equiv (f_n(\eta_1^{(n)}), \cdots, f_n(\eta_{n-1}^{(n)})), \qquad \bar{g}_n \equiv (g_n(\eta_1^{(n)}), \cdots, g_n(\eta_{n-1}^{(n)})).$$

Then, one has

$$(9.9.3)\qquad \begin{bmatrix} M_n & \mu I_n \\ -\mu I_n & M_n \end{bmatrix} \begin{bmatrix} \bar{p}_n \\ \bar{q}_n \end{bmatrix} = \begin{bmatrix} \bar{f}_n \\ \bar{g}_n \end{bmatrix},$$

where I_n denotes the $(n-1)\times(n-1)$ identity matrix and M_n is the reduced $(n-1)\times(n-1)$ matrix corresponding to problem (7.4.9) (see (7.4.11) for the case $n=3$). In the ultraspherical case with $-1<\alpha=\beta\leq 1$, one shows that the matrix in (9.9.3) has eigenvalues with positive real part (see section 8.2). In addition, we can invert the $(2n-2)\times(2n-2)$ linear system (9.9.3) by blocks, obtaining

$$(9.9.4)\qquad \begin{bmatrix} \bar{p}_n \\ \bar{q}_n \end{bmatrix} = \begin{bmatrix} (M_n^2+\mu^2 I_n)^{-1} & 0 \\ 0 & (M_n^2+\mu^2 I_n)^{-1} \end{bmatrix} \begin{bmatrix} M_n & -\mu I_n \\ \mu I_n & M_n \end{bmatrix} \begin{bmatrix} \bar{f}_n \\ \bar{g}_n \end{bmatrix}.$$

Therefore, we only need to compute the inverse of $M_n^2+\mu^2 I_n$.

9.10 Integral equations

In concluding this chapter, we say a few words about the possibility of using polynomials in the approximation of equations in integral form. A standard example is given by the following *singular integral equation:*

$$(9.10.1)\qquad A\,U(x) + \frac{B}{\pi}\int_{-1}^{1}\frac{U(s)}{s-x}\,ds = f(x), \qquad \forall x\in\,]-1,1[,$$

where $A,B\in\mathbf{R}$ are constants and $f:]-1,1[\to\mathbf{R}$ is a given function. The integral in (9.10.1) is a *principal value integral.* This is given by the limit

$$(9.10.2)\qquad \lim_{\epsilon\to 0^+}\left[\int_{-1}^{x-\epsilon}\frac{U(s)}{s-x}\,ds + \int_{x+\epsilon}^{1}\frac{U(s)}{s-x}\,ds\right], \qquad \forall x\in\,]-1,1[.$$

Problem (9.10.1) admits solutions satisfying appropriate integrability conditions. Theoretical results have been given in TRICOMI(1957), KANTOROVICH and KRYLOV(1964), POGORZELSKI(1966), HOCHSTADT(1973).

Galerkin and collocation type approximations of the solution U have been studied by many authors. In the latter approach, the integral is replaced by suitable quadrature sums based on Chebyshev polynomials. Consider the case $A = 0$ and $B = 1$. A classical scheme is obtained by defining $V(s) := U(s)\sqrt{1-s^2}$, $s \in]-1,1[$. For $n \geq 1$, recalling (3.4.1) and (3.4.6), we approximate the integral by the formula

(9.10.3) $$\int_{-1}^{1} \frac{V(s)}{s-x} \frac{ds}{\sqrt{1-s^2}} \approx \frac{\pi}{n} \sum_{j=1}^{n} \frac{V(\xi_j^{(n)})}{\xi_j^{(n)} - x}, \qquad x \neq \xi_j^{(n)}, \quad 1 \leq j \leq n,$$

where the $\xi_j^{(n)}$'s are the zeroes of T_n (see (3.1.4)).

Finally, for $n \geq 2$, V is approximated by the polynomial $p_n \in \mathbf{P}_{n-2}$ satisfying the set of equations

(9.10.4) $$\frac{1}{n} \sum_{j=1}^{n} \frac{p_n(\xi_j^{(n)})}{\xi_j^{(n)} - \eta_i^{(n)}} = f(\eta_i^{(n)}), \qquad 1 \leq i \leq n-1,$$

where the $\eta_j^{(n)}$'s are given in (3.1.11).

It is easy to set up the corresponding linear system. Instead, the theoretical analysis of convergence is not a trivial matter. Hints are provided in KRENK(1975). For other results and generalizations, we quote the following papers: DELVES(1968), ERDOGAN, GUPTA and COOK(1973), KRENK(1978), ELLIOTT(1981), IOAKIMIDIS(1981), MONEGATO(1982), ELLIOTT(1982).

* * * * * * * * * * *

The differential problems examined in this chapter are mainly of academic interest. Closer to physics applications are the examples of chapter twelve. It is our opinion, however, that a correct implementation of more realistic problems starts from a full understanding of the basic patterns.

For simplicity, boundary value problems in two or more variables are not considered. Actually, although this extension is sometimes straightforward (see chapter thirteen), most of the times it yields a rather involved theoretical analysis, even for simple model problems, which contrasts with the elementary exposition to which we have tried to adhere.

10
TIME-DEPENDENT PROBLEMS

The purpose of this chapter is to analyze the approximation, using spectral methods, of differential problems where the solution is time-dependent. We consider parabolic and hyperbolic partial differential equations in one space variable. Finite-differences are usually employed for the treatment of the time variable, hence part of the analysis will be devoted to this topic.

10.1 The Gronwall inequality

A basic result, known as *Gronwall lemma*, will be used in the following. Several versions exist, but we recall only one of them here.

Theorem 10.1.1 - *Let* $T > 0$ *and* $C \geq 0$. *Let* g *and* h *be two continuous functions in the interval* $[0, T]$, *such that*

$$g(t) \leq g(0) + \int_0^t [Cg(s) + h(s)]ds, \qquad \forall t \in [0, T]. \tag{10.1.1}$$

Then, we have

$$g(t) \leq e^{Ct}\left[g(0) + \int_0^t h(s)e^{-Cs}ds\right], \qquad \forall t \in [0, T]. \tag{10.1.2}$$

Proof - Using (10.1.1), we can easily check the inequality $G'(t) \leq 0,\ \forall t \in [0,T]$, where we defined

$$G(t) := e^{-Ct}\left[g(0) + \int_0^t [Cg(s) + h(s)]ds\right] - \int_0^t h(s)e^{-Cs}ds, \qquad t \in [0,T].$$

Therefore, the function G is not increasing. Finally, by (10.1.1), we get

(10.1.3) $$g(t)e^{-Ct} - \int_0^t h(s)e^{-Cs}ds \leq G(t) \leq G(0) = g(0), \qquad \forall t \in [0,T].$$

This implies (10.1.2).

■

10.2 Approximation of the heat equation

In the field of linear partial differential equations, the *heat equation* is a classical example. The solution $U \equiv U(x,t)$, physically interpreted as a temperature, depends on the space variable x and the time variable t. In the case of a heated bar $I =]-1,1[$ of homogeneous material, the temperature evolves in the time interval $[0,T]$, $T > 0$, according to the equation

(10.2.1) $$\frac{\partial U}{\partial t}(x,t) = \zeta \frac{\partial^2 U}{\partial x^2}(x,t), \qquad \forall x \in I, \ \forall t \in]0,T],$$

where $\zeta > 0$ is a constant (*thermal diffusivity*).

To pose the problem properly, we need an *initial condition*:

(10.2.2) $$U(x,0) \equiv U_0(x), \qquad \forall x \in \bar{I},$$

where $U_0 : \bar{I} \rightarrow \mathbf{R}$ is a given continuous function.

In addition, at any $t \in]0,T]$, suitable boundary conditions at the endpoints of the interval I are assumed. These are for instance

$$(10.2.3) \qquad U(-1,t) = \sigma_1(t), \quad U(1,t) = \sigma_2(t), \quad \forall t \in]0,T],$$

where $\sigma_1 : [0,T] \to \mathbf{R}$, $\sigma_2 : [0,T] \to \mathbf{R}$, are given continuous functions.

Equation (10.2.1) is of *parabolic* type. The literature on this subject is extensive and cannot be covered here. For a discussion of the general properties of parabolic equations, we only mention the following comprehensive books: COURANT and HILBERT (1953), SNEDDON (1957), GREENSPAN(1961), EPSTEIN(1962), SOBOLEV(1964), WEINBERGER(1965). The same books can be used as references for the other sections in this chapter. An approach more closely related to the heat equation is analyzed in WIDDER (1975), HILL and DEWYNNE(1987).

Under appropriate hypotheses on the data U_0, σ_1 and σ_2, existence and uniqueness of a solution satisfying (10.2.1), (10.2.2), (10.2.3) can be proven. In most cases, this can be expressed by a suitable series expansion.
We note that the *compatibility conditions*

$$(10.2.4) \qquad \sigma_1(0) = U_0(-1), \quad \sigma_2(0) = U_0(1),$$

are not required in general. However, though the solution U is very smooth for $t \in]0,T]$, if we drop conditions (10.2.4), heat propagation with infinite velocity is manifested at $t = 0$. This can adversely affect the numerical analysis. Thus, we assume henceforth that relations (10.2.4) are satisfied.
The substitution $V(x,t) := U(x,t) - \frac{1}{2}(1-x)\sigma_1(t) - \frac{1}{2}(1+x)\sigma_2(t)$, $x \in [-1,1]$, $t \in [0,T]$, leads to an equivalent formulation with homogeneous boundary conditions. Indeed, we now have

$$(10.2.5) \qquad \frac{\partial V}{\partial t}(x,t) = \zeta \, \frac{\partial^2 V}{\partial x^2}(x,t) + f(x,t), \quad \forall x \in I, \ \forall t \in]0,T],$$

$$(10.2.6) \qquad V(x,0) = U_0(x) - \tfrac{1}{2}(1-x)U_0(-1) - \tfrac{1}{2}(1+x)U_0(1), \quad \forall x \in \bar{I},$$

(10.2.7) $$V(-1,t) = V(1,t) = 0, \qquad \forall t \in [0,T],$$

where $f(x,t) := -\frac{1}{2}(1-x)\sigma_1'(t) - \frac{1}{2}(1+x)\sigma_2'(t)$, $x \in I$, $t \in]0,T]$. Here, we require that the functions σ_1 and σ_2 are differentiable.

We intend to use spectral methods in order to approximate the unknown V. To this end, for any $t \in [0,T]$, the function $V(\cdot,t) : I \to \mathbf{R}$ is approximated by an algebraic polynomial $p_n(\cdot,t) \in \mathbf{P}_n^0$ (see (6.4.11)), where the integer $n \geq 2$ has been fixed in advance. For example, we use the collocation method (see section 9.2). The approximate problem is

(10.2.8) $$\frac{\partial p_n}{\partial t}(\eta_i^{(n)},t) = \zeta\, \frac{\partial^2 p_n}{\partial x^2}(\eta_i^{(n)},t) + f(\eta_i^{(n)},t), \quad 1 \leq i \leq n-1, \quad \forall t \in]0,T].$$

As usual, the nodes $\eta_i^{(n)}$, $1 \leq i \leq n-1$, are the zeroes of the polynomial $\frac{d}{dx}P_n^{(\alpha,\beta)}$, $\alpha > -1$, $\beta > -1$. Initial and boundary conditions are respectively given by

(10.2.9) $$p_n(\eta_i^{(n)},0) = U_0(\eta_i^{(n)}) - \tfrac{1}{2}(1-\eta_i^{(n)})U_0(-1) - \tfrac{1}{2}(1+\eta_i^{(n)})U_0(1), \quad 0 \leq i \leq n,$$

(10.2.10) $$p_n(\eta_0^{(n)},t) = p_n(\eta_n^{(n)},t) = 0, \qquad \forall t \in [0,T].$$

At this point, our problem is transformed into a $(n+1)\times(n+1)$ linear system of ordinary differential equations. Actually, with the notations of sections 7.2 and 7.4, we can write for any $t \in]0,T]$ (we choose for instance $n = 4$):

(10.2.11) $$\frac{d}{dt}\begin{bmatrix} p_n(\eta_1^{(n)},t) \\ p_n(\eta_2^{(n)},t) \\ p_n(\eta_3^{(n)},t) \end{bmatrix} = \zeta \begin{bmatrix} \tilde{d}_{11}^{(2)} & \tilde{d}_{12}^{(2)} & \tilde{d}_{13}^{(2)} \\ \tilde{d}_{21}^{(2)} & \tilde{d}_{22}^{(2)} & \tilde{d}_{23}^{(2)} \\ \tilde{d}_{31}^{(2)} & \tilde{d}_{32}^{(2)} & \tilde{d}_{33}^{(2)} \end{bmatrix} \begin{bmatrix} p_n(\eta_1^{(n)},t) \\ p_n(\eta_2^{(n)},t) \\ p_n(\eta_3^{(n)},t) \end{bmatrix} + \begin{bmatrix} f(\eta_1^{(n)},t) \\ f(\eta_2^{(n)},t) \\ f(\eta_3^{(n)},t) \end{bmatrix}.$$

Together with condition (10.2.9), this differential system admits a unique solution. Moreover, following the remarks of section 8.2, the eigenvalues of the matrix in (10.2.11) are distinct, real and negative, under suitable assumptions on the parameters α and β.

The next step is to discretize (10.2.11) with respect to the variable t. This will be investigated in section 10.6. Let us first analyze the convergence of the solution of the *semi-discrete* problem (10.2.8), (10.2.9), (10.2.10) to the solution of problem (10.2.5), (10.2.6), (10.2.7). Arguing as in section 9.3, we can propose a variational formulation for equation (10.2.5) as follows:

$$\int_{-1}^{1} \frac{\partial V}{\partial t} \phi w \, dx = -\zeta \int_{-1}^{1} \frac{\partial V}{\partial x} \frac{\partial(\phi w)}{\partial x} \, dx + \int_{-1}^{1} f \phi w \, dx, \tag{10.2.12}$$

$$\forall \phi \in \mathbf{X}, \quad \forall t \in]0,T],$$

where the solution and the test functions both belong to the space $\mathbf{X} \equiv H^1_{0,w}(I)$ (see (5.7.6)). Existence and uniqueness of weak solutions of (10.2.12) are discussed in LIONS and MAGENES(1972) for the case $w \equiv 1$.

With the help of the quadrature formula (3.5.1), the approximating polynomials satisfy

$$\sum_{j=0}^{n} \left(\frac{\partial p_n}{\partial t} \phi \right) (\eta_j^{(n)}, t) \, \tilde{w}_j^{(n)} = -\zeta \int_{-1}^{1} \frac{\partial p_n}{\partial x} \frac{\partial(\phi w)}{\partial x} \, dx \tag{10.2.13}$$

$$+ \sum_{j=0}^{n} (f\phi)(\eta_j^{(n)}, t) \, \tilde{w}_j^{(n)}, \qquad \forall \phi \in \mathbf{P}_n^0, \quad \forall t \in]0,T].$$

Interesting properties are obtained from this equation. Consider $\nu := \alpha = \beta$ with $-1 < \nu \leq 1$. For any $t \in]0,T]$, setting $\phi(x) := p_n(x,t)$, $x \in I$, one gets

$$\frac{1}{2} \frac{d}{dt} \left(\sum_{j=0}^{n} [p_n(\eta_j^{(n)}, t)]^2 \, \tilde{w}_j^{(n)} \right) + \zeta \int_{-1}^{1} \frac{\partial p_n}{\partial x} \frac{\partial(p_n w)}{\partial x} \, dx \tag{10.2.14}$$

$$= \sum_{j=0}^{n} (f p_n)(\eta_j^{(n)}, t) \, \tilde{w}_j^{(n)} \leq \frac{1}{2} \sum_{j=0}^{n} [p_n(\eta_j^{(n)}, t)]^2 \, \tilde{w}_j^{(n)} + \frac{1}{2} \sum_{j=0}^{n} [f(\eta_j^{(n)}, t)]^2 \, \tilde{w}_j^{(n)}.$$

The last inequality in (10.2.14) is derived from the relation $ab \leq \frac{1}{2}(a^2 + b^2)$, $a, b \in \mathbf{R}$.

After integration with respect to the variable t, we have

$$(10.2.15)\qquad \|p_n(\cdot,t)\|^2_{w,n} + 2\zeta\int_0^t\left(\int_{-1}^1 \frac{\partial p_n}{\partial x}\frac{\partial(p_n w)}{\partial x}\,dx\right)ds$$

$$\le \|p_n(\cdot,0)\|^2_{w,n} + \int_0^t\left[\|p_n(\cdot,s)\|^2_{w,n} + \|\tilde{I}_{w,n}f(\cdot,s)\|^2_{w,n}\right]ds, \qquad \forall t\in]0,T].$$

The norm $\|\cdot\|_{w,n}$ is defined in (3.8.2) and the interpolation operator $\tilde{I}_{w,n}$ is defined in section 3.3. Finally, recalling lemma 8.2.1, we can eliminate the integral on the left-hand side of (10.2.15), since it turns out to be positive. Therefore, theorem 10.1.1 yields

$$(10.2.16)\qquad \|p_n(\cdot,t)\|^2_{w,n} \le e^t\left[\|p_n(\cdot,0)\|^2_{w,n} + \int_0^t \|\tilde{I}_{w,n}f(\cdot,s)\|^2_{w,n}\, e^{-s}ds\right],$$

$$\forall t\in]0,T].$$

Taking into account (10.2.9), one shows that the right-hand side in the above expression is bounded by a constant which does not depend on n. Thus, (10.2.16) shows that the norm $\|p_n(\cdot,t)\|_w$ (which is equivalent to the norm $\|p_n(\cdot,t)\|_{w,n}$ by theorem 3.8.2) is also bounded independently of n. This is interpreted as a *stability condition* (compare with (9.4.3)). Moreover, using again lemma 8.2.1, we can also give a uniform bound to the quantity $\int_0^t \|\frac{\partial p_n}{\partial x}(\cdot,s)\|^2_w ds$, $\forall t\in]0,T]$, by estimating the right-hand side of (10.2.15) with the help of (10.2.16).

The analysis of convergence is similar. For any $t\in]0,T]$, we define $\chi_n(t) := \hat{\Pi}^1_{0,w,n}V(\cdot,t)$ (see (6.4.13) and (9.4.5)). Thus, by (10.2.12) we can write

$$(10.2.17)\qquad \int_{-1}^1 \frac{\partial V}{\partial t}\phi w\,dx = -\zeta\int_{-1}^1 \frac{\partial \chi_n}{\partial x}\frac{\partial(\phi w)}{\partial x}\,dx$$

$$+ \sum_{j=0}^n (f\phi)(\eta_j^{(n)},t)\,\tilde{w}_j^{(n)}, \qquad \forall\phi\in\mathbf{P}_n^0,\ \forall t\in]0,T].$$

The sum on the right-hand side of (10.2.17) replaces the integral $\int_{-1}^1 f\phi w dx$, after noting that $f(\cdot,t)\in\mathbf{P}_1$, $\forall t\in]0,T]$. Subtracting equation (10.2.13) from equation (10.2.17) and taking $\phi := \chi_n - p_n$, we obtain by lemma 8.2.1

(10.2.18)
$$\int_{-1}^{1}\left[\frac{\partial V}{\partial t}(\chi_n - p_n)\right] w\ dx \ -\ \sum_{j=0}^{n}\left[\frac{\partial p_n}{\partial t}(\chi_n - p_n)\right](\eta_j^{(n)},t)\ \tilde{w}_j^{(n)}$$
$$= -\zeta\int_{-1}^{1}\frac{\partial(\chi_n - p_n)}{\partial x}\frac{\partial[(\chi_n - p_n)w]}{\partial x}\ dx \ \leq\ 0, \qquad \forall t\in]0,T].$$

Therefore

(10.2.19)
$$\frac{1}{2}\frac{d}{dt}\|\chi_n - p_n\|_{w,n}^2 \ \leq\ \sum_{j=0}^{n}\left[\frac{\partial \chi_n}{\partial t}(\chi_n - p_n)\right](\eta_j^{(n)},t)\ \tilde{w}_j^{(n)}$$
$$-\int_{-1}^{1}\frac{\partial V}{\partial t}(\chi_n - p_n)w\ dx \ =\ \left[(\frac{\partial \chi_n}{\partial t},\chi_n - p_n)_{w,n}\ -\ (\frac{\partial \chi_n}{\partial t},\chi_n - p_n)_w\right]$$
$$+\,(\frac{\partial(\chi_n - V)}{\partial t},\chi_n - p_n)_w, \qquad \forall t\in]0,T].$$

The estimate of the error $\|V-p_n\|_w \leq \|V-\chi_n\|_w+\|\chi_n-p_n\|_w$ is now a technical matter. The difference between the inner products $(\cdot,\cdot)_{w,n}$ and $(\cdot,\cdot)_w$ can be bounded with the help of (3.8.15) (see also section 9.4). In addition, χ_n converges to V for $n\to+\infty$, by virtue of (9.4.7). Theorem 10.1.1 is used as done in the proof of stability. The reader can easily provide the details.

It turns out that, for any $t\in[0,T]$, the error decays for $n\to+\infty$, with a rate only depending on the regularity of V. However, using inequality (10.1.2), the multiplicative term e^t appears in the right-hand side of the estimates. Therefore, for large times, the error bound is not in general very meaningful for practical applications. Refinements are possible, for example when $\sigma_1\equiv\sigma_2\equiv 0$. In this case, the solution $V(\cdot,t)$ and the polynomial $p_n(\cdot,t)$ are known to decay exponentially to zero for $t\to+\infty$.

The convergence analysis of the collocation method for the heat equation is developed in CANUTO and QUARTERONI(1981). The same conclusions hold for the tau and Galerkin methods. These techniques can be extended to other parabolic operators or to different boundary conditions, but very few theoretical results are available. For homogeneous Neumann boundary conditions, an analysis of stability in the Chebyshev case is provided in GOTTLIEB, HUSSAINI and ORSZAG(1984). A numerical example for a nonlinear parabolic equation is presented in section 12.1.

A special case worth mentioning is $I \equiv \mathbf{R}$ in (10.2.1) and (10.2.2). Of course, we no longer have boundary conditions. These are supplanted by the condition that U tends to zero at infinity with a specified minimum rate. We derive a collocation type approximation in this case. According to ESCOBEDO and KAVIAN(1987), first set $V(x,t) := U(\sqrt{4\zeta}xe^{2\zeta t}, e^{4\zeta t} - 1)$, $x \in \mathbf{R}$, $t \in]0, \hat{T}]$, where $\hat{T} := \frac{1}{4\zeta}\ln(T+1)$. Then, it is clear that V satisfies the equation

$$(10.2.20) \qquad \frac{\partial V}{\partial t}(x,t) = \zeta\left(\frac{\partial^2 V}{\partial x^2}(x,t) + 2x\frac{\partial V}{\partial x}(x,t)\right), \qquad \forall x \in \mathbf{R}, \ \forall t \in]0, \hat{T}],$$

$$(10.2.21) \qquad V(x,0) = U(\sqrt{4\zeta}x, 0) = U_0(\sqrt{4\zeta}x), \qquad \forall x \in \mathbf{R}.$$

Assuming that the solution V decays as fast as e^{-x^2} at infinity, we approximate by Hermite functions (see section 6.7). Let $\xi_i^{(n)}$, $1 \leq i \leq n$, be the zeroes of H_n. For any $t \in]0, \hat{T}]$, we want to find $P_n(\cdot, t) \in \mathbf{S}_{n-1}$ such that

$$(10.2.22) \qquad \frac{\partial P_n}{\partial t}(\xi_i^{(n)}, t) = \zeta\left(\frac{\partial^2 P_n}{\partial x^2}(\xi_i^{(n)}, t) + 2\xi_i^{(n)}\frac{\partial P_n}{\partial x}(\xi_i^{(n)}, t)\right), \qquad 1 \leq i \leq n,$$

$$(10.2.23) \qquad P_n(\xi_i^{(n)}, 0) = U_0(\sqrt{4\zeta}\xi_i^{(n)}), \qquad 1 \leq i \leq n.$$

We can formulate the same problem in the space of polynomials by setting $p_n := P_n e^{x^2}$ (see section 7.4). This is equivalent to a system of ordinary differential equations. The corresponding matrix has non-positive eigenvalues, as pointed out in section 8.2, which is very important for proving convergence of P_n to V for $n \to +\infty$. Actually, recalling formula (3.4.1), the proof follows the same arguments as the case when I is bounded. The theory is presented in FUNARO and KAVIAN(1988). Finally, we can get an approximation of the unknown U with the change of variables: $x \to x[4\zeta(t+1)]^{-1/2}$, $t \to \frac{1}{4\zeta}\ln(t+1)$.

We remark that the direct approximation of the unknown U by the Hermite-collocation method applied to equation (10.2.1) with $I \equiv \mathbf{R}$, is *unstable*. This is due to the fact that the quantity $\int_{\mathbf{R}} \psi'(\psi v)' dx$ is not in general positive for $\psi \in H_v^1(\mathbf{R})$, where $v(x) := e^{x^2}$, $x \in \mathbf{R}$.

We now write $P_n(x,t) = \sum_{m=0}^{n-1} c_m(t)H_m(x)e^{-x^2}$, where the $c_m(t)$'s are the Fourier coefficients of P_n at $t \in]0,\hat{T}]$. Substituting P_n in (10.2.22) and recalling (1.7.1), we obtain

$$\text{(10.2.24)} \qquad \frac{d}{dt}c_m(t) = -2\zeta(m+1)c_m(t), \qquad 0 \le m \le n-1, \quad \forall t \in]0,\hat{T}].$$

Now, we can find the explicit expression of the coefficients. We note that, from (10.2.23), the initial datum is $P_n(x,0) = (I^*_{v,n}U_0)(\sqrt{4\zeta}x)$, $v(x) := e^{x^2}$, $x \in \mathbf{R}$ (see (6.7.6)).

Another scheme is obtained with the initial condition $P_n(x,0) = (\Pi^*_{v,n}U_0)(\sqrt{4\zeta}x)$, $x \in \mathbf{R}$ (see (6.7.3)). This amounts to assigning the values $c_m(0)$, $0 \le m \le n-1$, as the Fourier coefficients of U_0. As the reader can easily check, after this modification, the solution to (10.2.22) turns out to be the approximation of V by the Galerkin method. Indeed, P_n also satisfies

$$\text{(10.2.25)} \qquad \int_{\mathbf{R}} \frac{\partial P_n}{\partial t}\phi v \, dx = -\zeta \int_{\mathbf{R}} \frac{\partial(P_n v)}{\partial x}\frac{\partial \phi}{\partial x} \, dx, \qquad \forall \phi \in \mathbf{S}_{n-1}, \quad \forall t \in]0,\hat{T}].$$

10.3 Approximation of linear first-order problems

We now focus the attention to another time-dependent partial differential equation. The unknown $U : [-1,1] \times [0,T] \to \mathbf{R}$, satisfies

$$\text{(10.3.1)} \qquad \frac{\partial U}{\partial t}(x,t) = -\zeta \, \frac{\partial U}{\partial x}(x,t), \qquad \forall x \in]-1,1], \quad \forall t \in]0,T],$$

$$\text{(10.3.2)} \qquad U(x,0) = U_0(x), \qquad \forall x \in [-1,1],$$

$$\text{(10.3.3)} \qquad U(-1,t) = \sigma(t), \qquad \forall t \in]0,T],$$

where $\zeta > 0$, $U_0 : [-1,1] \to \mathbf{R}$, $\sigma : [0,T] \to \mathbf{R}$, are given. We assume that U_0 and σ are continuous and $\sigma(0) = U_0(-1)$.

All the standard spectral schemes can be considered to approximate U. Nevertheless, the theoretical analysis of convergence is more difficult than the analysis of the problem in section 10.2. We discuss some examples.

Take $\sigma \equiv 0$. In this case, we can explicitly compute the solution, given by $U(x,t) = U_0(x - \zeta t)$ if $-1 < x - \zeta t \leq 1$, $U(x,t) = 0$ if $x - \zeta t \leq -1$. Consider the collocation method. We must find $p_n(\cdot,t) \in \mathbf{P}_n$, $n \geq 1$, $t \in]0,T]$, such that

$$\text{(10.3.4)} \qquad \frac{\partial p_n}{\partial t}(\eta_i^{(n)},t) = -\zeta \frac{\partial p_n}{\partial x}(\eta_i^{(n)},t), \qquad 1 \leq i \leq n, \quad \forall t \in]0,T],$$

$$\text{(10.3.5)} \qquad p_n(\eta_i^{(n)},0) = U_0(\eta_i^{(n)}), \qquad 0 \leq i \leq n,$$

$$\text{(10.3.6)} \qquad p_n(\eta_0^{(n)},t) = 0, \qquad \forall t \in]0,T].$$

Again, these equations can be reduced to a $n \times n$ linear system of ordinary differential equations. For instance, using the notations of section 7.2, in the case $n = 3$, we have

$$\text{(10.3.7)} \qquad \frac{d}{dt}\begin{bmatrix} p_n(\eta_1^{(n)},t) \\ p_n(\eta_2^{(n)},t) \\ p_n(\eta_3^{(n)},t) \end{bmatrix} = -\zeta \begin{bmatrix} \tilde{d}_{11}^{(1)} & \tilde{d}_{12}^{(1)} & \tilde{d}_{13}^{(1)} \\ \tilde{d}_{21}^{(1)} & \tilde{d}_{22}^{(1)} & \tilde{d}_{23}^{(1)} \\ \tilde{d}_{31}^{(1)} & \tilde{d}_{32}^{(1)} & \tilde{d}_{33}^{(1)} \end{bmatrix} \begin{bmatrix} p_n(\eta_1^{(n)},t) \\ p_n(\eta_2^{(n)},t) \\ p_n(\eta_3^{(n)},t) \end{bmatrix}, \qquad t \in]0,T].$$

The eigenvalues of the matrix in (10.3.7) (see also (7.4.3)) have been studied in section 8.1. We will return to this point in section 10.6.

In the Legendre case ($w \equiv 1$), we can provide a simple analysis of stability. In fact, formula (3.5.1) allows us to write for any $t \in]0,T]$

$$\text{(10.3.8)} \qquad \frac{d}{dt}\|p_n(\cdot,t)\|_{w,n}^2 = 2\sum_{j=0}^{n}\left(\frac{\partial p_n}{\partial t}p_n\right)(\eta_j^{(n)},t)\ w_j^{(n)} =$$

$$= -2\zeta \sum_{j=0}^{n} \left(\frac{\partial p_n}{\partial x} p_n \right) (\eta_j^{(n)}, t) \, w_j^{(n)} \;=\; -2\zeta \int_{-1}^{1} \frac{\partial p_n}{\partial x} p_n \, dx \;=\; -\zeta p_n^2(1,t) \leq 0.$$

This shows that the norm $\|p_n(\cdot,t)\|_{w,n}$, $t \in [0,T]$, is bounded by $\|\tilde{I}_{w,n} U_0\|_{w,n}$. This last term is bounded with respect to n, when U_0 is sufficiently regular. A proof of convergence is given, using similar arguments, by estimating the error $p_n - U_n$, where $U_n \in \mathbf{P}_n$, $n \geq 1$, is a suitable projection of the solution U.

Unfortunately, the same techniques are not effective when different weight functions w are considered. As pointed out in GOTTLIEB and ORSZAG(1977), p.89, in the Chebyshev case ($w(x) = 1/\sqrt{1-x^2}$, $x \in]-1,1[$), we cannot expect convergence of p_n to U in the norm $\|\cdot\|_w$. In the papers of GOTTLIEB and TURKEL(1985) and SALOMONOFF and TURKEL(1989), convergence estimates for Chebyshev-collocation approximations are proven in a norm weighted by the function $w(x) := \sqrt{(1-x)/(1+x)}$, $x \in]-1,1[$, but the theoretical analysis is harder than in the Legendre case.

A different treatment of the boundary conditions is possible. We modify (10.3.6) as follows:

$$\text{(10.3.9)} \qquad \frac{\partial p_n}{\partial t}(\eta_0^{(n)}, t) \;=\; -\zeta \, \frac{\partial p_n}{\partial x}(\eta_0^{(n)}, t) \;-\; \gamma \, p_n(\eta_0^{(n)}, t), \qquad \forall t \in]0,T],$$

where $\gamma > 0$. Basically, we are trying to force the polynomial p_n to satisfy, at the point $x = -1$, both the boundary constraint and the differential equation (the same trick used in section 9.4 for the Neumann problem). This leads to a $(n+1) \times (n+1)$ differential system, where the corresponding matrix is considered in section 7.4 (see (7.4.6) for the case $n = 3$). The eigenvalues of this matrix are examined in section 8.1. When γ is proportional to n^2, a proof of convergence can be given. The Legendre and Chebyshev cases are examined in FUNARO and GOTTLIEB(1989) and FUNARO and GOTTLIEB(1988) respectively.

Let us now introduce the tau method for $\sigma \equiv 0$. This is obtained by projecting equation (10.3.1) onto the space $\mathbf{P}_{n-1}$, $n \geq 1$. Then, we are concerned with finding $p_n(\cdot,t) \in \mathbf{P}_n$, $t \in [0,T]$, such that

$$\text{(10.3.10)} \qquad \left[\Pi_{w,n-1} \frac{\partial p_n}{\partial t} \right] (x,t) \;=\; -\zeta \, \frac{\partial p_n}{\partial x}(x,t), \qquad \forall x \in]-1,1], \quad \forall t \in]0,T].$$

According to theorem 6.2.2, we have

$$(10.3.11)\qquad \int_{-1}^{1}\left(\frac{\partial p_n}{\partial t}-\Pi_{w,n-1}\frac{\partial p_n}{\partial t}\right)\phi w\ dx \;=\; 0 \;=\; \frac{d}{dt}\int_{-1}^{1}(p_n-\Pi_{w,n-1}p_n)\phi w\ dx,$$

$$\forall\phi\in\mathbf{P}_{n-1},\quad \forall t\in]0,T],$$

which shows that the operator $\partial/\partial t$ commutes with the operator $\Pi_{w,n-1}$. Therefore, we can rewrite (10.3.10) as

$$(10.3.12)\qquad \left[\frac{\partial}{\partial t}(\Pi_{w,n-1}p_n)\right](x,t) \;=\; -\zeta\,\frac{\partial p_n}{\partial x}(x,t),\qquad \forall x\in]-1,1],\quad \forall t\in]0,T].$$

In addition, we require that $p_n(-1,t) = 0$, $\forall t\in]0,T]$, and we impose the initial condition

$$(10.3.13)\qquad p_n(\cdot,0) \;=\; r\Pi_{w,n-1}(U_0 r^{-1}),\qquad \forall x\in[-1,1],$$

where $r(x) := (1+x)$, $x\in[-1,1]$. In this way, we also get $p_n(-1,0)=0$.

Let $c_k(t)$, $0\le k\le n$, $t\in[0,T]$, be the Fourier coefficients of $p_n(\cdot,t)$ with respect to the basis $u_k\equiv P_k^{(\alpha,\beta)}$, $k\in\mathbf{N}$, $\alpha>-1$, $\beta>-1$. Therefore, $p_n=\sum_{k=0}^{n}c_k u_k$. Moreover, by (10.3.12), p_n satisfies the set of equations (see section 7.1):

$$(10.3.14)\qquad \frac{d}{dt}c_k(t) \;=\; -\zeta\; c_k^{(1)}(t),\qquad 0\le k\le n-1,\qquad \forall t\in]0,T].$$

This is equivalent to a $n\times n$ linear system of ordinary differential equations in the unknowns c_k, $0\le k\le n-1$, where the initial values $c_k(0)$, $0\le k\le n-1$, are obtained from (10.3.13) by virtue of the results of section 2.3. We eliminate c_n by expressing it as a linear combination of the other coefficients with the help of the relation

$$(10.3.15)\qquad p_n(-1,t) \;=\; \sum_{k=0}^{n} c_k(t)u_k(-1) \;=\; 0,\qquad \forall t\in[0,T].$$

Another characterization is obtained from (10.3.10) by noting that p_n satisfies

$$(10.3.16)\qquad \frac{\partial p_n}{\partial t}(x,t) \;=\; -\zeta\,\frac{\partial p_n}{\partial x}(x,t) \;+\; c_n'(t)u_n(x),\qquad \forall x\in]-1,1],\ \forall t\in]0,T].$$

In fact, the last term in (10.3.16) disappears when we apply the projector $\Pi_{w,n-1}$.

Thus, we can finally view $p_n(\cdot,t) \in \mathbf{P}_n$, with $p_n(-1,t) = 0$, $\forall t \in [0,T]$, as the solution of the collocation scheme

$$(10.3.17) \qquad \frac{\partial p_n}{\partial t}(\xi_i^{(n)},t) \;=\; -\zeta\,\frac{\partial p_n}{\partial x}(\xi_i^{(n)},t), \qquad 1 \le i \le n, \quad \forall t \in]0,T],$$

where the $\xi_i^{(n)}$'s are the zeroes of u_n (see section 3.1). By defining $q_n := p_n/r$, where $r(x) := (1+x)$, $x \in [-1,1]$, we have $q_n(\cdot,t) \in \mathbf{P}_{n-1}$, $\forall t \in [0,T]$. Using (10.3.17), q_n also satisfies

$$(10.3.18) \quad \frac{\partial q_n}{\partial t}(\xi_i^{(n)},t) = -\zeta\left(\frac{\partial q_n}{\partial x}(\xi_i^{(n)},t) + \frac{q_n(\xi_i^{(n)},t)}{r(\xi_i^{(n)})}\right), \quad 1 \le i \le n, \quad \forall t \in]0,T].$$

This is equivalent to a $n \times n$ linear system of differential equations in the unknowns $q_n(\xi_i^{(n)},\cdot)$, $1 \le i \le n$. The entries of the corresponding matrix are (see (7.2.5)): $-\zeta[d_{ij}^{(1)} + \delta_{ij}/(1+\xi_i^{(n)})]$, $1 \le i \le n$, $1 \le j \le n$.

We can prove stability when $\alpha > -1$ and $-1 < \beta \le 1$. We first note that $\frac{1-x}{1+x}p_n = -c_n u_n + \{\textit{lower degree terms}\}$. Then, (9.3.16) yields

$$(10.3.19) \qquad \frac{d}{dt}\left(\int_{-1}^{1} \tfrac{1-x}{1+x}\, p_n^2 w\, dx \;+\; c_n^2(t)\|u_n\|_w^2\right)$$

$$= 2\int_{-1}^{1} \tfrac{1-x}{1+x}\,\frac{\partial p_n}{\partial t} p_n w\, dx \;+\; 2c_n'(t)c_n(t)\|u_n\|_w^2$$

$$= -2\zeta\int_{-1}^{1} \tfrac{1-x}{1+x}\,\frac{\partial p_n}{\partial x} p_n w\, dx \;+\; 2c_n'(t)\int_{-1}^{1} \tfrac{1-x}{1+x}\, u_n p_n w\, dx \;+\; 2c_n'(t)c_n(t)\|u_n\|_w^2$$

$$= -2\zeta\int_{-1}^{1} \tfrac{1-x}{1+x}\,\frac{\partial p_n}{\partial x} p_n w\, dx \;=\; \zeta\int_{-1}^{1} p_n^2\,\frac{d}{dx}\left(\tfrac{1-x}{1+x}w\right) dx \;\le\; 0, \qquad \forall t \in]0,T].$$

The last integral is negative by virtue of the restrictions on α and β. This shows that a certain weighted norm of $p_n(\cdot,t)$ is bounded by the initial data for any $t \in]0,T]$. Further details and improvements are given in GOTTLIEB and TADMOR(1990).

A proof of stability in a norm which also controls the derivative $\partial p_n/\partial x$ has been outlined in GOTTLIEB and ORSZAG(1977), section 8, for the Chebyshev case. The same proof can be extended to the other Jacobi cases, provided $-1 < \beta \le 0$ and $\alpha > -1$.

Once the stability of the scheme is achieved, error estimates are obtained by applying the same proof to the error $p_n - U_n$, where $U_n \in \mathbf{P}_n$, $n \geq 1$, is a suitable projection of U. The same argument was used in section 10.2 for the heat equation (see also CANUTO, HUSSAINI, QUARTERONI and ZANG(1988), chapter 10). Other spectral type approximations and theoretical results relative to equation (10.3.1) are considered in GOTTLIEB and ORSZAG(1977), section 8, GOTTLIEB(1981), CANUTO and QUARTERONI (1982b), MERCIER(1982), MERCIER(1989), TAL-EZER(1986b), DUBINER(1987).

Similar results are available for the equation

$$\frac{\partial U}{\partial t}(x,t) = \zeta \frac{\partial U}{\partial x}(x,t), \qquad \forall x \in [-1,1[, \ \forall t \in]0,T], \tag{10.3.20}$$

when the boundary condition is imposed at the point $x = 1$.

The reader should pay a little more care to the theoretical analysis of partial differential equations, before trying experiments on general problems, such as the following one:

$$\frac{\partial U}{\partial t}(x,t) = A(x, U(x,t)) \frac{\partial U}{\partial x}(x,t), \qquad x \in]-1,1[, \quad t \in]0,T], \tag{10.3.21}$$

where $A :]-1,1[\times \mathbf{R} \to \mathbf{R}$ is given.

An analysis for equations such as (10.3.21) is carried out in SMOLLER(1983), RHEE, ARIS and AMUNDSON(1986), KREISS and LORENZ(1989). It is known that the solution of (10.3.21) maintains a constant value along the so-called *characteristic curves*. These are parallel straight-lines in the (x,t) plane with slope $1/\zeta$ for equation (10.3.1), and slope $-1/\zeta$ for equation (10.3.20). In the former case, the solution $U(x,\cdot)$, $x \in [-1,1]$, shifts on the right-hand side during the time evolution. With terminology deriving from fluid physics, the point $x = -1$ is the *inflow boundary*, while $x = 1$ is the *outflow boundary*. The situation is reversed for equation (10.3.20). The characteristic curves are no longer straight-lines when A is not a constant. Even in the case of smooth initial conditions and smooth boundary data, the solution can loose regularity and *shocks* can be generated when two or more characteristic curves intersect. The numerical analysis becomes a delicate issue in this situation, especially for spectral methods which are particularly sensitive to the smoothness of the solution. Despite much effort devoted to this subject, many problems remain. Polynomial approximations by spectral methods of equation (10.3.21), when $A(x, U) := x$ or $A(x, U) := -x$, have been considered in GOTTLIEB and ORSZAG(1977), section 7. Nonlinear examples will be examined in the coming section.

10.4 Nonlinear time-dependent problems

An interesting family of nonlinear first order problems is represented by *conservation equations*. They are written as

(10.4.1)
$$\frac{\partial U}{\partial t}(x,t) = \frac{\partial F(U)}{\partial x}(x,t), \qquad x \in]-1,1[, \quad t \in]0,T],$$

where $F : \mathbf{R} \to \mathbf{R}$ is a given function. If F is differentiable, then (10.4.1) is equivalent to taking $A(x, U) := F'(U)$ in (10.3.21). These equations are consequence of conservation of energy, mass or momentum, in a physics phenomenon. Boundary conditions have to be imposed at the inflow boundaries, which are determined according to the direction of the characteristic curves. We note that now the slope of a characteristic curve at a given point (x, t) depends on the unknown U. The case corresponding to $F(U) := -\frac{1}{2}U^2$, $U \in \mathbf{R}$ (hence $A(x, U) = -U$), is sufficient to illustrate the possible ways in which the solution can develop. A theoretical discussion is too lengthy to include here. We suggest the books of SMOLLER(1983), chapter 15, and KREISS and LORENZ(1989), chapter 4, for a more in-depth analysis.

A large variety of finite-difference schemes has been proposed to solve the above equations numerically. These are sometimes very sophisticated in order to treat solutions with sharp derivatives or shocks. Among many celebrated papers, we refer to SOD(1985) for a general introductive overview. There is less published material available in the field of spectral methods. Algorithms are proposed for the approximation of periodic solutions by trigonometric polynomials, but some of the techniques can be adapted to algebraic polynomials. If the solution displays a smooth behavior, a straightforward application of the usual methods (tau and collocation) generally provides good results. For the treatment of other situations, a lot of extra care is needed. Many research groups are working with great effort on this subject, following different ideas and devising new schemes. A detailed analysis of these techniques is an ambitious project that we prefer to avoid in this book. We will only outline the main patterns. For a closer approach, an update review and numerous references are given in CANUTO, HUSSAINI, QUARTERONI and ZANG(1988), chapters 8 and 12.

One strategy commonly used in computations is *filtering*. The procedure consists in appropriately damping the higher Fourier coefficients of the approximating polynomial, to control oscillations near the points of discontinuity of the solution. A drawback to using *filters* is that such a smoothing affects the convergence, and reduces the accuracy of the scheme.

Another approach is to apply domain decomposition methods (see chapter eleven), either to improve the resolution in the critical zones, or to allow discontinuities of the approximating functions in accordance with those of the solution. These methods have been employed for instance in KOPRIVA(1986), MACARAEG and STREETT(1986). The location of the singular points can be estimated by a procedure called *shock-fitting*.

Numerical codes based on *cell-averaging* offer another promising field of research. Results for Chebyshev approximations are provided in CAI, GOTTLIEB and HARTEN (1990).

As far as the theory is concerned, very little is known about these techniques, especially in the case of algebraic polynomials.

Let us examine some other equations. An interesting variation of (10.4.1) is

$$(10.4.2) \qquad \frac{\partial U}{\partial t}(x,t) = \left(\epsilon \frac{\partial^2 U}{\partial x^2} + \frac{\partial F(U)}{\partial x}\right)(x,t), \qquad x \in]-1,1[, \quad t \in]0,T],$$

where $\epsilon > 0$ ia a given parameter.

For $F(U) := -\frac{1}{2}U^2$, $U \in \mathbf{R}$, (10.4.2) is known as *Burgers equation*. Due to the presence of the second-order derivative (which, in analogy with fluid dynamics, corresponds to *viscosity* in the physical model problem), solutions of (10.4.2) are smoother than solutions corresponding to (10.4.1). This definitely helps the theoretical analysis. In addition, one expects that solutions of (10.4.2) for a small ϵ are in some way close to solution of (10.4.1). We refer to SMOLLER, p.257, for a theoretical explanation of this fact. For boundary conditions of the form $U(\pm 1, t) = 0$, $t \in]0,T]$, approximations of the Burgers equation by Galerkin and collocation methods are respectively considered in MADAY and QUARTERONI(1981), and MADAY and QUARTERONI(1982), for the Chebyshev and Legendre cases. The analysis is carried out for the *steady* equation, which means that U does not depend on t (see (9.8.5)). Other results are provided in BRESSAN and QUARTERONI(1986a).

Finally, results are also available for the *Korteveg-De Vries equation* (see PAVONI (1988), BRESSAN and PAVONI(1990)):

(10.4.3) $$\frac{\partial U}{\partial t}(x,t) = \left(\epsilon \frac{\partial^3 U}{\partial x^3} - U\frac{\partial U}{\partial x}\right)(x,t), \qquad x \in]-1,1[, \quad t \in]0,T],$$

where $\epsilon \neq 0$ is a given constant.

10.5 Approximation of the wave equation

Another classical time-dependent problem is given by the *wave equation*. Here, the unknown is the function $U : [-1,1] \times [0,T] \to \mathbf{R}$ satisfying

(10.5.1) $$\frac{\partial^2 U}{\partial t^2}(x,t) = \zeta^2 \frac{\partial^2 U}{\partial x^2}(x,t), \qquad x \in]-1,1[, \quad t \in]0,T],$$

$$(10.5.2) \qquad U(\pm 1, t) = 0, \qquad \forall t \in]0, T],$$

$$(10.5.3) \qquad U(x, 0) = U_0(x), \qquad \frac{\partial U}{\partial t}(x, 0) = \hat{U}_0(x), \qquad \forall x \in]-1, 1[,$$

where $\zeta > 0$ and $U_0 :]-1,1[\to \mathbf{R}$, $\hat{U}_0 :]-1,1[\to \mathbf{R}$ are given initial data. Equation (10.5.1) is of ***hyperbolic*** type. The solution U represents the displacement of a vibrating string of length equal 2, fixed at the endpoints. Other oscillatory phenomena are described by the wave equation. Examples are collected in BALDOCK and BRIDGEMAN (1981).

From the numerical point of view, we can apply the same techniques of section 10.2 to obtain the semi-discrete spectral approximation of problem (10.5.1), (10.5.2), (10.5.3). This time, after discretization in the variable x, we get an initial-value second-order differential system in the variable t.

In the Legendre case, stability results and error estimates can be obtained from a weak formulation of the problem. The reader is addressed to LIONS and MAGENES(1972), chapter 5, theorem 2.1, for results concerning this aspect of the theory. To our knowlwdge, no analysis is currently available for the other Jacobi cases.

Another approach consists in writing the solution as a suitable superposition of two waves travelling in opposite directions. Actually, one checks that the general solution of (10.5.1) is given by $U(x,t) = \tau(x - \zeta t) + \upsilon(x + \zeta t)$, where τ and υ can be uniquely determined by the initial data (10.5.3) (***d'Alembert solution***). Thus, U can be decomposed into the sum of two functions, which are solutions of equations (10.3.1) and (10.3.20) respectively, with some prescribed initial guess. Namely, we have $U = V + W$, where V and W solve the following ***linear symmetric hyperbolic system***:

$$(10.5.4) \qquad \frac{\partial}{\partial t}\begin{pmatrix} V(x,t) \\ W(x,t) \end{pmatrix} = \begin{pmatrix} -\zeta & 0 \\ 0 & \zeta \end{pmatrix} \frac{\partial}{\partial x}\begin{pmatrix} V(x,t) \\ W(x,t) \end{pmatrix}, \qquad x \in]-1,1[, \quad t \in]0,T].$$

Since (10.5.4) is in a diagonal form, V and W are also called *characteristic variables.* From (10.5.2), the two unknowns are *coupled* via the boundary relations

$$(10.5.5) \qquad \begin{cases} V(-1,t) = L\, W(-1,t) \\ W(1,t) = R\, V(1,t) \end{cases} \qquad \forall t \in]0,T],$$

where $R = L = -1$.

The physical interpretation of formula (10.5.5) is that the travelling waves are reflected at the endpoints of the vibrating string. First-order systems, such as (10.5.4), are often encountered in fluid dynamics models (see KREISS and LORENZ(1989), section 7.6).

For spectral type approximations, results are generally obtained for hyperbolic systems written in terms of characteristic variables which are combined together at the boundary points by (10.5.5), with R and L satisfying $|RL| < 1$ (note that for the wave equation one has instead $|RL| = 1$). In this situation, the amplitude of the oscillations decays for $t \to +\infty$ (*energy dissipation* due to friction effects).

The collocation method is obtained in the following way. We are concerned with finding $p_n(\cdot,t) \in \mathbf{P}_n$, $q_n(\cdot,t) \in \mathbf{P}_n$, $n \geq 1$, $t \in]0,T]$, such that

$$(10.5.6) \qquad \begin{cases} \dfrac{\partial p_n}{\partial t}(\eta_i^{(n)},t) = -\zeta\, \dfrac{\partial p_n}{\partial x}(\eta_i^{(n)},t) & 1 \leq i \leq n, \quad t \in]0,T], \\[2ex] \dfrac{\partial q_n}{\partial t}(\eta_i^{(n)},t) = \zeta\, \dfrac{\partial q_n}{\partial x}(\eta_i^{(n)},t) & 0 \leq i \leq n-1, \quad t \in]0,T]. \end{cases}$$

At the boundaries, we require that

$$(10.5.7) \qquad \begin{cases} p_n(-1,t) = L\, q_n(-1,t) \\ q_n(1,t) = R\, p_n(1,t) \end{cases} \qquad \forall t \in]0,T].$$

When R and L in (10.5.7) satisfy $|LR| < 1$, the polynomials $p_n(\cdot,t)$ and $q_n(\cdot,t)$ decay to zero when $t \to +\infty$ (see LUSTMAN(1986)). It is an easy exercise to set up the $2n \times 2n$ system of ordinary differential equations corresponding to (10.5.6). We note that two unknowns can be eliminated by virtue of (10.5.7). For $n = 2$, we get for example

$$
(10.5.8)\qquad \frac{d}{dt}\begin{bmatrix} p_n(\eta_1^{(n)},t) \\ p_n(\eta_2^{(n)},t) \\ q_n(\eta_0^{(n)},t) \\ q_n(\eta_1^{(n)},t) \end{bmatrix} = \zeta \begin{bmatrix} -\tilde{d}_{11}^{(1)} & -\tilde{d}_{12}^{(1)} & -L\tilde{d}_{10}^{(1)} & 0 \\ -\tilde{d}_{21}^{(1)} & -\tilde{d}_{22}^{(1)} & -L\tilde{d}_{20}^{(1)} & 0 \\ 0 & R\tilde{d}_{02}^{(1)} & \tilde{d}_{00}^{(1)} & \tilde{d}_{01}^{(1)} \\ 0 & R\tilde{d}_{12}^{(1)} & \tilde{d}_{10}^{(1)} & \tilde{d}_{11}^{(1)} \end{bmatrix} \begin{bmatrix} p_n(\eta_1^{(n)},t) \\ p_n(\eta_2^{(n)},t) \\ q_n(\eta_0^{(n)},t) \\ q_n(\eta_1^{(n)},t) \end{bmatrix},
$$

$$t \in]0,T].$$

We note that the determinant of the matrix of the system vanishes when $RL = 1$. In fact, polynomials of degree zero satisfying (10.5.7) are eigenfunctions relative to the eigenvalue zero. Unfortunately, we are not aware of stability and convergence results for this approximation scheme.

Approximations of equation (10.5.4) by polynomials in $\mathbf{P}_n$, using the collocation method at the points $\eta_i^{(n+1)}$, $1 \leq i \leq n$ are considered in GOTTLIEB, LUSTMAN and TADMOR(1987a) and GOTTLIEB, LUSTMAN and TADMOR(1987b). The boundary conditions are treated as in (10.5.7), where $|RL| < 1$. In this case, a theoretical analysis is given for the ultraspherical case when $\nu := \alpha = \beta$ satisfies $-1 < \nu \leq 0$. In FUNARO and GOTTLIEB(1989), the boundary relations (10.5.7) are modified as

$$
(10.5.9)\qquad \begin{cases} \dfrac{\partial p_n}{\partial t}(-1,t) = -\zeta\, \dfrac{\partial p_n}{\partial x}(-1,t) - \gamma[p_n - L\, q_n](-1,t) \\[2ex] \dfrac{\partial q_n}{\partial t}(1,t) = \zeta\, \dfrac{\partial q_n}{\partial x}(1,t) - \gamma[q_n - R\, p_n](1,t) \end{cases}
$$

$$\forall t \in]0,T],$$

where $\gamma > 0$ is a constant. As in (10.3.9) the differential equation is also considered at the points $x = \pm 1$. The new collocation scheme is now equivalent to a $(2n+2)\times(2n+2)$ differential system. For $n = 2$, we have $\forall t \in]0,T]$

(10.5.10)

$$\frac{d}{dt}\begin{bmatrix} p_n(\eta_0^{(n)},t) \\ p_n(\eta_1^{(n)},t) \\ p_n(\eta_2^{(n)},t) \\ q_n(\eta_0^{(n)},t) \\ q_n(\eta_1^{(n)},t) \\ q_n(\eta_2^{(n)},t) \end{bmatrix} = \zeta \begin{bmatrix} -\tilde{d}_{00}^{(1)}-\gamma & -\tilde{d}_{01}^{(1)} & -\tilde{d}_{02}^{(1)} & \gamma L/\zeta & 0 & 0 \\ -\tilde{d}_{10}^{(1)} & -\tilde{d}_{11}^{(1)} & -\tilde{d}_{12}^{(1)} & 0 & 0 & 0 \\ -\tilde{d}_{20}^{(1)} & -\tilde{d}_{21}^{(1)} & -\tilde{d}_{22}^{(1)} & 0 & 0 & 0 \\ 0 & 0 & 0 & \tilde{d}_{00}^{(1)} & \tilde{d}_{01}^{(1)} & \tilde{d}_{02}^{(1)} \\ 0 & 0 & 0 & \tilde{d}_{10}^{(1)} & \tilde{d}_{11}^{(1)} & \tilde{d}_{12}^{(1)} \\ 0 & 0 & \gamma R/\zeta & \tilde{d}_{20}^{(1)} & \tilde{d}_{21}^{(1)} & \tilde{d}_{22}^{(1)}-\gamma \end{bmatrix} \begin{bmatrix} p_n(\eta_0^{(n)},t) \\ p_n(\eta_1^{(n)},t) \\ p_n(\eta_2^{(n)},t) \\ q_n(\eta_0^{(n)},t) \\ q_n(\eta_1^{(n)},t) \\ q_n(\eta_2^{(n)},t) \end{bmatrix}$$

Convergence estimates for the error $\sqrt{|R|\|p_n - V\|_w^2 + |L|\|q_n - W\|_w^2}$, are given in the Legendre case ($w \equiv 1$) when $|RL| < 1$ and $\gamma := \frac{1}{2}\zeta n(n+1)/\sqrt{|RL|}$.
For the proof of stability, we argue as follows. Let us assume for simplicity that $L > 0$ and $R > 0$ with $LR < 1$. By (3.5.6), we first note that $\tilde{w}_0^{(n)} = \tilde{w}_n^{(n)} = \zeta/\gamma\sqrt{RL}$. Then, formula (3.5.1) with $w \equiv 1$ yields $\forall t \in]0,T]$

(10.5.11)
$$\frac{d}{dt}\left(R\sum_{j=0}^{n} p_n^2(\eta_j^{(n)},t)\ \tilde{w}_j^{(n)} \ +\ L\sum_{j=0}^{n} q_n^2(\eta_j^{(n)},t)\ \tilde{w}_j^{(n)} \right)$$

$$= \ -2\zeta R\int_{-1}^{1} \frac{\partial p_n}{\partial x} p_n\ dx \ +\ 2\zeta L\int_{-1}^{1} \frac{\partial q_n}{\partial x} q_n\ dx$$

$$-\ 2R\gamma\tilde{w}_0^{(n)}\Big[p_n(-1,t) - Lq_n(-1,t)\Big]p_n(-1,t) - 2L\gamma\tilde{w}_n^{(n)}\Big[q_n(1,t) - Rp_n(1,t)\Big]q_n(1,t)$$

$$= \ \Big[-\zeta Rp_n^2(1,t) \ +\ 2\gamma RL\tilde{w}_n^{(n)}p_n(1,t)q_n(1,t) \ +\ L(\zeta - 2\gamma\tilde{w}_n^{(n)})q_n^2(1,t)\Big]$$

$$+\ \Big[-\zeta Lq_n^2(-1,t) \ +\ 2\gamma RL\tilde{w}_0^{(n)}p_n(-1,t)q_n(-1,t) \ +\ R(\zeta - 2\gamma\tilde{w}_0^{(n)})p_n^2(-1,t)\Big] \ =$$

$$= \zeta\Big[- Rp_n^2(1,t) \ + \ 2\sqrt{RL}p_n(1,t)q_n(1,t) \ + \ (L - 2\sqrt{L/R})q_n^2(1,t)\Big]$$

$$+ \zeta\Big[- Lq_n^2(-1,t) \ + \ 2\sqrt{RL}p_n(-1,t)q_n(-1,t) \ + \ (R - 2\sqrt{R/L})p_n^2(-1,t)\Big] \ \leq 0.$$

The last terms in (10.5.11) are negative in view of the inequalities

$$2\sqrt{RL}ab \ \leq \ Ra^2 + Lb^2 \ \leq \ Ra^2 + \left(2\sqrt{L/R} - L\right) b^2,$$

$$2\sqrt{RL}ab \ \leq \ La^2 + Rb^2 \ \leq \ La^2 + \left(2\sqrt{R/L} - R\right) b^2,$$

where $a, b \in \mathbf{R}, \ \ R > 0, \ \ L > 0, \ \ RL < 1.$

By (3.8.6), we obtain a bound to the norm $\sqrt{R\|p_n(\cdot,t)\|_w^2 + L\|q_n(\cdot,t)\|_w^2}$, $t \in]0,T]$, in terms of the initial data.

We also outline the proof of convergence. One introduces two additional polynomials $\hat{p}_n(\cdot,t) \in \mathbf{P}_n$, $\hat{q}_n(\cdot,t) \in \mathbf{P}_n$, $n \geq 1$, $t \in]0,T]$. These are solution of

$$\text{(10.5.12)} \qquad \begin{cases} \dfrac{\partial \hat{p}_n}{\partial t}(\eta_i^{(n)},t) \ = \ -\zeta \ \dfrac{\partial \hat{p}_n}{\partial x}(\eta_i^{(n)},t) \qquad 1 \leq i \leq n, \qquad t \in]0,T], \\[2ex] \dfrac{\partial \hat{q}_n}{\partial t}(\eta_i^{(n)},t) \ = \ \zeta \ \dfrac{\partial \hat{q}_n}{\partial x}(\eta_i^{(n)},t) \qquad 0 \leq i \leq n-1, \qquad t \in]0,T], \end{cases}$$

with

$$\text{(10.5.13)} \qquad \begin{cases} \dfrac{\partial \hat{p}_n}{\partial t}(-1,t) \ = \ -\zeta \ \dfrac{\partial \hat{p}_n}{\partial x}(-1,t) \ - \ \gamma[\hat{p}_n \ - \ V](-1,t) \\[2ex] \dfrac{\partial \hat{q}_n}{\partial t}(1,t) \ = \ \zeta \ \dfrac{\partial \hat{q}_n}{\partial x}(1,t) \ - \ \gamma[\hat{q}_n \ - \ W](1,t) \end{cases}$$

$$\forall t \in]0,T].$$

The functions $\hat{p}_n$ and $\hat{q}_n$ are *decoupled.* Therefore, we can estimate the errors $\|(\hat{p}_n - V)(\cdot,t)\|_w$ and $\|(\hat{q}_n - W)(\cdot,t)\|_w$, $t \in]0,T]$, by virtue of the results of section 10.3. On the other hand, by subtracting the equations (10.5.12) from the equations (10.5.6), we get an equivalent collocation scheme in the unknowns $r_n := p_n - \hat{p}_n$ and $s_n := q_n - \hat{q}_n$, with the following boundary relations, obtained by combining (10.5.5), (10.5.9) and (10.5.13):

$$
(10.5.14) \quad \begin{cases} \dfrac{\partial r_n}{\partial t}(-1,t) = -\zeta \, \dfrac{\partial r_n}{\partial x}(-1,t) - \gamma[r_n - Ls_n](-1,t) + \gamma L[\hat{q}_n - W](-1,t) \\ \\ \dfrac{\partial s_n}{\partial t}(1,t) = \zeta \, \dfrac{\partial s_n}{\partial x}(1,t) - \gamma[s_n - Rr_n](1,t) + \gamma R[\hat{p}_n - V](1,t) \end{cases}
$$

$$\forall t \in]0,T].$$

Adapting the proof of stability to this new system, we can bound $\|r_n(\cdot,t)\|_w$, $\|s_n(\cdot,t)\|_w$, $t \in]0,T]$, by the initial data and by the errors $|\hat{p}_n - V|(1,t)$, $|\hat{q}_n - W|(-1,t)$, $t \in]0,T]$. The final error estimate follows from the triangle inequalities

$$\|p_n - V\|_w \le \|r_n\|_w + \|\hat{p}_n - V\|_w, \qquad \|q_n - W\|_w \le \|s_n\|_w + \|\hat{q}_n - W\|_w.$$

We finally observe that the numerical experiments are in general more accurate using (10.5.9) in place of (10.5.7), though the expression of the constant γ is known only in the Legendre case.

Other polynomial approximations are examined in GOTTLIEB, GURZBURGER and TURKEL(1982) and TAL-EZER(1986b). Techniques based on spectral methods are under development for the numerical discretization of nonlinear hyperbolic systems.

10.6 Time discretization

In the previous sections we analyzed polynomial approximations, with respect to the space variable x, of various time-dependent partial differential equations. In order to implement our algorithms we also need to discretize the time variable t. Although polynomials can be used to get a global approximation in the time interval $[0,T]$, $T > 0$ (see MORCHOISNE(1979), TAL-EZER(1986a) and TAL-EZER(1989)), finite-differences are generally preferred. Much has been written on this subject. We refer the reader to

GEAR(1971), LAPIDUS and SEINFELD(1971), JAIN(1984). Here, we only present some basic algorithms.

Our purpose here is to solve first-order differential systems of the form

$$(10.6.1)\qquad \begin{cases} \dfrac{d}{dt}\bar{p}(t) = D\bar{p}(t) + \bar{f}(t), & t \in]0,T], \\ \bar{p}(0) = \bar{p}_0, \end{cases}$$

where $\bar{p}(t) \in \mathbf{R}^n$, $t \in [0,T]$, is the unknown. In (10.6.1), D is a $n \times n$ matrix, $\bar{p}_0 \in \mathbf{R}^n$ is an initial guess and $\bar{f}(t)$ is a given vector of $\mathbf{R}^n$, $\forall t \in]0,T]$. Systems like the one considered above were derived in sections 10.2, 10.3 and 10.5.

The most straightforward time-discretization method is the *Euler method.* We subdivide the interval $[0,T]$ in $m \geq 1$ equal parts of size $h := T/m > 0$. Then, for any $m \geq 1$, we construct the vector $\bar{p}_h \equiv \bar{p}_h^{(m)} \in \mathbf{R}^n$, according to the recursion formula

$$(10.6.2)\qquad \begin{cases} \bar{p}_h^{(j)} := (I + hD)\bar{p}_h^{(j-1)} + h\bar{f}(t_{j-1}) & 1 \leq j \leq m, \\ \bar{p}_h^{(0)} := \bar{p}_0, \end{cases}$$

where I denotes the $n \times n$ identity matrix and $t_j := jh$, $0 \leq j \leq m$. We assume that D admits a diagonal form and we denote by λ_i, $1 \leq i \leq n$, its eigenvalues. Then, we have the following result.

Theorem 10.6.1 - *Let the eigenvalues of the matrix D satisfy $\mathrm{Re}\lambda_i < 0$, $1 \leq i \leq n$. Then, if $\bar{p}(t)$, $t \in [0,T]$, is the solution of problem (10.6.1), and $\bar{p}_h$ is obtained by (10.6.2), it is possible to find a constant $C > 0$ such that*

$$(10.6.3)\qquad \|\bar{p}_h - \bar{p}(T)\|_{\mathbf{R}^n} \leq Ch \sup_{t\in]0,T[} \left\| \frac{d^2\bar{p}}{dt^2}(t) \right\|_{\mathbf{R}^n},$$

for any h satisfying

$$(10.6.4)\qquad 0 < h < \tilde{h} := \min_{1\leq i\leq n} \left\{ \frac{-2\mathrm{Re}\lambda_i}{|\lambda_i|^2} \right\}.$$

■

It is well-known that, when condition (10.6.4) is violated, severe instabilities can occur when applying the algorithm (10.6.2). Actually, (10.6.4) is equivalent to requiring that the spectral radius $\rho(I + hD)$ is less than 1 (see section 7.6). This stresses the importance of working with matrices having eigenvalues with a negative real part. All the examples analyzed in the previous sections lead to systems of ordinary differential equations where the corresponding matrices fulfill such a requirement (see chapter eight). Moreover, in some cases (see (10.3.8), (10.3.19) and (10.5.11)), we were able to prove an additional property, that is, that a certain norm in $\mathbf{R}^n$ of the vector $\bar{p}(t)$ is a decreasing function of $t \in [0, T]$. This automatically excludes the existence of eigenvalues with a positive real part, otherwise the components of the vector $\bar{p}$ along the directions of the corresponding eigenvectors would be amplified during the time evolution. Due to (10.6.3), the error committed at time T decays linearly with the size of the interval h. This means that the method is *first-order accurate*. We observe that the constant C in (10.6.3) depends on n.

To illustrate the above, we provide the results of a numerical test. Consider equation (10.2.1) with $\zeta = 1$ and $T = 1$. Initial and boundary conditions are such that the solution is $U(x,t) = e^{-t}\cos x + e^{-4t}\sin x$, $x \in [-1, 1]$, $t \in [0, 1]$. By changing the variable as suggested in section 10.2, the new unknown V satisfies (10.2.5), (10.2.6) and (10.2.7). Then, for $n \geq 2$, p_n denotes the polynomial approximation of V by the collocation method (10.2.8) at the Chebyshev nodes given in (3.1.11). The relative differential system (such as (10.2.11) for $n = 4$) is discretized by the Euler method. For any $m \geq 1$, at the end of the iterative process, we obtain a polynomial $p_{n,m} \in \mathbf{P}_n^0$ identified by its value at the points $\eta_j^{(n)}$, $1 \leq j \leq n-1$. In table 10.6.1, we report the error

$$E_{n,m} := \left(\int_{-1}^{1} \left[p_{n,m}(x) - (\tilde{I}_{w,n}V)(x,1)\right]^2 \frac{dx}{\sqrt{1-x^2}} \right)^{\frac{1}{2}}, \qquad n \geq 2, \quad m \geq 1.$$

From relation (3.8.14), we can evaluate $E_{n,m}$ by a quadrature formula based on the collocation points (see also the numerical examples of section 9.2). The quantity $E_{n,m}$ is the sum of two contributions. One is the error relative to the approximation in the space variable, which decays exponentially when n grows (compare for instance

$E_{4,3200}$ with $E_{6,3200}$). The other is the error in the time variable, which only behaves like $h := 1/m$. Indeed, except for the case $n = 4$, $E_{n,m}$ is halved when m is doubled. The reader will note that $E_{n,m}$ does not decrease for $n \geq 6$. This is because the *time-step* h is not small enough to compete with the good accuracy in space, thus, the global error is dominated by the time-discretization error. The situation is different for the case $n = 4$, where the space resolution is poor and the results cannot improve by reducing the time-step. In our example, relation (10.6.4) is not satisfied when $n = 10$ and $m = 200$. This is the reason why the error $E_{10,200}$ is rather high.

m	$E_{4,m}$	$E_{6,m}$	$E_{8,m}$	$E_{10,m}$
200	$.5782 \times 10^{-3}$	$.2426 \times 10^{-3}$	$.2427 \times 10^{-3}$	$.3458 \times 10^{28}$
400	$.5546 \times 10^{-3}$	$.1212 \times 10^{-3}$	$.1213 \times 10^{-3}$	$.1213 \times 10^{-3}$
800	$.5522 \times 10^{-3}$	$.6069 \times 10^{-4}$	$.6066 \times 10^{-4}$	$.6066 \times 10^{-4}$
1600	$.5534 \times 10^{-3}$	$.3055 \times 10^{-4}$	$.3033 \times 10^{-4}$	$.3032 \times 10^{-4}$
3200	$.5545 \times 10^{-3}$	$.1576 \times 10^{-4}$	$.1517 \times 10^{-4}$	$.1516 \times 10^{-4}$

Table 10.6.1 - Errors relative to the Chebyshev collocation approximation of problem (10.2.5), (10.2.6), (10.2.7).

The above experiment shows that, in order to retain the good approximation properties of spectral methods, the time-step should be very small. This results in a large number of steps. Although the FFT (see section 4.3) speeds up matrix-vector multiplications in the Chebyshev case, the global computational cost can still be very high. An alternative is to use more accurate time-discretization methods, such as ***Runge-Kutta***, ***Adams-Bashforth*** or ***Du Fort-Frankel*** methods (see JAIN(1984)). By these techniques, the error decays as a fixed integer power of the time-step. For instance, experiments based on equation (10.3.20), using a fourth-order Runge-Kutta method, are discussed in SOLOMONOFF and TURKEL(1989). Like the Euler method, the schemes mentioned above are *explicit*. This implies that they can be used only if h satisfies an inequality of the type (10.6.4) (***absolute stability condition***). Indications for the choice of the parameter h are given in GOTTLIEB and TADMOR(1990) for spectral approximations of first-order partial differential equations. Unfortunately, in most of the practical applications, these stability conditions are considered to be very restrictive, especially when D is related to the space-discretization of second-order differential operators. In this case, according to (8.3.6), the maximum time-step $\tilde{h}$ in (10.6.4) is required to be proportional to $1/n^4$. Indeed, this limitation is quite severe, even when n is not too large. For first-order problems, KOSLOFF and TAL-EZER(1989) propose a Chebyshev-like spectral method with a weaker restriction on $\tilde{h}$.

No restrictions are in general required for ***implicit*** methods, such as the ***backward Euler method***. This is obtained by modifying (10.6.2) to be

$$
\text{(10.6.5)} \qquad \begin{cases} \bar{p}_h^{(j)} := \bar{p}_h^{(j-1)} + hD\bar{p}_h^{(j)} + h\bar{f}(t_j) & 1 \le j \le m, \\ \bar{p}_h^{(0)} := \bar{p}_0. \end{cases}
$$

The scheme (10.6.5) is ***inconditionally stable*** (i.e. no limitations on h are required), provided the eigenvalues of D have a negative real part. In addition, we still have (10.6.3), i.e., the method is first-order accurate. At each step, the new vector $\bar{p}_h^{(j)}$, $1 \le j \le m$, is computed by solving a $n \times n$ linear system whose matrix is $I - hD$. We refer to section 7.6 for the numerical treatment of this system. The algorithm is now more costly, but we are allowed to choose a larger time-step. This results in a loss of accuracy, which is moderate, however, for solutions that have a slow time variation.

Further theoretical results and experiments are discussed in MERCIER(1982) and MERCIER(1989) for first-order problems, and CANUTO and QUARTERONI(1987) for hyperbolic systems. Interesting comments are given in TREFETHEN and TRUMMER(1987) and TREFETHEN(1988), where numerical tests show the sensitivity to rounding errors of certain explicit methods, when coupled with spectral approximations.

Similar techniques apply to nonlinear equations. For example, the Burgers equation (see section 10.4) can be discretized by introducing a sequence of polynomials $p_n^{(j)} \in \mathbf{P}_n^0$,

$0 \le j \le m$, according to the recursion formula

$$(10.6.6)\quad p_n^{(j)}(\eta_i^{(n)}) := p_n^{(j-1)}(\eta_i^{(n)}) + \epsilon h \left[\frac{d^2}{dx^2} p_n^{(j)}\right](\eta_i^{(n)}) - h \left[p_n^{(j-1)} \frac{d}{dx} p_n^{(j-1)}\right](\eta_i^{(n)}),$$

$$1 \le i \le n-1, \quad 0 \le j \le m.$$

The initial polynomial is $p_n^{(0)}(\eta_i^{(n)}) := U_0(\eta_i^{(n)})$, $1 \le i \le n-1$. The derivatives at the nodes are evaluated with the usual arguments (see section 7.2). Then, the final polynomial $p_n^{(m)}$ represents an approximation of the exact solution $U(\cdot, T)$. To avoid severe time-step restrictions the scheme is implicit. Since a full implicit scheme forces us to solve a set of nonlinear equations at any step, the nonlinear term is treated explicitly. A theoretical analysis of convergence of the method for the Chebyshev case is carried out in BRESSAN and QUARTERONI(1986a).

The literature on this subject offers an extensive collection of (more or less efficient) algorithms. An overview of the time-discretization techniques most used in spectral methods can be found in GOTTLIEB and ORSZAG(1977), chapter 9, TURKEL(1980), CANUTO, HUSSAINI, QUARTERONI and ZANG(1988), chapter 4, BOYD(1989), chapter 8. A deeper analysis is too technical at this point. Nevertheless, this short introduction should be sufficient for running the first experiments and interpreting the results.

11 DOMAIN-DECOMPOSITION METHODS

For several reasons, when approximating differential equations numerically, it is often convenient to decompose the domain, where the solution is defined, into different subsets. Then, independently in each subdomain, one computes polynomial approximations by the techniques introduced in the previous chapters. The goal now is to find a suitable way to match the different pieces to obtain a discretization of the global solution.

11.1 Introductory remarks

Recently, *domain-decomposition* (or *multidomain*) methods have become a fundamental subject of research in spectral methods. This is especially true for boundary-value problems in two or more space variables, when the solutions are defined in domains with a complicated geometry. Actually, spectral-type techniques are well suited for very simple domains, obtainable by cartesian products of intervals (see chapter thirteen). It is clear that, when the given domain does not conform with these requirements, setting up a spectral approximation scheme is not a straightforward procedure. From this point of view, this is a severe drawback in comparison with other more flexible methods, such as finite-differences or finite element methods.

When possible, the initial domain is divided into subsets, each one having an elementary shape. Then, the approximation of the global solution results from combining the different discrete solutions relative to each subdomain. The way of coupling the various schemes will be discussed in the coming sections.

In this book we treat only one-dimensional problems. In this context, our domains, being intervals of $\mathbf{R}$, are simple right from the start. Nevertheless, even for equations in one variable, there are several situations where a domain-decomposition method is preferable to a direct computation in the whole domain. This is the case for instance for solutions displaying different regularity behaviors in different parts of the domain. A global high-degree polynomial approximation is generally more expensive and less accurate than using high-degree polynomials only in the regions where the degree of smoothness is lower. In this way, we can treat sharp gradients or shocks when we can guess their location. Moreover, different techniques can be used in each subdomain (see the example of section 12.2). The partitioning of the domain is often suggested by the problem itself, as a result of matching different type of equations. In addition, we stress that a preliminary theoretical analysis in one variable is often necessary before experimenting with more complicated situations.

We denote by I an open interval of $\mathbf{R}$. Multidomain methods are usually divided into two categories: *non-overlapping* and *overlapping*. In the first case I is decomposed into a finite number $m \geq 1$ of open intervals S_k, $1 \leq k \leq m$, in such a way that $\bar{I} = \bigcup_{1 \leq k \leq m} \bar{S}_k$, and $S_{k_1} \cap S_{k_2} = \emptyset$, if $k_1 \neq k_2$. In the second case, we assume that there exist two integers k_1 and k_2 such that $S_{k_1} \cap S_{k_2} \neq \emptyset$.

11.2 Non-overlapping multidomain methods

Let $m \geq 1$ be an integer and $s_k \in \mathbf{R}$, $0 \leq k \leq m$, an increasing set of real numbers. We subdivide the interval $I :=]s_0, s_m[$ into the subdomains $S_k :=]s_{k-1}, s_k[$, $1 \leq k \leq m$.

Consider first the second-order boundary-value problem (see also (9.1.4)):

$$\text{(11.2.1)} \qquad \begin{cases} -U'' = f & \text{in } I, \\ U(s_0) = \sigma_1, & U(s_m) = \sigma_2, \end{cases}$$

where $\sigma_1, \sigma_2 \in \mathbf{R}$, and $f : I \to \mathbf{R}$ is a given continuous function.
Given the decomposition of the domain I, (11.2.1) is equivalent to finding m functions $U_k : \bar{S}_k \to \mathbf{R}$, $1 \leq k \leq m$, such that

$$\text{(11.2.2)} \qquad \begin{cases} -U_k'' = f & \text{in } S_k, \quad 1 \leq k \leq m, \\ U_k(s_k) = U_{k+1}(s_k) & 1 \leq k \leq m-1, \\ U_k'(s_k) = U_{k+1}'(s_k) & 1 \leq k \leq m-1, \\ U_1(s_0) = \sigma_1, \quad U_m(s_m) = \sigma_2. \end{cases}$$

Of course, U_k turns out to be the restriction of U to the set $\bar{S}_k$, $1 \leq k \leq m$.
Instead of approximating the solution of (11.2.1) by a unique polynomial in the whole interval I as suggested in section 9.4, we look for a continuous approximating function $\pi_n : \bar{I} \to \mathbf{R}$, such that $p_{n,k}(x) := \pi_n(x)$, $x \in \bar{S}_k$, is a polynomial in $\mathbf{P}_n$ converging for $n \to +\infty$ to the corresponding function U_k, $1 \leq k \leq m$.

Let us examine for instance the collocation method. First, we map the points $\eta_j^{(n)} \in [-1,1]$, $0 \leq j \leq n$, in each interval $\bar{S}_k$, $1 \leq k \leq m$. This is done by defining the new set of nodes

$$\text{(11.2.3)} \qquad \theta_j^{(n,k)} := \tfrac{1}{2}[(s_k - s_{k-1})\eta_j^{(n)} + s_k + s_{k-1}] \qquad 0 \leq j \leq n, \quad 1 \leq k \leq m.$$

We note that $\theta_n^{(n,k)} = \theta_0^{(n,k+1)} = s_k$, $1 \leq k \leq m-1$.
Then, following ORSZAG(1980), we define the polynomials $p_{n,k}$, $1 \leq k \leq m$, to be solutions of the set of equations

$$\text{(11.2.4)} \qquad -p_{n,k}''(\theta_i^{(n,k)}) = f(\theta_i^{(n,k)}) \qquad 1 \leq i \leq n-1, \quad 1 \leq k \leq m,$$

(11.2.5) $$p_{n,k}(s_k) = p_{n,k+1}(s_k) \qquad 1 \leq k \leq m-1,$$

(11.2.6) $$p'_{n,k}(s_k) = p'_{n,k+1}(s_k) \qquad 1 \leq k \leq m-1,$$

(11.2.7) $$p_{n,1}(s_0) = \sigma_1, \qquad p_{n,m}(s_m) = \sigma_2.$$

Basically, in (11.2.4) we collocated the differential equation at the nodes inside S_k, $1 \leq k \leq m$, and imposed the boundary conditions (11.2.7). Relations (11.2.5) and (11.2.6) require the continuity of π_n and its derivative in the interval I, i.e., $\pi_n \in C^1(I)$. It is clear that (11.2.4)-(11.2.7) is equivalent to a $m(n+1) \times m(n+1)$ linear system. To determine the corresponding matrix, we argue as in sections 7.2 and 7.4. For instance, the entries of the derivative matrix at the collocation nodes $\theta_j^{(n,k)}$, $0 \leq j \leq n$, are given by $\left\{2\tilde{d}_{ij}^{(1)}/(s_k - s_{k-1})\right\}_{\substack{0 \leq i \leq n \\ 0 \leq j \leq n}}$, $1 \leq k \leq m$.

In the ultraspherical case, we can replace (11.2.6) by

(11.2.8) $$\frac{-1}{s_{k+1} - s_{k-1}} \Big[(s_k - s_{k-1})p''_{n,k} + (s_{k+1} - s_k)p''_{n,k+1}\Big](s_k)$$
$$+ \gamma_k \Big[p'_{n,k} - p'_{n,k+1}\Big](s_k) = f(s_k) \qquad 1 \leq k \leq m-1,$$

where γ_k, $1 \leq k \leq m$, are suitable constants. In this case we only have $\pi_n \in C^0(I)$. The use of formula (11.2.8) was proposed in FUNARO(1986) for Legendre nodes and FUNARO(1988) for Chebyshev nodes. This new condition, which is closely related to (9.4.23), results from a variational formulation of problem (11.2.1). When the constants γ_k in (11.2.8) are chosen appropriately, numerical results are in general more accurate than those obtained with condition (11.2.6). In both cases, we expect convergence of the approximating function π_n to the exact solution U when $n \to +\infty$. A convergence analysis can be developed following the variational approach described in sections 9.3 and 9.4. To illustrate the basic principles, we show the theory in a simple case. We recall that there is no loss of generality if we assume $\sigma_1 = \sigma_2 = 0$ in (11.2.1).

Theorem 11.2.1 - *Let* U_k, $1 \leq k \leq m$, *be the solutions of problem* (11.2.2) *with* $\sigma_1 = \sigma_2 = 0$. *Let* $\eta_j^{(n)}$, $0 \leq j \leq n$, *in* (11.2.3), *be the nodes of the Legendre* ($w \equiv 1$) *Gauss-Lobatto formula* (3.5.1). *Let* $p_{n,k}$, $1 \leq k \leq m$, *be the polynomials satisfying* (11.2.4), (11.2.5), (11.2.7) *with* $\sigma_1 = \sigma_2 = 0$, *and* (11.2.8) *with* $\gamma_k := \frac{n(n+1)}{s_{k+1}-s_{k-1}}$, $1 \leq k \leq m-1$. *Then, we have*

$$(11.2.9)\quad \lim_{n\to+\infty}\left(\int_{s_{k-1}}^{s_k}(U_k - p_{n,k})^2 dx \; + \; \int_{s_{k-1}}^{s_k}(U_k' - p_{n,k}')^2 dx\right)^{\frac{1}{2}} = 0, \qquad 1 \leq k \leq m.$$

Proof - For any $1 \leq k \leq m$, define the weights $\tilde{w}_j^{(n,k)} := \frac{1}{2}(s_k - s_{k-1})\tilde{w}_j^{(n)}$, $0 \leq j \leq n$ (see (3.5.6)). Therefore, by (3.5.1) with $w \equiv 1$, we get the integration formula

$$(11.2.10)\qquad \int_{s_{k-1}}^{s_k} p \, dx = \sum_{j=0}^{n} p(\theta_j^{(n,k)}) \, \tilde{w}_j^{(n,k)} \qquad \forall p \in \mathbf{P}_{2n-1}, \; 1 \leq k \leq m.$$

Let $\mathbf{X}_n$ denote the space of continuous functions ϕ in I, such that $\phi(s_0) = \phi(s_m) = 0$ and such that the restriction of ϕ to any interval $\bar{S}_k$ is a polynomial of degree n. Then, by noting that $\gamma_k[\tilde{w}_n^{(n,k)} + \tilde{w}_0^{(n,k+1)}] = 1$, $1 \leq k \leq m-1$, the function $\pi_n \in \mathbf{X}_n$ satisfies

$$(11.2.11)\qquad \int_I \pi_n' \phi' \, dx = \sum_{k=1}^{m} \int_{S_k} \pi_n' \phi' \, dx$$

$$= -\sum_{k=1}^{m} \int_{S_k} \pi_n'' \phi \, dx + \sum_{k=1}^{m-1} \left[\lim_{x\to s_k^-} (\pi_n' \phi)(x) - \lim_{x\to s_k^+} (\pi_n' \phi)(x) \right]$$

$$= -\sum_{k=1}^{m} \int_{s_{k-1}}^{s_k} p_{n,k}'' \phi \, dx + \sum_{k=1}^{m-1} \left[p_{n,k}' \phi - p_{n,k+1}' \phi \right](s_k)$$

$$= -\sum_{k=1}^{m} \left(\sum_{i=1}^{n} [p_{n,k}'' \phi](\theta_i^{(n,k)}) \tilde{w}_i^{(n,k)} \right) + \sum_{k=1}^{m-1} \gamma_k (\tilde{w}_n^{(n,k)} + \tilde{w}_0^{(n,k+1)}) \left[p_{n,k}' \phi - p_{n,k+1}' \phi \right](s_k)$$

$$= \sum_{k=1}^{m} \left(\sum_{i=1}^{n} [f\phi](\theta_i^{(n,k)}) \, \tilde{w}_i^{(n,k)} \right) =: F_n(\phi), \qquad \forall \phi \in \mathbf{X}_n.$$

Thus, our collocation scheme is now in a variational form. On the other hand, U satisfies $\int_I U'\psi' dx = \int_I f\psi dx, \quad \forall\psi \in \mathbf{X}, \ \psi(s_0) = \psi(s_m) = 0$, where $\mathbf{X}$ is the Hilbert space with norm $\|\psi\|_{\mathbf{X}} = \left(\int_I \psi^2 dx + \int_I [\psi']^2 dx\right)^{1/2}$. Due to theorem 9.4.1, it is sufficient to estimate the quantity $\left|\int_I f\phi dx - F_n(\phi)\right|$, $\forall\phi \in \mathbf{X}_n$, to conclude that the error $|U - \pi_n|$ tends to zero in the norm of $\mathbf{X}$. To this end, for any S_k, we can argue as in (9.4.12), (9.4.13), which lead to (11.2.9).

■

With some technical modifications, we can use the same proof to obtain a convergence result for the scheme (11.2.4), (11.2.5), (11.2.6), (11.2.7), in the Legendre case. For other sets of collocation nodes the theory becomes more difficult. The idea is to recast the collocation scheme into a variational framework which then reduces to a problem like (9.4.16) in each subdomain.

We get similar conclusions when different degrees n_k, $1 \le k \le m$, are assigned to the approximating polynomials. Now, we can either use condition (11.2.6) as is, or condition (11.2.8) after suitable modification. For example, in the Legendre case, starting from the variational formulation (11.2.11), we deduce that the following relations are satisfied at the *interface* nodes:

$$(11.2.12) \qquad -\left[\tilde{w}_{n_k}^{(n_k,k)} + \tilde{w}_0^{(n_{k+1},k+1)}\right]^{-1} \left[\tilde{w}_{n_k}^{(n_k,k)} p''_{n_k,k} + \tilde{w}_0^{(n_{k+1},k+1)} p''_{n_{k+1},k+1}\right](s_k)$$

$$+ \gamma_k \left[p'_{n_k,k} - p'_{n_{k+1},k+1}\right](s_k) = f(s_k) \qquad 1 \le k \le m-1,$$

where $\gamma_k := 1 \Big/ \left(\tilde{w}_{n_k}^{(n_k,k)} + \tilde{w}_0^{(n_{k+1},k+1)}\right)$, $1 \le k \le m-1$.

In the applications, we can prescribe the polynomial degrees n_k, depending on the regularity of U in the different subintervals S_k. We show an example for $I =]-1,1[$. In (11.2.1) we take $\sigma_1 = \sigma_2 = 0$ and f such that $U(x) := \sin\left(\frac{12\pi}{5x+7}\right)$, $x \in \bar{I}$. The function U is plotted in figure 11.2.1. In the first test, we approximate U by a single polynomial p_n with the standard collocation method at the Legendre points (see (9.2.15)). In table 11.2.1, we give the error $E_n := \left(\sum_{j=1}^{n} (p_n - U)^2(\eta_j^{(n)})\tilde{w}_j^{(n)}\right)^{\frac{1}{2}}$ for different values of n.

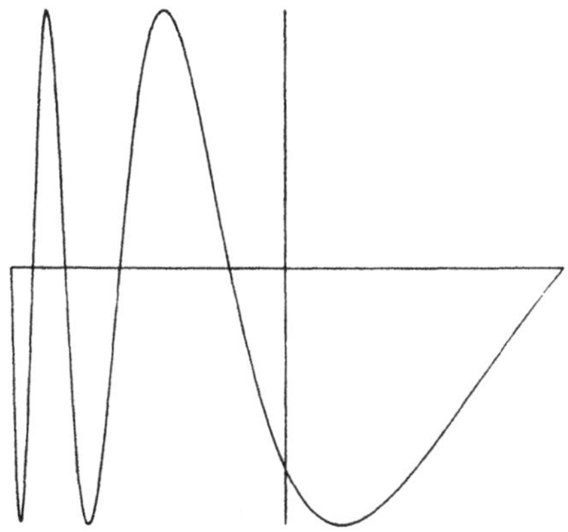

Figure 11.2.1 - Plot of the function $U(x)$, $x \in [-1,1]$.

n	E_n
12	.557106
16	$.4418 \times 10^{-1}$
20	$.4262 \times 10^{-2}$
24	$.6324 \times 10^{-3}$
28	$.4265 \times 10^{-4}$
32	$.7923 \times 10^{-5}$

Table 11.2.1 - Errors for a single-domain approximation of problem (11.2.1).

In the second experiment, we consider two domains, namely $S_1 =]s_0, s_1[=]-1, 0[$ and $S_2 =]s_1, s_2[=]0, 1[$. In each interval S_k, $1 \leq k \leq 2$, we approximate the solution by a polynomial $p_{n_k,k} \in \mathbf{P}_{n_k}$, with the collocation method at the Legendre nodes. At the point $s_1 = 0$ we impose $p_{n_1,1}(s_1) = p_{n_2,2}(s_1)$ and condition (11.2.12) for $m = 2$. We report in table 11.2.2 the error $E_{n_1,n_2} := \left(\sum_{k=1}^{2} \sum_{j=1}^{n_k} (p_{n_k,k} - U)^2(\theta_j^{(n_k,k)}) \tilde{w}_j^{(n_k,k)}\right)^{\frac{1}{2}}$ for various choices of the parameters n_1 and n_2.

n_1	$E_{n_1,2}$	$E_{n_1,4}$	$E_{n_1,6}$	$E_{n_1,8}$
12	.130794	$.1817 \times 10^{-1}$	$.1839 \times 10^{-1}$	$.1839 \times 10^{-1}$
16	.133438	$.1645 \times 10^{-2}$	$.1022 \times 10^{-2}$	$.1021 \times 10^{-2}$
20	.133418	$.1274 \times 10^{-2}$	$.5449 \times 10^{-4}$	$.2638 \times 10^{-4}$
24	.133418	$.1274 \times 10^{-2}$	$.4772 \times 10^{-4}$	$.2089 \times 10^{-5}$
28	.133418	$.1274 \times 10^{-2}$	$.4769 \times 10^{-4}$	$.9319 \times 10^{-6}$

Table 11.2.2 - Errors for a multidomain approximation of problem (11.2.1).

Due to the different behavior of U in the two subdomains, the best results are obtained with an uneven distribution of nodes ($n_1 > n_2$). A correct balance of the two parameters guarantees performances as good as the ones of the single-domain approximation, with the same degrees of freedom and a lower computational cost. In fact, the matrices involved in the multidomain approach are ***block-diagonal***, with a number of blocks equal to the number of subdomains and a bandwidth proportional to the largest degree used. We can take advantage of this structure for a more efficient numerical implementation. The details are discussed in section 11.3. In (11.2.23), we explicitly write the system from the above example with $n_1 = n_2 = 2$ (hence $\gamma_1 = 3$):

(11.2.13)

$$\begin{bmatrix} 1 & 0 & 0 & 0 & 0 & 0 \\ -4\tilde{d}_{10}^{(2)} & -4\tilde{d}_{11}^{(2)} & -4\tilde{d}_{12}^{(2)} & 0 & 0 & 0 \\ 0 & 0 & 1 & -1 & 0 & 0 \\ \tau_1 & \tau_2 & \tau_3 & \tau_4 & \tau_5 & \tau_6 \\ 0 & 0 & 0 & -4\tilde{d}_{10}^{(2)} & -4\tilde{d}_{11}^{(2)} & -4\tilde{d}_{12}^{(2)} \\ 0 & 0 & 0 & 0 & 0 & 1 \end{bmatrix} \begin{bmatrix} p_{n_1,1}(\theta_0^{(n_1,1)}) \\ p_{n_1,1}(\theta_1^{(n_1,1)}) \\ p_{n_1,1}(\theta_2^{(n_1,1)}) \\ p_{n_2,2}(\theta_0^{(n_2,2)}) \\ p_{n_2,2}(\theta_1^{(n_2,2)}) \\ p_{n_2,2}(\theta_2^{(n_2,2)}) \end{bmatrix} = \begin{bmatrix} \sigma_1 \\ f(\theta_1^{(n_1,1)}) \\ 0 \\ f(s_1) \\ f(\theta_1^{(n_2,2)}) \\ \sigma_2 \end{bmatrix}$$

where $\tau_1 := -2\tilde{d}_{20}^{(2)} + 2\gamma_1 \tilde{d}_{20}^{(1)}$, $\tau_2 := -2\tilde{d}_{21}^{(2)} + 2\gamma_1 \tilde{d}_{21}^{(1)}$, $\tau_3 := -2\tilde{d}_{22}^{(2)} + 2\gamma_1 \tilde{d}_{22}^{(1)}$, $\tau_4 := -2\tilde{d}_{00}^{(2)} - 2\gamma_1 \tilde{d}_{00}^{(1)}$, $\tau_5 := -2\tilde{d}_{01}^{(2)} - 2\gamma_1 \tilde{d}_{01}^{(1)}$, $\tau_6 := -2\tilde{d}_{02}^{(2)} - 2\gamma_1 \tilde{d}_{02}^{(1)}$. We can remove one unknown by eliminating the third row and summing up the third and the fourth columns. Similarly, condition (11.2.6) is imposed by defining $\tau_1 := 2\tilde{d}_{20}^{(1)}$, $\tau_2 := 2\tilde{d}_{21}^{(1)}$, $\tau_3 := 2\tilde{d}_{22}^{(1)}$, $\tau_4 := -2\tilde{d}_{00}^{(1)}$, $\tau_5 := -2\tilde{d}_{01}^{(1)}$, $\tau_6 := -2\tilde{d}_{02}^{(1)}$, and by setting to zero the fourth entry of the right-hand side vector.

In the Chebyshev case, relation (11.2.6) is more suitable for implementation since it is difficult to obtain the exact expression of γ_k in (11.2.8), or the counterpart of formula (11.2.12). As an alternative, we can use the condition

$$(11.2.14) \qquad p'_{n_k,k}(s_{k-1}) - p'_{n_{k+1},k+1}(s_{k+1}) = \int_{s_{k-1}}^{s_{k+1}} f\, dx \qquad 1 \le k \le m-1,$$

which is suggested by the identity

$$(11.2.15) \quad U'(s_{k-1}) - U'(s_{k+1}) = -\int_{s_{k-1}}^{s_{k+1}} U''\, dx = \int_{s_{k-1}}^{s_{k+1}} f\, dx \quad 1 \le k \le m-1.$$

The integral in (11.2.14) can be computed by quadrature, for instance by the Clenshaw-Curtis formula once the values of f at the nodes are known (see section 3.7). This approach was suggested in MACARAEG and STREETT(1986) and produces results comparable to those based on relation (11.2.6). Going back to the example previously illustrated, (11.2.14) with $m = 2$ is obtained by setting in (11.2.13): $\tau_1 := 2\tilde{d}_{00}^{(1)}$, $\tau_2 := 2\tilde{d}_{01}^{(1)}$, $\tau_3 := 2\tilde{d}_{02}^{(1)}$, $\tau_4 := -2\tilde{d}_{20}^{(1)}$, $\tau_5 := -2\tilde{d}_{21}^{(1)}$, $\tau_6 := -2\tilde{d}_{22}^{(1)}$, and by replacing the fourth entry of the right-hand side vector by $\int_{-1}^{1} f dx$.

The spectral element method (see PATERA(1984) and section 9.6) uses in the Chebyshev case a variational formulation similar to that in (11.2.11) (we recall that this corresponds to an approximation of problem (11.2.1) with $\sigma_1 = \sigma_2 = 0$). In any interval $\bar{S}_k$, $1 \le k \le m$, we consider the expansion $p_{n,k} = \sum_{j=0}^{n} p_{n,k}(\theta_j^{(n,k)})\tilde{l}_j^{(n,k)}$, where

$$(11.2.16) \quad \tilde{l}_j^{(n,k)}(x) := \tilde{l}_j^{(n)}((2x - s_k - s_{k-1})/(s_k - s_{k-1})), \qquad \forall x \in \bar{S}_k, \ \ 0 \le j \le n.$$

The nodes are obtained from (11.2.3), where $\eta_j^{(n)}$, $0 \le j \le n$, are the Gauss-Lobatto Chebyshev points (see (3.1.11)). The associated Lagrange polynomials $\tilde{l}_j^{(n)}$, $0 \le j \le n$, are defined in (3.2.10). Next, we modify the right-hand side of (11.2.11) by defining

$$(11.2.17) \qquad F_n(\phi) := \sum_{k=1}^{m} \left(\sum_{i=1}^{n} f(\theta_i^{(n,k)}) \int_{S_k} \tilde{l}_i^{(n,k)} \phi\, dx \right), \qquad \forall \phi \in \mathbf{X}_n.$$

The set of equations is obtained by testing on the functions $\phi \equiv \phi_{k,j} \in \mathbf{X}_n$, such that

$$\phi_{k,j}(x) := \begin{cases} \tilde{l}_j^{(n,k)}(x) & \text{if } x \in \bar{S}_k \\ 0 & \text{if } x \in \bar{I} - \bar{S}_k \end{cases} \qquad 1 \le j \le n-1, \ \ 1 \le k \le m,$$

and on the set of functions $\phi \equiv \phi_k \in \mathbf{X}_n$, such that

$$\phi_k(x) := \begin{cases} \tilde{l}_n^{(n,k)}(x) & \text{if } x \in \bar{S}_k \\ \tilde{l}_0^{(n,k+1)}(x) & \text{if } x \in \bar{S}_{k+1} \\ 0 & \text{elsewhere} \end{cases} \qquad 1 \le k \le m-1.$$

The cumbersome construction of the corresponding $(nm-1)\times(nm-1)$ linear system is left to the reader. Since we are using Chebishev nodes instead of Legendre nodes, we can apply the FFT (see section 4.3) in our computations. Besides, the proof of convergence follows from theorem 9.4.1 and by showing that the term $\left| \int_I f\phi dx - F_n(\phi)\right|$, where $\phi \in \mathbf{X}_n$, tends to zero. Different polynomial degrees may be also considered in each domain.

We can use the same techniques introduced above when the approximating polynomials are represented in the frequency space. For example, the multidomain tau method is obtained by seeking $p_{n,k} = \sum_{j=0}^n c_{j,k} u_{j,k}$, $1 \le k \le m$, where $u_{j,k}(x) := u_j\big((2x - s_k - s_{k-1})/(s_k - s_{k-1})\big)$, $x \in \bar{S}_k$, $1 \le k \le m$, $j \in \mathbf{N}$, and the u_j's are the classical orthogonal polynomials in $[-1,1]$. Then, in analogy with (11.2.4)-(11.2.7), we require that

(11.2.18) $$-\, c_{j,k}^{(2)} = f_{j,k} \qquad 1 \le k \le m,$$

(11.2.19) $$\sum_{j=0}^n c_{j,k}\, u_{j,k}(s_k) = \sum_{j=0}^n c_{j,k+1}\, u_{j,k+1}(s_k) \qquad 1 \le k \le m-1,$$

(11.2.20) $$\sum_{j=0}^n c_{j,k}\, u'_{j,k}(s_k) = \sum_{j=0}^n c_{j,k+1}\, u'_{j,k+1}(s_k) \qquad 1 \le k \le m-1,$$

(11.2.21) $$\sum_{j=0}^n c_{j,1}\, u_{j,1}(s_0) = \sigma_1, \qquad \sum_{j=0}^n c_{j,m}\, u_{j,m}(s_m) = \sigma_2.$$

In (11.2.18), the $c_{j,k}^{(2)}$'s are the Fourier coefficients of the polynomial $p''_{n,k}$, which, thanks to the results of section 7.1, can be expressed in terms of the $c_{j,k}$'s. For any $1 \le k \le m$, the quantities $f_{j,k}$, $j \in \mathbf{N}$, are the Fourier coefficients of f in $\bar{S}_k$ with respect to $u_{j,k}$, $j \in \mathbf{N}$. In the Legendre case, (11.2.18)-(11.2.21) are equivalent to the variational problem (11.2.11), for a suitable right-hand side function F.

We finally note that the Legendre-Galerkin method is obtained by modifying the right-hand side of (11.2.11) by $F(\phi) := \int_I f\phi dx$, $\forall \phi \in \mathbf{X}_n$.

Of course, different discretizations can be used in each subdomain. An example of a problem defined in an unbounded domain, solved by coupling Legendre and Laguerre approximations, is presented in section 12.2. The coupling of finite element and spectral methods is considered in BERNARDI, MADAY and SACCHI-LANDRIANI(1989) and BERNARDI, DEBIT and MADAY(1990).

First-order problems are approached in a similar way. For example, the differential equation (see also (9.1.3))

$$\text{(11.2.22)} \qquad \begin{cases} U' + AU = f & \text{in }]s_0, s_m], \\ U(s_0) = \sigma, \end{cases}$$

can be written as

$$\text{(11.2.23)} \qquad \begin{cases} U_k' + AU_k = f & \text{in }]s_{k-1}, s_k[, \quad 1 \le k \le m-1, \\ U_m' + AU_m = f & \text{in }]s_{m-1}, s_m], \\ U_k(s_k) = U_{k+1}(s_k) & 1 \le k \le m-1, \\ U_1(s_0) = \sigma, \end{cases}$$

where U_k is the restriction of U to the set $[s_{k-1}, s_k]$, $1 \le k \le m$.
A multidomain collocation scheme is obtained by finding the polynomials $p_{n,k} \in \mathbf{P}_n$, $1 \le k \le m$, such that

$$\text{(11.2.24)} \qquad (p'_{n,k} + Ap_{n,k})(\theta_i^{(n,k)}) = f(\theta_i^{(n,k)}) \quad 1 \le i \le n-1, \ 1 \le k \le m-1,$$

$$\text{(11.2.25)} \qquad (p'_{n,m} + Ap_{n,m})(\theta_i^{(n,m)}) = f(\theta_i^{(n,m)}) \quad 1 \le i \le n,$$

(11.2.26) $$p_{n,k}(s_k) = p_{n,k+1}(s_k) \qquad 1 \le k \le m-1,$$

(11.2.27) $$p_{n,1}(s_0) = \sigma.$$

Another approximation is obtained by replacing condition (11.2.26) by

(11.2.28) $$\frac{1}{s_{k+1}-s_{k-1}}\Big[(s_k - s_{k-1})(p'_{n,k} + Ap_{n,k}) + (s_{k+1}-s_k)(p'_{n,k+1} + Ap_{n,k+1})\Big](s_k)$$
$$+ \gamma_k\big[p_{n,k} - p_{n,k+1}\big](s_k) = f(s_k) \quad 1 \le k \le m-1,$$

where γ_k, $1 \le k \le m$, are suitable constants. We note that using (11.2.28) the global approximating function in $[s_0, s_m]$ is not continuous. Nevertheless, we still observe a spectral rate of convergence.

Conditions at the interface points for linear or nonlinear first-order time-dependent problems (see sections 10.3, 10.4 and 10.5) are considered for instance in PATERA(1984), KOPRIVA(1986), MACARAEG and STREETT(1986), CANUTO and QUARTERONI(1987), FUNARO(1990b). Other time-dependent equations are examined in PAVONI(1988) and BRESSAN and PAVONI(1990). These techniques are under development and very little is known about the theory. Finally, the matching of equations having different orders in the different domains is examined in GASTALDI and QUARTERONI(1989).

11.3 Solution techniques

The matrices related to the numerical approximation of a differential problem by domain-decomposition can be partitioned into diagonal blocks, coupled through the conditions imposed at the interface points. The number of blocks and their size is arbitrary. Some authors are inclined to use many subdomains and low polynomial degrees. As in the finite element method, they achieve convergence to the exact solution by letting the size of these domains tend to zero. This is the case of the *h-p*-version of the finite element method (see BABUŠKA, SZABO and KATZ(1981), BABUŠKA and SURI(1987)). In contrast, the common practice in spectral methods is to use large domains and high polynomial degrees. Whether the first approach is better than the latter depends on the circumstances.

The use of domain-decomposition methods gives us the possibility of developing useful iterative algorithms by taking advantage of the structure of the matrices. At each iteration, the linear equations in the system are decoupled by blocks, which are solved independently. Of course, such a feature is particularly well adapted to computers with a parallel architecture. We illustrate the basic ideas with an example. We study the approximation π_n of problem (11.2.1), given by the multidomain collocation scheme (11.2.4)-(11.2.7). We first introduce some notations. Let y_k, $0 \le k \le m$, be a set of real values with $y_0 := \sigma_1$ and $y_m := \sigma_2$. For any $1 \le k \le m$, we define the polynomial $q_{n,k} \in \mathbf{P}_n$ satisfying the set of equations

$$
(11.3.1) \qquad \begin{cases} -q''_{n,k}(\theta_i^{(n,k)}) = f(\theta_i^{(n,k)}) & 1 \le i \le n-1, \\ q_{n,k}(s_{k-1}) = y_{k-1}, \quad q_{n,k}(s_k) = y_k. \end{cases}
$$

Next, we define the mapping $\Gamma : \mathbf{R}^{m-1} \to \mathbf{R}^{m-1}$, $\Gamma \equiv (\Gamma_1, \cdots, \Gamma_{m-1})$, whose components are given by

$$
(11.3.2) \qquad \Gamma_k(y_1, \cdots, y_{m-1}) := q'_{n,k}(s_k) - q'_{n,k+1}(s_k), \qquad 1 \le k \le m-1.
$$

It is clear that π_n is characterized by the relation

$$
(11.3.3) \qquad \Gamma_k(\pi_n(s_1), \cdots, \pi_n(s_{m-1})) = 0, \qquad 1 \le k \le m-1.
$$

From the values $\pi_n(s_k)$, $1 \le k \le m-1$, it is easy to recover the entire function π_n by noting that $q_{n,k}$ in (11.3.1) coincides with $p_{n,k}$ when $y_{k-1} = \pi_n(s_{k-1})$ and $y_k = \pi_n(s_k)$. The problem is consequently reduced to finding the $m-1$ zeroes of Γ. This can be done by an iterative procedure. Using an explicit method, at each step we solve in parallel $m-1$ linear systems of dimension $(n+1) \times (n+1)$ corresponding to (11.3.1). This suggests to constructing once for all, for any $1 \le k \le m$, an operator, that for any pair of input values (y_{k-1}, y_k) in (11.3.1) provides the output values $(q'_{n,k}(s_{k-1}), q'_{n,k}(s_k))$. For $f \equiv \mathbf{0}$, $\sigma_1 = \sigma_2 = 0$, the mapping Γ is linear and corresponds to the *Schur complement* of the matrix associated with (11.2.4)-(11.2.7), with respect to the unknowns corresponding to the interface points (see DUFF, ERISMAN and REID(1986), p.60). Several iterative techniques are available for the numerical determination of the zeroes of Γ. Convergence results in an abstract context are given in AGOSHKOV(1988). The *conjugate gradient* method, which is theoretically examined in the case of finite element multidomain approximations in BJØRSTAD and WIDLUND(1986), is studied in CHAN and GOOVAERTS(1989) for spectral Legendre approximations. We note that the rate of convergence of the method depends on the ratio between the maximum and the minumum size of the domains.

Variants of the algorithm proposed above have been analyzed recently. An iterative approach is obtained by alternating Dirichlet and Neumann boundary conditions in the solution of the problems in the different subdomains. The solution in a given domain, recovered by imposing Dirichlet (respectively Neumann) boundary conditions, provides the Neumann (respectively Dirichlet) data to be used in the contiguous domains. Although only half of the blocks can be simultaneously processed at any step, a speed up of the convergence is observed when using an appropriate *relaxation* scheme. The procedure has been investigated in ZANOLLI(1987), FUNARO, QUARTERONI and ZANOLLI(1988), and interpreted in QUARTERONI and SACCHI-LANDRIANI(1988) as a preconditioned iterative method for the Schur complement of the matrix corresponding to (11.2.4)-(11.2.7). Other results are provided in ZAMPIERI(1989). A similar algorithm for the discretization of first-order time-dependent differential problems is analyzed in QUARTERONI(1990).

The interest in applying domain-decomposition methods, for the solution of partial differential equations, has increased in recent years, together with the development of parallel computers. Therefore, many other sophisticated techniques have been experimented. The reader will find a collection of papers and references in CHAN, GLOWINSKI, PERIAUX and WIDLUND(1989) and CHAN, GLOWINSKI, PERIAUX and WIDLUND(1990).

11.4 Overlapping multidomain methods

To avoid complex situations and cumbersome notation, we confine ourselves to the case of two subdomains only. These are given by $S_1 :=]s_0, s_2[$ and $S_2 :=]s_1, s_3[$, where $s_0 < s_1 < s_2 < s_3$. The set $S_1 \cap S_2 =]s_1, s_2[\neq \emptyset$ is the *overlapping region*. Let us consider problem (11.2.1)($m = 3$), which is equivalent to writing

$$(11.4.1)\qquad \begin{cases} -U_1'' = f & \text{in } S_1, \\ U_1(s_0) = \sigma_1, & U_1(s_2) = U_2(s_2), \end{cases} \qquad \begin{cases} -U_2'' = f & \text{in } S_2, \\ U_2(s_1) = U_1(s_1), & U_1(s_3) = \sigma_2, \end{cases}$$

where U_1 and U_2 are the restrictions of U to the intervals $\bar{S}_1$ and $\bar{S}_2$ respectively. The functions U_1 and U_2 coincide in $S_1 \cap S_2$.

With the intent of using the collocation method, in analogy with (11.2.3), we first define the collocation nodes

$$(11.4.2)\quad \theta_j^{(n,k)} := \frac{1}{2}\Big[(s_{k+1} - s_{k-1})\eta_j^{(n)} + s_{k+1} + s_{k-1}\Big] \qquad 0 \le j \le n, \quad 1 \le k \le 2.$$

For any $n \geq 2$, we would now like to determine two polynomials $p_{n,k} \in \mathbf{P}_n$, $1 \le k \le 2$, such that

$$(11.4.3)\qquad \begin{cases} -p''_{n,1}(\theta_i^{(n,1)}) = f(\theta_i^{(n,1)}) & 1 \le i \le n-1, \\ p_{n,1}(s_0) = \sigma_0, \quad p_{n,1}(s_2) = p_{n,2}(s_2), \end{cases}$$

$$(11.4.4)\qquad \begin{cases} -p''_{n,2}(\theta_i^{(n,2)}) = f(\theta_i^{(n,2)}) & 1 \le i \le n-1, \\ p_{n,2}(s_1) = p_{n,1}(s_1), \quad p_{n,2}(s_3) = \sigma_2. \end{cases}$$

The value $p_{n,1}(s_1)$ is obtained from the values $p_{n,1}(\theta_i^{(n,1)})$, $0 \le i \le n$, by interpolation (see formula (3.2.7)). The same argument holds for the value $p_{n,2}(s_2)$. The first step is to show that $\lim_{n\to+\infty} p_{n,k} = U_k$ in $\bar{S}_k$, $1 \le k \le 2$. This result is proven in CANUTO and FUNARO(1988) in the Legendre case for any choice of the s_k's, and in the Chebyshev case when the size of the overlapping region is sufficiently large. Note that, for any $n \geq 2$, the polynomials $p_{n,1}$ and $p_{n,2}$ do not coincide in $S_1 \cap S_2$. The construction of the linear system relative to (11.4.3)-(11.4.4), and the generalization to the case of more domains, is left to the reader.

Finally, as in the previous section, we suggest an iterative algorithm for the solution of the linear system corresponding to (11.4.3)-(11.4.4). For any $y \in \mathbf{R}$, we construct two polynomials $q_{n,k} \in \mathbf{P}_n$, $1 \leq k \leq 2$, satisfying

$$\text{(11.4.5)} \qquad \begin{cases} -q''_{n,1}(\theta_i^{(n,1)}) = f(\theta_i^{(n,1)}) & 1 \leq i \leq n-1, \\ q_{n,1}(s_0) = \sigma_0, \quad q_{n,1}(s_2) = y, \end{cases}$$

$$\text{(11.4.6)} \qquad \begin{cases} -q''_{n,2}(\theta_i^{(n,2)}) = f(\theta_i^{(n,2)}) & 1 \leq i \leq n-1, \\ q_{n,2}(s_1) = q_{n,1}(s_1), \quad q_{n,2}(s_3) = \sigma_2. \end{cases}$$

Thus, we define the mapping $\Gamma : \mathbf{R} \to \mathbf{R}$ such that $\Gamma(y) := q_{n,2}(s_2)$. Therefore, $q_{n,k} \equiv p_{n,k}$ in $\bar{S}_k$, $1 \leq k \leq 2$, if and only if $\Gamma(y) = y$. The aim is now to find the *fixed point* y^* of the function Γ. This can be determined by the recursion formula $y_{m+1} = \Gamma(y_m)$, $m \in \mathbf{N}$, where y_0 is given. At any step, we have to solve the collocation problems (11.4.5) and (11.4.6). Such a procedure is known as the *Schwarz alternating method* (see SCHWARZ(1890), Vol.2). In CANUTO and FUNARO(1988), the authors prove that $\lim_{n\to+\infty} y_m = y^*$, for Legendre and Chebyshev approximations. In particular, in the Legendre case, we get the relation

$$\text{(11.4.7)} \quad |y_m - y^*| \leq \frac{\kappa^m}{1-\kappa}\,|y_1 - y_0| \qquad m \in \mathbf{N}, \quad \text{where} \quad \kappa := \frac{(s_1 - s_0)(s_3 - s_2)}{(s_2 - s_0)(s_3 - s_1)}.$$

Note that the parameter $\kappa < 1$ approaches 1 when the size of the overlapping region tends to zero.

A presentation of the Schwarz alternating method in a very general framework is given in DRYJA and WIDLUND(1990), where a unifying theory provides a link between overlapping and non-overlapping multidomain methods.

12
EXAMPLES

In this chapter, we show how to apply spectral methods to the approximation of four upgraded model problems, without pretending that our approach is competitive or improves upon other known numerical algorithms. No theoretical convergence analysis is provided for the schemes proposed here.

12.1 A free boundary problem

In our first example, given $T > 0$, we search for two functions $U : [-1,1] \times [0,T] \to \mathbf{R}$, and $\Gamma : [0,T] \to [0,1]$, coupled by the following set of equations:

(12.1.1) $$\frac{\partial U}{\partial t}(x,t) = \frac{\partial^2 U}{\partial x^2}(x,t), \qquad \forall x \in]-1,\Gamma(t)[, \quad \forall t \in]0,T],$$

(12.1.2) $$\frac{\partial U}{\partial x}(-1,t) = 0, \qquad U(\Gamma(t),t) = 0, \quad \forall t \in]0,T],$$

(12.1.3) $$U(x,t) = 0, \qquad \forall x \in]\Gamma(t),1], \quad \forall t \in]0,T],$$

(12.1.4) $$U(x,0) = -1, \quad \forall x \in [-1,0[, \qquad U(x,0) = 0, \quad \forall x \in [0,1],$$

$$(12.1.5) \qquad \Gamma(0) = 0, \qquad \lambda\,\Gamma'(t) = \frac{\partial U}{\partial x}(\Gamma(t),t), \quad \forall t \in]0,T].$$

In (12.1.5), λ is a positive given constant.

Let us provide a physical interpretation. We assume that the segment $[-1,1]$ represents a one-dimensional bar of some prescribed material. Initially, the interval $[-1,0[$ is in the *solid state*, while the interval $[0,1]$ is in the *liquid state*. After appropriate scaling, the function U represents the temperature of the bar, in such a way that $U = 0$ corresponds to the *solidification temperature* of the liquid. Below this level the material changes its physical state. The initial configuration is expressed by conditions (12.1.4). Then, as the time passes, due to the lower temperature of the solid part, the liquid begins a solidification process and the boundary between the solid and the liquid regions is determined by the function Γ. For any $t \in]0,T]$, in the interval $]-1,\Gamma(t)[$ the temperature is governed by the heat equation (12.1.1) (see also section 10.2). In the remaining portion of the bar, the temperature is assumed to be zero (condition (12.1.3)). Relations (12.1.2) are the boundary conditions of equation (12.1.1). In particular, at the endpoint $x = -1$ the *heat flux* is set to zero. Thus, the bar is *thermally insulated*, i.e., there is no heat tranfer across that point. Finally, the differential relation in (12.1.5) describes the evolution of Γ. This is known as the *Stefan condition* and states that, for any $t \in]0,T]$, the moving point $\Gamma(t)$ has a velocity proportional to the heat flux at $(\Gamma(t),t)$. The parameter λ is the *Stefan constant* and it is proportional to the *latent heat* of the material. For high values of λ solidification is slow, and vice versa. For simplicity, we do not take into account the effects of heat conduction inside the portion of liquid material.

According to the literature, (12.1.1)-(12.1.5) is a *single-phase Stefan problem* and represents an example of *free boundary problem* (see for instance FRIEDMAN(1959), RUBINSTEIN(1971), FASANO and PRIMICERIO(1972), ELLIOTT and OCKENDON(1982), HILL and DEWYNNE(1987), chapter 7). A theoretical analysis shows that U and Γ are uniquely determined and Γ, as expected, is an increasing function.

In addition, one can deduce that

(12.1.6)
$$\frac{d}{dt}\left(\lambda\,\Gamma(t)\;-\;\int_{-1}^{\Gamma(t)} U(x,t)\;dx\right)$$

$$=\;\lambda\Gamma'(t)\;-\;\int_{-1}^{\Gamma(t)}\frac{\partial U}{\partial t}(x,t)\;dx\;-\;\Gamma'(t)\;U(\Gamma(t),t)$$

$$=\;\frac{\partial U}{\partial x}(\Gamma(t),t)\;-\;\int_{-1}^{\Gamma(t)}\frac{\partial^2 U}{\partial x^2}(x,t)\;dx\;=\;\frac{\partial U}{\partial x}(-1,t)\;=\;0,\qquad \forall t\in]0,T].$$

This shows that the quantity $\lambda\Gamma(t)-\int_{-1}^{\Gamma(t)}U(x,t)dx$ is constant for $t\in]0,T]$ (*heat balance condition*).

We now propose an approximation scheme. We use spectral methods for the space variable and finite-differences for the time variable. We divide the interval $[0,T]$ in $m\geq 1$ parts of size $h:=1/m$. Then, the function Γ is approximated at the points $t_k:=kh,\;\;0\leq k\leq m$, by some values $\gamma_k\in\mathbf{R},\;0\leq k\leq m$, to be determined later. We assume that the γ_k's are increasing and $\gamma_0=0$. At the time $t_k,\;0\leq k\leq m$, the function U is approximated in the interval $[-1,\gamma_k]$ by a polynomial $p_n^{(k)}\in\mathbf{P}_n$. On the other hand, according to (12.1.3), the approximating function is equal to zero in the interval $]\gamma_k,1]$. The polynomial $p_n^{(k)},\;0\leq k\leq m$, is defined by its value at the $n+1$ points

(12.1.7)
$$\theta_j^{(n,k)}\;:=\tfrac{1}{2}(\eta_j^{(n)}+1)(1+\gamma_k)-1,\qquad 0\leq j\leq n,\;\;0\leq k\leq m.$$

Here, $\eta_0^{(n)}=-1,\;\eta_n^{(n)}=1$, and $\eta_j^{(n)},\;1\leq j\leq n-1$, are the zeroes of the derivative of the Legendre polynomial of degree n (see section 3.1). In practice, the Legendre Gauss-Lobatto nodes are mapped to the interval $[-1,\gamma_k]$ (see (11.2.3)).
From (3.5.1) with $w\equiv 1$, we get the quadrature formula

(12.1.8)
$$\int_{-1}^{\gamma_k} p\;dx\;=\;\sum_{j=0}^{n}\;p(\theta_j^{(n,k)})\;\tilde{w}_j^{(n,k)}\qquad \forall p\in\mathbf{P}_{2n-1},\;\;0\leq k\leq m,$$

where the new weights are defined by $\tilde{w}_j^{(n,k)} := \frac{1}{2}(1+\gamma_k)\tilde{w}_j^{(n)}$, $0 \le j \le n$, $0 \le k \le m$ (see (3.5.6)).

We are now ready to describe the algorithm. In view of (12.1.4), the first polynomial is determined by the relations

$$p_n^{(0)}(\theta_j^{(n,0)}) := -1, \quad 0 \le j \le n-1, \qquad p_n^{(0)}(\theta_n^{(n,0)}) = 0. \tag{12.1.9}$$

To proceed we need to establish the new position of the free boundary. Following (12.1.5) we define

$$\gamma_{k+1} := \gamma_k + \frac{h}{\lambda}\left[\frac{d}{dx}p_n^{(k)}\right](\gamma_k), \qquad 0 \le k \le m-1. \tag{12.1.10}$$

The next step is to adapt the polynomial $p_n^{(k)}$ to the stretched interval $[-1, \gamma_{k+1}]$. To this purpose we define an auxiliary polynomial $q_n^{(k)} \in \mathbf{P}_n$ satisfying for $0 \le j \le n$

$$\begin{cases} q_n^{(k)}(\theta_j^{(n,k+1)}) = p_n^{(k)}(\theta_j^{(n,k+1)}) & \text{if } \theta_j^{(n,k+1)} \in [-1, \gamma_k[, \\ q_n^{(k)}(\theta_j^{(n,k+1)}) = 0 & \text{if } \theta_j^{(n,k+1)} \in [\gamma_k, \gamma_{k+1}]. \end{cases} \tag{12.1.11}$$

The computation of $q_n^{(k)}$ can be done by interpolation. In fact, in analogy with (3.2.7), we write

$$p_n^{(k)}(x) = \sum_{j=0}^{n} p_n^{(k)}(\theta_j^{(n,k)})\, \tilde{l}_j^{(n,k)}(x), \qquad \forall x \in [-1, \gamma_k], \tag{12.1.12}$$

where $\tilde{l}_j^{(n,k)} \in \mathbf{P}_n$, $0 \le j \le n$, are the Lagrange polynomials with respect to the points $\theta_i^{(n,k)}$, $0 \le i \le n$. It is an easy matter to check that (see also (11.2.16))

$$\tilde{l}_j^{(n,k)}(x) = \tilde{l}_j^{(n)}\left(2\tfrac{x+1}{1+\gamma_k} - 1\right), \qquad \forall x \in [-1, \gamma_k]. \tag{12.1.13}$$

Thus, by virtue of (3.2.8) with $\alpha = \beta = 0$, we can determine the values of $q_n^{(k)}$ at the points $\theta_j^{(n,k+1)}$, $0 \le j \le n$.

Then, we construct the polynomial $p_n^{(k+1)}$ by discretizing the heat equation. Thus, we consider the set of equations

$$(12.1.14)\qquad \begin{cases} p_n^{(k+1)}(\theta_i^{(n,k+1)}) = \left[q_n^{(k)} + h\frac{d^2}{dx^2}p_n^{(k+1)}\right](\theta_i^{(n,k+1)}), \qquad 1 \le i \le n, \\ \\ p_n^{(k+1)}(-1) = \left[q_n^{(k)} + h\frac{d^2}{dx^2}p_n^{(k+1)} + h[\tilde{w}_0^{(n,k+1)}]^{-1}\ \frac{d}{dx}p_n^{(k+1)}\right](-1). \end{cases}$$

Equation (12.1.1) has been collocated at all the ponts $\theta_i^{(n,k+1)}$, $0 \le i \le n$, and for $i = 0$ we imposed the boundary condition as suggested in (9.4.23). This choice will allow us to obtain a discrete counterpart of relation (12.1.6).

Basically, we used the backward Euler method (see (10.6.5)). Hence, the scheme is implicit and $p_n^{(k+1)}$ is recovered by solving a linear system. To determine the corresponding matrix, we first note that, by (12.1.12) and (12.1.13), one has

$$(12.1.15)\qquad \left[\frac{d}{dx}p_n^{(k+1)}\right](\theta_i^{(n,k+1)}) = \sum_{j=0}^{n} p_n^{(k+1)}(\theta_j^{(n,k+1)})\left[\frac{d}{dx}\tilde{l}_j^{(n,k+1)}\right](\theta_i^{(n,k+1)})$$

$$= \frac{2}{1+\gamma_{k+1}}\sum_{j=1}^{n}\tilde{d}_{ij}^{(1)}\ p_n^{(k+1)}(\theta_j^{(n,k+1)}), \qquad 0 \le i \le n,$$

where we use the notations of section 7.2. Therefore, the matrix takes the form

$$I_n - \frac{4h}{(1+\gamma_{k+1})^2}\tilde{D}_n^2 - \frac{2h}{\tilde{w}_0^{(n,k+1)}(1+\gamma_{k+1})}\begin{bmatrix} \tilde{d}_{00}^{(1)} & \cdot & \cdot & \cdot & \tilde{d}_{0n}^{(1)} \\ 0 & \cdot & \cdot & \cdot & 0 \\ \cdot & & & & \cdot \\ \cdot & & & & \cdot \\ 0 & \cdot & \cdot & \cdot & 0 \end{bmatrix},$$

where I_n is the $(n+1)\times(n+1)$ identity matrix and $\tilde{D}_n^2$ is the second derivative matrix at the collocation points (see section 7.2).

Once we obtain $p_n^{(k+1)}$, we can finally update the value of the free boundary location in (12.1.10) with the help of (12.1.15) for $i = n$.

Provided h is sufficiently small, this algoritm gives stable and consistent results. The restriction on h is severe ($h \approx 1/n^4$) due to the jump in the initial condition, but can be relaxed ($h \approx 1/n^2$) when the heat flux at the free boundary starts decreasing in magnitude.

Numerical results for $T = 1$, $\lambda = 5$, $n = 6$, $h = 1/800$, are now presented to illustrate the algorithm. The polynomials $p_n^{(k)}$, $k = 2,\ 10,\ 100,\ 200,\ 400,\ 800$, are plotted in figure 12.1.1. Some terms of the sequence $\{\gamma_k\}_{0\leq k\leq 800}$ are given in figure 12.1.2.

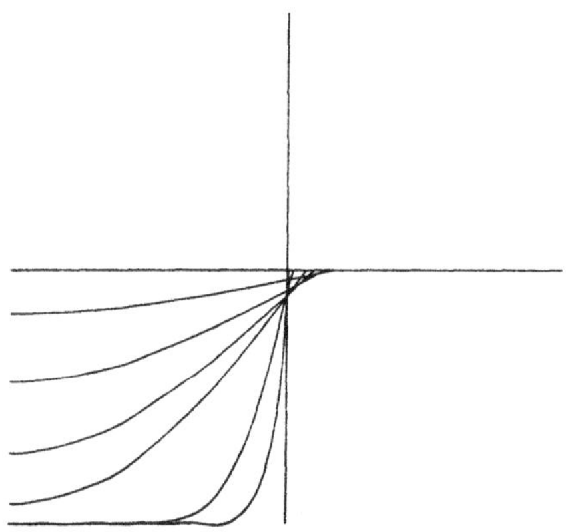

Figure 12.1.1 - Heat propagation for the Stefan problem.

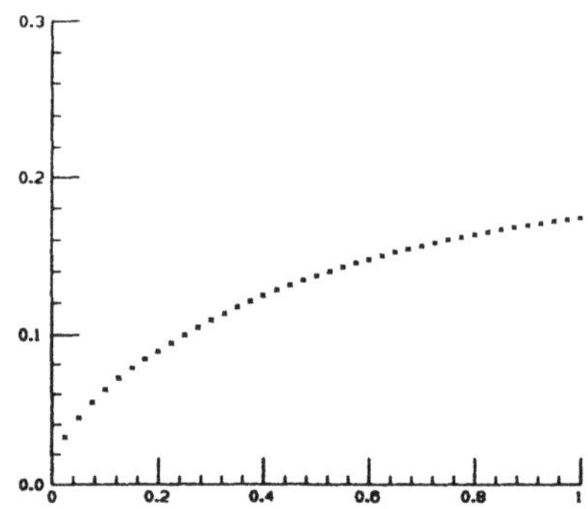

Figure 12.1.2 - Approximation of the free boundary.

As predicted by asympotic estimates we have $\Gamma(t) \approx \sigma\sqrt{t}$, where $\sigma > 0$ is related to λ. As the reader can see in figure 12.1.2, this behavior is maintained by the approximate free boundary.

We conclude by noting that a relation similar to (12.1.6) holds for the discretizing functions. Actually, using (12.1.8), we have

(12.1.16)
$$\frac{d}{dt}\left(\lambda\,\Gamma(t)\ -\ \int_{-1}^{\Gamma(t)} U(x,t)\,dx\right)_{t=t_k}$$

$$\approx \lambda\,\frac{\gamma_{k+1}-\gamma_k}{h}\ -\ \frac{1}{h}\int_{-1}^{\gamma_k}[p_n^{(k)}-q_n^{(k-1)}](x)\,dx$$

$$= \left[\frac{d}{dx}p_n^{(k)}\right](\gamma_k)\ -\ \frac{1}{h}\sum_{j=0}^{n}[p_n^{(k)}-q_n^{(k-1)}](\theta_j^{(n,k)})\,\tilde{w}_j^{(n,k)}$$

$$= \left[\frac{d}{dx}p_n^{(k)}\right](\gamma_k)\ -\ \sum_{j=0}^{n}\left[\frac{d^2}{dx^2}p_n^{(k)}\right](\theta_j^{(n,k)})\,\tilde{w}_j^{(n,k)}\ -\ \left[\frac{d}{dx}p_n^{(k)}\right](-1)$$

$$= \left[\frac{d}{dx}p_n^{(k)}\right](\gamma_k)\ -\ \int_{-1}^{\gamma_k}\left[\frac{d^2}{dx^2}p_n^{(k)}\right](x)\,dx\ -\ \left[\frac{d}{dx}p_n^{(k)}\right](-1)\ =\ 0,\qquad 1\le k\le m-1.$$

A version of the scheme similar to the one proposed here has been tried for the two-phase Stefan problem in RØNQUIST and PATERA(1987). Different discretizations with finite-differences for both the space and the time variable are compared in FURZELAND(1980). For a survey of numerical methods for free boundary problems we refer the reader to NOCHETTO(1990).

12.2 An example in an unbounded domain

We study in this section the approximation of a second-order differential equation defined in $I\equiv]0,+\infty[$. Let $f:I\rightarrow\mathbf{R}$ be a given function. We are concerned with finding the solution $U:\bar{I}\rightarrow\mathbf{R}$ to the problem

(12.2.1)
$$\begin{cases} -(e^x\,U'(x))' \ =\ f(x) & x\in I,\\ U(0)\ =\ 0, \quad \lim_{x\to+\infty}U(x)\ =\ 0. \end{cases}$$

We assume for example that f is such that the unique solution of (12.2.1) is $U(x) = e^{-x}\sin\frac{7x}{1+x^2}$, $x\in\bar{I}$, which is shown in figure 12.2.1 for $x\in[0,4]$.

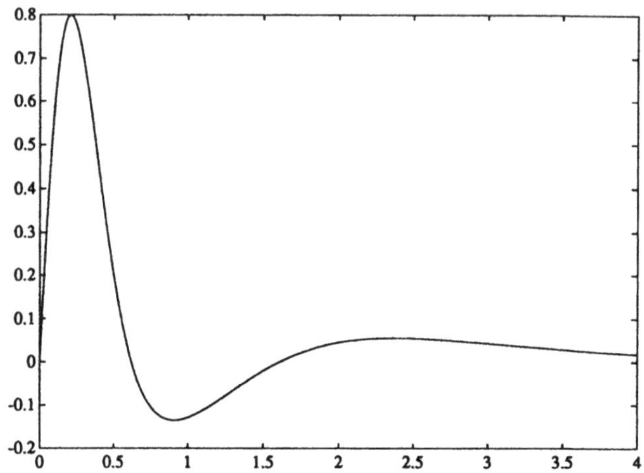

Figure 12.2.1 - The function $U(x) = e^{-x} \sin \frac{7x}{1+x^2}$, $x \in [0,4]$.

In our first experiment, we discretize the equation in a bounded subset of I, i.e., the interval $]0,2[$. For any $n \geq 2$, we use the collocation method at the nodes $\hat{\eta}_j^{(n)} := \eta_j^{(n)} + 1$, $0 \leq j \leq n$, obtained by shifting the Chebyshev Gauss-Lobatto points in $[-1,1]$ (see (3.1.11)). Thus, we want to find a polynomial $p_n \in \mathbf{P}_n$ such that

$$(12.2.2) \qquad \begin{cases} -p_n''(\hat{\eta}_i^{(n)}) \; - \; p_n'(\hat{\eta}_i^{(n)}) \; = \; \exp(-\hat{\eta}_i^{(n)}) \; f(\hat{\eta}_i^{(n)}), \qquad 1 \leq i \leq n-1, \\ p_n(0) \; = \; 0, \qquad\qquad p_n(2) \; = \; 0. \end{cases}$$

To determine a unique solution of (12.2.2), we force the homogeneous Dirichlet boundary condition at the point $x = 2$. This condition is suggested by the behavior prescribed for U at infinity. We note that $U(2) \approx -.04533$, so that the limit $\lim_{n \to +\infty} p_n$ does not coincide with the function U. We examine for instance the approximation of the quantity $U'(0) = 7$. We give in table 12.2.1 the error $E_n := |U'(0) - p_n'(0)|$, $n \geq 2$, for different n. As expected E_n does not decay to zero, but approaches a value of the same order of magnitude of the error $|U(2) - p_n(2)|$. Of course, we can improve on these results by increasing the size of the computational domain.

n	E_n
4	2.88462
8	.164540
12	.087322
16	.049646
20	.052595

Table 12.2.1 - Errors for the approximation of problem (12.2.1) by the scheme (12.2.2).

Another approach is to use Laguerre functions (see section 6.7) for obtaining a global approximation of U in the whole domain I (see section 9.5). It is recommended not to use high degree Laguerre polynomials, due to the effect of rounding errors in the computations. Thus, we adopt a non-overlapping multidomain method (see section 11.2). For any $n \geq 2$ and $m \geq 1$, we approximate the solution of (12.2.1) in $S_1 :=]0, 2[$ by a polynomial $p_n \in \mathbf{P}_n$, and in $S_2 :=]2, +\infty[$ by a Laguerre function $Q_m \in \mathbf{S}_{m-1}$. At the interface point $x = 2$ we assume the continuity of the approximating function and its derivative. Then, we define the nodes $\tilde{\eta}_j^{(m)} := \eta_j^{(m)} + 2$, $0 \leq j \leq m-1$, where $\eta_0^{(m)} := 0$ and $\eta_j^{(m)}$, $1 \leq j \leq m-1$, are the zeroes of $\frac{d}{dx} L_m^{(\alpha)}$ for $\alpha = 0$. The following collocation scheme is considered:

$$
(12.2.3) \qquad \begin{cases} -p_n''(\hat{\eta}_i^{(n)}) \ - \ p_n'(\hat{\eta}_i^{(n)}) \ = \ \exp(-\hat{\eta}_i^{(n)}) \ f(\hat{\eta}_i^{(n)}), & 1 \leq i \leq n-1, \\ p_n(0) \ = \ 0, \quad p_n(2) \ = \ Q_m(2), \quad p_n'(2) \ = \ Q_m'(2), & \\ -Q_m''(\tilde{\eta}_i^{(m)}) \ - \ Q_m'(\tilde{\eta}_i^{(m)}) \ = \ \exp(-\tilde{\eta}_i^{(m)}) \ f(\tilde{\eta}_i^{(m)}), & 1 \leq i \leq m-1. \end{cases}
$$

The substitution $q_m(x) := Q_m(x) e^{x-2}$, $\forall x \in \bar{S}_2$, leads to $q_m \in \mathbf{P}_{m-1}$, and (12.2.3) becomes

$$
(12.2.4) \qquad \begin{cases} -p_n''(\hat{\eta}_i^{(n)}) \ - \ p_n'(\hat{\eta}_i^{(n)}) \ = \ \exp(-\hat{\eta}_i^{(n)}) \ f(\hat{\eta}_i^{(n)}), & 1 \leq i \leq n-1, \\ p_n(0) \ = \ 0, \quad p_n(2) \ = \ q_m(2), \quad p_n'(2) \ = \ q_m'(2) - q_m(2), & \\ -q_m''(\tilde{\eta}_i^{(m)}) \ + \ q_m'(\tilde{\eta}_i^{(m)}) \ = \ e^{-2} \ f(\tilde{\eta}_i^{(m)}), & 1 \leq i \leq m-1. \end{cases}
$$

Now the error $E_{n,m} := |U'(0) - p'_n(0)|$, depends also on the discretization parameter m. One can prove that $\lim_{n,m\to+\infty} E_{n,m} = 0$. We give in table 12.2.2 some computed values. As the reader will note, with nearly the same number of collocation nodes as the previous algorithm, one may obtain better results.

n	$E_{n,3}$	$E_{n,4}$	$E_{n,5}$
8	.087622	.115119	.126074
12	.057295	.029799	.018844
16	.031526	.004030	.006925

Table 12.2.2 - Errors for the approximation of problem (12.2.1) by the scheme (12.2.4).

12.3 The nonlinear Schrödinger equation

Many phenomena in plasma physics, quantum physics and optics are described by the *nonlinear Schrödinger equation*, which is introduced for instance in TRULLINGER, ZAKHAROV and POKROVSKY(1986). After dimensional scaling the equation takes the form

$$(12.3.1)\qquad \mathbf{i}\,\frac{\partial U}{\partial t}(x,t) \;+\; \zeta\,\frac{\partial^2 U}{\partial x^2}(x,t) \;+\; \epsilon\,|U(x,t)|^2 U(x,t) \;=\; 0 \quad x \in I,\ t \in]0,T],$$

where $\mathbf{i} = \sqrt{-1}$ is the complex unity, $\zeta > 0$ is proportional to the so called *dispersion parameter*, $\epsilon \in \mathbf{R}$ and $T > 0$. The unknown $U : I \times [0,T] \to \mathbf{C}$ is a function assuming complex values. Generally we have $I \equiv \mathbf{R}$ and U is required to decay to zero at infinity. For simplicity, we take $I =]-1,1[$ and impose the homogeneous boundary conditions $U(\pm 1,t)=0,\ t\in[0,T]$. Finally, an initial condition $U(x,t)=U_0(x),\ x\in\bar{I}$, is provided. A conservation property is immediately obtained by noting that

$$(12.3.2)\qquad \frac{d}{dt}\int_I |U|^2(x,t)\,dx \;=\; \int_I \left(U\frac{\partial \bar{U}}{\partial t}+\bar{U}\frac{\partial U}{\partial t}\right)(x,t)\,dx$$

$$= \mathbf{i}\int_I \left(\zeta\, U\frac{\partial^2 \bar{U}}{\partial x^2}+\epsilon\,|U|^4-\zeta\,\bar{U}\frac{\partial^2 U}{\partial x^2}-\epsilon\,|U|^4\right)(x,t)\,dx \;=\; 0, \qquad \forall t\in]0,T],$$

where the overbar denotes the complex conjugate and we used integration by parts. This means that the quantity $\int_I |U|^2(x,t)dx$ is constant with respect to t. Another quantity which does not vary with respect to t is $\int_I\left(\zeta\left|\frac{\partial U}{\partial x}\right|^2-\frac{\epsilon}{2}|U|^4\right)(x,t)dx$.

We denote by V and W the real and imaginary parts of U respectively, hence $U=V+\mathbf{i}W$. Thus, we rewrite (12.3.1) as a system

$$(12.3.3)\qquad \begin{cases} -\dfrac{\partial W}{\partial t}(x,t) = -\zeta\,\dfrac{\partial^2 V}{\partial x^2}(x,t)-\epsilon\,|U(x,t)|^2 V(x,t) \\[2ex] \dfrac{\partial V}{\partial t}(x,t) = -\zeta\,\dfrac{\partial^2 W}{\partial x^2}(x,t)-\epsilon\,|U(x,t)|^2 W(x,t) \end{cases} \qquad x\in I,\ \ t\in]0,T],$$

with

$$(12.3.4)\qquad V(\pm 1,t) \;=\; W(\pm 1,t) \;=\; 0, \qquad t\in[0,T].$$

The initial conditions are $V(x,0)=V_0(x),\ W(x,0)=W_0(x),\ x\in\bar{I},\ \ U_0=V_0+\mathbf{i}W_0$.

For the approximation we use the collocation method based on the Legendre Gauss-Lobatto points for the variable x, and an implicit finite-difference scheme for the variable t (see section 10.6). The initial polynomials in $\mathbf{P}_n^0$, $n\geq 2$ (see (6.4.11)) are such that $p_n^{(0)} := \tilde{I}_{w,n}V_0$ and $q_n^{(0)} := \tilde{I}_{w,n}W_0$, where the interpolation operator is defined in section 3.3 ($w\equiv 1$). We subdivide the interval $[0,T]$ in $m\geq 1$ equal parts of size $h:=T/m$. The successive polynomials $p_n^{(k)},\ q_n^{(k)}\in\mathbf{P}_n^0,\ 1\leq k\leq m$, are determined at the nodes $\eta_i^{(n)},\ 1\leq i\leq n-1$, by the formulas

$$(12.3.5)\quad \begin{cases} \dfrac{q_n^{(k)} - q_n^{(k-1)}}{h} = -\dfrac{\zeta}{2}[p_n^{(k)} + p_n^{(k-1)}]'' - \dfrac{\epsilon}{4}[\Theta_n^{(k)} + \Theta_n^{(k-1)}][p_n^{(k)} + p_n^{(k-1)}] \\ \\ \dfrac{p_n^{(k)} - p_n^{(k-1)}}{h} = -\dfrac{\zeta}{2}[q_n^{(k)} + q_n^{(k-1)}]'' - \dfrac{\epsilon}{4}[\Theta_n^{(k)} + \Theta_n^{(k-1)}][q_n^{(k)} + q_n^{(k-1)}] \end{cases}$$

where $\Theta_n^{(k)} := [p_n^{(k)}]^2 + [q_n^{(k)}]^2$, $0 \le k \le m$. The polynomials $p_n^{(m)}$ and $q_n^{(m)}$ are expected to be the approximations of $V(\cdot, T)$ and $W(\cdot, T)$ respectively. Using the notations of section 9.9, (12.3.5) is equivalent to

$$(12.3.6)\quad \begin{bmatrix} M_n & \mu I_n \\ -\mu I_n & M_n \end{bmatrix} \begin{bmatrix} \bar{p}_n^{(k)} \\ \bar{q}_n^{(k)} \end{bmatrix} = -\begin{bmatrix} M_n & -\mu I_n \\ \mu I_n & M_n \end{bmatrix} \begin{bmatrix} \bar{p}_n^{(k-1)} \\ \bar{q}_n^{(k-1)} \end{bmatrix}$$

$$+ \frac{\epsilon}{2\zeta} \begin{bmatrix} \Delta_n^{(k)} & 0 \\ 0 & \Delta_n^{(k)} \end{bmatrix} \begin{bmatrix} \bar{p}_n^{(k)} + \bar{p}_n^{(k-1)} \\ \bar{q}_n^{(k)} + \bar{q}_n^{(k-1)} \end{bmatrix} \qquad 1 \le k \le m,$$

where $\Delta_n^{(k)} := \mathrm{diag}\{(\Theta_n^{(k)} + \Theta_n^{(k-1)})(\eta_1^{(n)}), \cdots, (\Theta_n^{(k)} + \Theta_n^{(k-1)})(\eta_{n-1}^{(n)})\}$ and $\mu := 2/\zeta h$. The inverse of the matrix on the left-hand side of (12.3.6) can be evaluated once and for all with the help of (9.9.4).

For any $1 \le k \le m$, we can solve the nonlinear equation (12.3.6) iteratively. One approach is to construct a sequence of vectors according to the prescription

$$(12.3.7)\quad \begin{bmatrix} \bar{p}_n^{(k,j)} \\ \bar{q}_n^{(k,j)} \end{bmatrix} := \begin{bmatrix} M_n & \mu I_n \\ -\mu I_n & M_n \end{bmatrix}^{-1} \left\{ -\begin{bmatrix} M_n & -\mu I_n \\ \mu I_n & M_n \end{bmatrix} \begin{bmatrix} \bar{p}_n^{(k-1)} \\ \bar{q}_n^{(k-1)} \end{bmatrix} \right.$$

$$\left. + \frac{\epsilon}{2\zeta} \begin{bmatrix} \Delta_n^{(k,j-1)} & 0 \\ 0 & \Delta_n^{(k,j-1)} \end{bmatrix} \begin{bmatrix} \bar{p}_n^{(k,j-1)} + \bar{p}_n^{(k-1)} \\ \bar{q}_n^{(k,j-1)} + \bar{q}_n^{(k-1)} \end{bmatrix} \right\} \qquad j \ge 1,$$

where $\Delta_n^{(k,j)} := \mathrm{diag}\{(\Theta_n^{(k,j)} + \Theta_n^{(k-1)})(\eta_1^{(n)}), \cdots, (\Theta_n^{(k,j)} + \Theta_n^{(k-1)})(\eta_{n-1}^{(n)})\}$ with $\Theta_n^{(k,j)} := [p_n^{(k,j)}]^2 + [q_n^{(k,j)}]^2$, $j \in \mathbf{N}$, and $\bar{p}_n^{(k,0)} := \bar{p}_n^{(k-1)}$, $\bar{q}_n^{(k,0)} := \bar{q}_n^{(k-1)}$.

Therefore, one obtains

$$\lim_{j\to+\infty}\begin{bmatrix}\bar{p}_n^{(k,j)}\\ \\ \bar{q}_n^{(k,j)}\end{bmatrix}=\begin{bmatrix}\bar{p}_n^{(k)}\\ \\ \bar{q}_n^{(k)}\end{bmatrix},\qquad 1\le k\le m. \tag{12.3.8}$$

The scheme considered here was suggested in DELFOUR, FORTIN and PAYRE(1981) for finite-difference approximations. It is second-order accurate, i.e., the error in the variable t decays as h^2. In addition, it preserves certain physical quantities. For example, using formula (3.5.1) for $w \equiv 1$, we can prove the discrete counterpart of (12.3.2), i.e., that the sum $\sum_{i=0}^{n}\Theta_n^{(k)}(\eta_i^{(n)})\,\tilde{w}_i^{(n)}$ does not change with k.

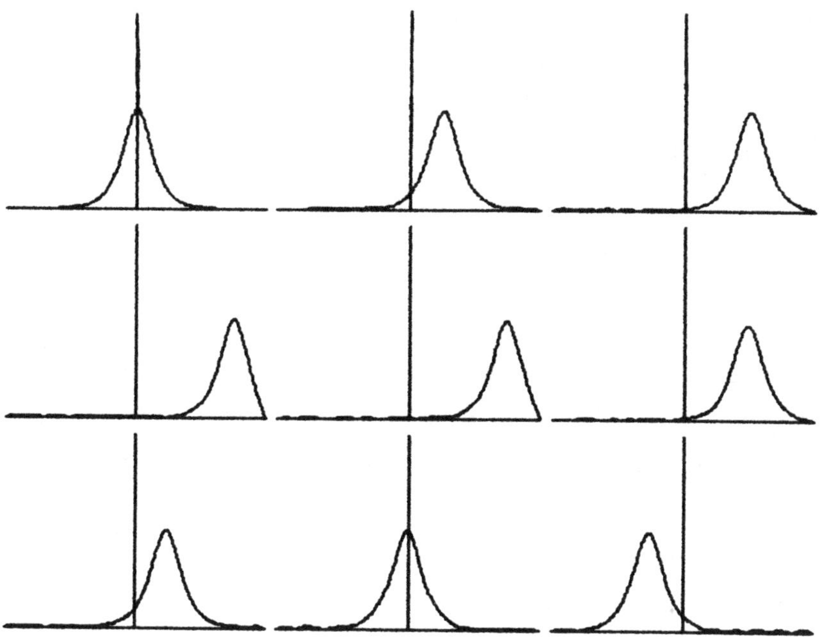

Figure 12.3.1 - Evolution of an approximate solution of equation (12.3.1).

We give results from a numerical experiment corresponding to the following data: $\zeta = 1$, $\epsilon = 200$, $T = .32$, $n = 40$, $m = 64$. The initial guess is $U_0(x) := \exp(\mathbf{i}\pi x) \big/ \cosh\left(\sqrt{\frac{\epsilon}{2\zeta}}x\right)$, $x \in I$, $U_0(\pm 1) = 0$. The function $\sqrt{\Theta_n^{(k)}}$, plotted in figure 12.3.1 for $k = 0, 8, 16, 24, 32, 40, 48, 56, 64$, approximates a travelling *soliton*.

Approximations of the nonlinear Schrödinger equation in $\mathbf{R}$, by a non-overlapping multidomain method using Laguerre functions as in the previous section, are presented in DE VERONICO(1991), and DE VERONICO, FUNARO and REALI(1991).

12.4 Zeroes of Bessel functions

Bessel functions frequently arise when solving boundary-value partial differential equations by the method of separation of variables. For the sake of simplicity, we only consider Bessel functions J_k with a positive integer index $k \geq 1$. These are determined, up to a multiplicative constant, by the Sturm-Liouville problem (see section 1.1):

$$(12.4.1) \qquad \begin{cases} -(x\ J_k'(x))' - \left(x - \dfrac{k^2}{x}\right) J_k(x) = 0, & x > 0, \\ J_k(0) = 0. \end{cases}$$

Many properties are known for this family of functions (they can be found, for example, in WATSON(1966)). In particular, we have $|J_k(x)| \leq 1/\sqrt{2}$, $k \geq 1$, $x > 0$.

An interesting issue is to evaluate the zeroes of Bessel functions. It turns out that J_k, $k \geq 1$, has an infinite sequence of positive real zeroes $z_j^{(k)}$, $j \geq 1$, that we assume to be in increasing order. We describe here one of the techniques available for the computation of the $z_j^{(k)}$'s. We take for instance $k = 1$. For $\lambda > 0$, we define $U(x) := J_1(\sqrt{\lambda}x)$, $x > 0$. After substitution in (12.4.1), we find the differential equation

$$(12.4.2) \qquad \begin{cases} -\ U''(x) - \dfrac{1}{x}U'(x) + \dfrac{1}{x^2}U(x) = \lambda\ U(x), & x > 0, \\ U(0) = 0. \end{cases}$$

By imposing the condition $U(1) = 0$, it is possible to show that the eigenvalue problem (12.4.2) has a countable number of positive real eigenvalues λ_m, $m \geq 1$. Therefore, it is clear that $z_j^{(1)} = \sqrt{\lambda_j}$, $j \geq 1$. Our goal is to find an approximation to these eigenvalues. According to the results of section 8.6, for $n \geq 2$, we discretize (12.4.2) by the eigenvalue problem

$$(12.4.3) \quad \begin{cases} -p''_{n,m}(\tilde{\eta}_i^{(n)}) - [\tilde{\eta}_i^{(n)}]^{-1} p'_{n,m}(\tilde{\eta}_i^{(n)}) \\ \qquad + [\tilde{\eta}_i^{(n)}]^{-2} p_{n,m}(\tilde{\eta}_i^{(n)}) = \lambda_{n,m}\, p_{n,m}(\tilde{\eta}_i^{(n)}) \quad 1 \leq i \leq n-1, \\ \\ p_{n,m}(0) = \lambda_{n,m}\, p_{n,m}(0), \qquad p_{n,m}(1) = \lambda_{n,m}\, p_{n,m}(1), \end{cases}$$

$$0 \leq m \leq n,$$

where $p_{n,m} \in \mathbf{P}_n$, $0 \leq m \leq n$, and the collocation points are recovered from the Chebyshev Gauss-Lobatto nodes by setting $\tilde{\eta}_i^{(n)} := \frac{1}{2}(\eta_i^{(n)}+1)$, $0 \leq i \leq n$ (see (3.1.11)). The $(n+1) \times (n+1)$ matrix relative to (12.4.3) can be written in the standard way (see chapter seven). Two of the corresponding $n+1$ eigenvalues $\lambda_{n,m}$, $0 \leq m \leq n$, are equal to 1 (say $\lambda_{n,0} = \lambda_{n,n} = 1$). The others approach the exact eigenvalues λ_m, $m \geq 1$. In table 12.4.1, we report the zeroes $z_m^{(1)}$ and the computed eigenvalues for $1 \leq m \leq 5$ (the first five zeroes of J_1 are tabulated in LUKE(1969), Vol.2, p.233). In figure 12.4.1, we plot J_1 in the window $[0, 18] \times [-1, 1]$. An approximation of the zeroes of J_k, $k \geq 2$, is obtained with the same arguments.

m	$z_m^{(1)}$	$\sqrt{\lambda_{8,m}}$	$\sqrt{\lambda_{12,m}}$	$\sqrt{\lambda_{16,m}}$
1	3.831705970207	3.831705650608	3.831705970200	3.831705970207
2	7.015586669815	7.015773798574	7.015586712768	7.015586669820
3	10.17346813506	10.16416616370	10.17347111164	10.17346813668
4	13.32369193631	13.50667201292	13.32368681006	13.32369172635
5	16.47063005087	15.87375296304	16.46311800444	16.47062464105

Table 12.4.1 - Approximation of the zeroes of J_1.

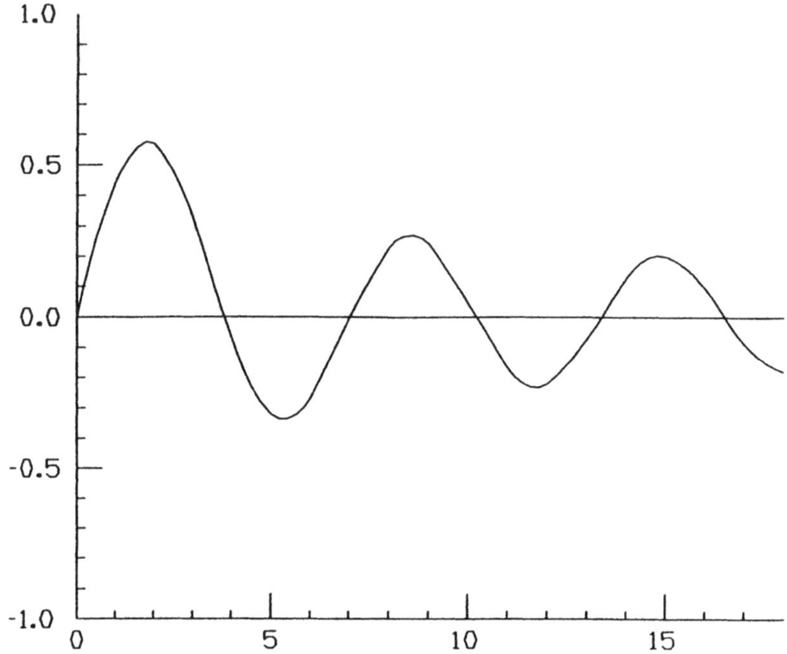

Figure 12.4.1 - Plot of $J_1(x)$, $x \in [0, 18]$.

The proof of the orthogonality relation (which somehow justifies the interest for the Bessel zeroes):

$$\int_0^1 x\ J_k(z_i^{(k)}x)\ J_k(z_j^{(k)}x)\ dx \ = \ 0, \quad \text{if } i \neq j, \quad k \geq 1, \tag{12.4.4}$$

is left to the reader.

13 AN EXAMPLE IN TWO DIMENSIONS

To illustrate the construction of polynomial approximations of the solution of a partial differential equation, we examine a simple boundary-value problem in two dimensions.

13.1 Poisson's equation

Let Ω be the open square $]-1,1[\times]-1,1[$ in $\mathbf{R}^2$. We denote by $\partial\Omega$ the boundary of Ω. Given the function $f:\Omega\to\mathbf{R}$, we want to find the solution $U:\bar{\Omega}\to\mathbf{R}$ of the boundary-value problem

$$(13.1.1)\qquad \begin{cases} -\Delta U := -\left(\dfrac{\partial^2 U}{\partial x^2}+\dfrac{\partial^2 U}{\partial y^2}\right) = f & \text{in } \Omega. \\ U \equiv 0 \qquad \text{on } \partial\Omega, \end{cases}$$

The partial differential equation in (13.1.1) is known as *Poisson's equation*. The symbol Δ is called the *Laplace operator*. There are many applications of equation (13.1.1) in the field of physics. For example, if Ω is a plane region of *dielectric* material surrounded by a grounded *conductor*, and f is the *charge density* in Ω, then the solution U of (13.1.1) is proportional to the *electrostatic potential* on Ω.

Numerous theoretical results pertaining to (13.1.1) are available. For a preliminary introduction, we mention for instance COURANT and HILBERT(1953), SNEDDON(1957), SOBOLEV(1964), WEINBERGER(1965). Here, we only state the following theorem.

Theorem 13.1.1 - *Let f be a continuous function in Ω satisfying $\int_\Omega f^2 dxdy < +\infty$, then there exists a unique solution $U \in C^0(\bar{\Omega}) \cap C^1(\Omega)$ of* (13.1.1).

■

The aim of the following sections is to suggest a technique for approximating the solution of Poisson's equation by means of algebraic polynomials in two variables.

13.2 Approximation by the collocation method

In the following, for any $n \in \mathbf{N}$, $\mathbf{P}_n^\star$ denotes the space of polynomials in two variables which degree is less or equal to n for each variable. Thus, the dimension of the linear space $\mathbf{P}_n^\star$ is $(n+1)^2$. For $n \geq 2$, we also define

$$\text{(13.2.1)} \qquad \mathbf{P}_n^{\star,0} := \Big\{ p \in \mathbf{P}_n^\star \Big| \ p \equiv 0 \ \text{ on } \partial\Omega \Big\}.$$

The dimension of $\mathbf{P}_n^{\star,0}$ is $(n-1)^2$.

We analyze the approximation of problem (13.1.1) by the collocation method. For any $n \geq 2$, let $\eta_j^{(n)}$, $0 \leq j \leq n$, be the nodes in $[-1,1]$ associated to the Jacobi Gauss-Lobatto formula (3.5.1). Then, in $\bar{\Omega}$ we consider the set of points $\aleph_n := \bigcup_{\substack{0\leq i\leq n \\ 0\leq j\leq n}} \{(\eta_i^{(n)}, \eta_j^{(n)})\}$. These nodes are displayed in figure 13.2.1 for $n = 8$ in the case $\alpha = \beta = -\frac{1}{2}$. We note that, if a polynomial in $\mathbf{P}_n^\star$ vanishes on $\aleph_n \cap \partial\Omega$, then it belongs to $\mathbf{P}_n^{\star,0}$. Moreover, any polynomial in $\mathbf{P}_n^{\star,0}$ is uniquely determined by the values attained at the points in $\aleph_n \cap \Omega$.

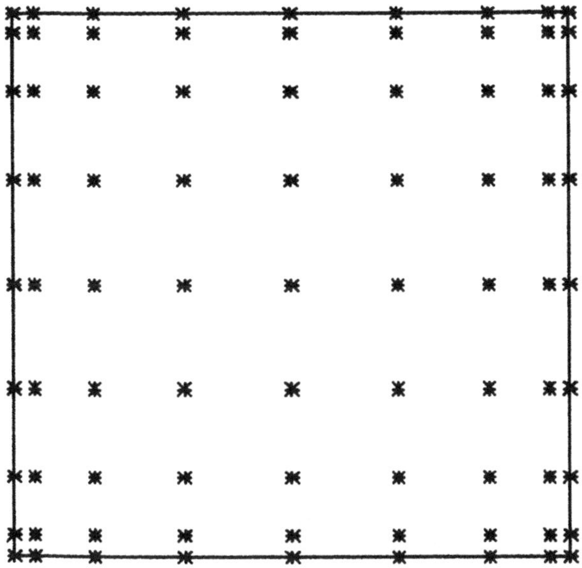

Figure 13.2.1 - The set $\aleph_8 \subset \bar{\Omega}$ for $\alpha = \beta = -\frac{1}{2}$.

Thus, we are concerned with finding a polynomial $p_n \in \mathbf{P}_n^{\star,0}$ such that

(13.2.2) $$-\Delta p_n(\eta_i^{(n)}, \eta_j^{(n)}) = f(\eta_i^{(n)}, \eta_j^{(n)}), \qquad 1 \le i \le n-1, \ 1 \le j \le n-1.$$

The above set of equations is equivalent to a $(n-1)^2 \times (n-1)^2$ linear system. We examine for simplicity the case $n = 4$, although the generalization to other values of n is straightforward. Let us define the vectors

$$\bar{p}_n \equiv \Big(p_n(\eta_1^{(n)}, \eta_1^{(n)}), p_n(\eta_2^{(n)}, \eta_1^{(n)}), p_n(\eta_3^{(n)}, \eta_1^{(n)}), p_n(\eta_1^{(n)}, \eta_2^{(n)}), p_n(\eta_2^{(n)}, \eta_2^{(n)}),$$
$$p_n(\eta_3^{(n)}, \eta_2^{(n)}), p_n(\eta_1^{(n)}, \eta_3^{(n)}), p_n(\eta_2^{(n)}, \eta_3^{(n)}), p_n(\eta_3^{(n)}, \eta_3^{(n)})\Big),$$

$$\bar{f}_n \equiv \Big(f(\eta_1^{(n)}, \eta_1^{(n)}), f(\eta_2^{(n)}, \eta_1^{(n)}), f(\eta_3^{(n)}, \eta_1^{(n)}), f(\eta_1^{(n)}, \eta_2^{(n)}), f(\eta_2^{(n)}, \eta_2^{(n)}),$$
$$f(\eta_3^{(n)}, \eta_2^{(n)}), f(\eta_1^{(n)}, \eta_3^{(n)}), f(\eta_2^{(n)}, \eta_3^{(n)}), f(\eta_3^{(n)}, \eta_3^{(n)})\Big).$$

Then, we can write

$$-D_n^\star\, \bar{p}_n \;=\; \bar{f}_n, \tag{13.2.3}$$

where, according to the notations of section 7.2, the matrix $D_n^\star$ for $n = 4$ turns out to be

$$\begin{bmatrix}
2\tilde{d}^{(2)}_{11} & \tilde{d}^{(2)}_{12} & \tilde{d}^{(2)}_{13} & \tilde{d}^{(2)}_{12} & 0 & 0 & \tilde{d}^{(2)}_{13} & 0 & 0 \\
\tilde{d}^{(2)}_{21} & \tilde{d}^{(2)}_{11}+\tilde{d}^{(2)}_{22} & \tilde{d}^{(2)}_{23} & 0 & \tilde{d}^{(2)}_{12} & 0 & 0 & \tilde{d}^{(2)}_{13} & 0 \\
\tilde{d}^{(2)}_{31} & \tilde{d}^{(2)}_{32} & \tilde{d}^{(2)}_{11}+\tilde{d}^{(2)}_{33} & 0 & 0 & \tilde{d}^{(2)}_{12} & 0 & 0 & \tilde{d}^{(2)}_{13} \\
\tilde{d}^{(2)}_{21} & 0 & 0 & \tilde{d}^{(2)}_{11}+\tilde{d}^{(2)}_{22} & \tilde{d}^{(2)}_{12} & \tilde{d}^{(2)}_{13} & \tilde{d}^{(2)}_{23} & 0 & 0 \\
0 & \tilde{d}^{(2)}_{21} & 0 & \tilde{d}^{(2)}_{21} & 2\tilde{d}^{(2)}_{22} & \tilde{d}^{(2)}_{23} & 0 & \tilde{d}^{(2)}_{23} & 0 \\
0 & 0 & \tilde{d}^{(2)}_{21} & \tilde{d}^{(2)}_{31} & \tilde{d}^{(2)}_{32} & \tilde{d}^{(2)}_{22}+\tilde{d}^{(2)}_{33} & 0 & 0 & \tilde{d}^{(2)}_{23} \\
\tilde{d}^{(2)}_{31} & 0 & 0 & \tilde{d}^{(2)}_{32} & 0 & 0 & \tilde{d}^{(2)}_{11}+\tilde{d}^{(2)}_{33} & \tilde{d}^{(2)}_{12} & \tilde{d}^{(2)}_{13} \\
0 & \tilde{d}^{(2)}_{31} & 0 & 0 & \tilde{d}^{(2)}_{32} & 0 & \tilde{d}^{(2)}_{21} & \tilde{d}^{(2)}_{22}+\tilde{d}^{(2)}_{33} & \tilde{d}^{(2)}_{23} \\
0 & 0 & \tilde{d}^{(2)}_{31} & 0 & 0 & \tilde{d}^{(2)}_{32} & \tilde{d}^{(2)}_{31} & \tilde{d}^{(2)}_{32} & 2\tilde{d}^{(2)}_{33}
\end{bmatrix}$$

Such a matrix can be decomposed as $D_n^\star \;=\; \mathcal{A}_n + \mathcal{K}_n \mathcal{A}_n \mathcal{K}_n$, where $\mathcal{A}_n$ $(n = 4)$ is the block diagonal matrix

$$\mathcal{A}_n := \begin{bmatrix} \hat{\mathcal{A}}_n & 0 & 0 \\ 0 & \hat{\mathcal{A}}_n & 0 \\ 0 & 0 & \hat{\mathcal{A}}_n \end{bmatrix}, \quad \text{with} \quad \hat{\mathcal{A}}_n := \begin{bmatrix} \tilde{d}^{(2)}_{11} & \tilde{d}^{(2)}_{12} & \tilde{d}^{(2)}_{13} \\ \tilde{d}^{(2)}_{21} & \tilde{d}^{(2)}_{22} & \tilde{d}^{(2)}_{23} \\ \tilde{d}^{(2)}_{31} & \tilde{d}^{(2)}_{32} & \tilde{d}^{(2)}_{33} \end{bmatrix}, \tag{13.2.4}$$

and $\mathcal{K}_n$ $(n = 4)$ is the *permutation* matrix (note that $\mathcal{K}_n = \mathcal{K}_n^{-1}$):

$$(13.2.5) \qquad \mathcal{K}_n := \begin{bmatrix} 1 & 0 & 0 & 0 & 0 & 0 & 0 & 0 & 0 \\ 0 & 0 & 0 & 1 & 0 & 0 & 0 & 0 & 0 \\ 0 & 0 & 0 & 0 & 0 & 0 & 1 & 0 & 0 \\ 0 & 1 & 0 & 0 & 0 & 0 & 0 & 0 & 0 \\ 0 & 0 & 0 & 0 & 1 & 0 & 0 & 0 & 0 \\ 0 & 0 & 0 & 0 & 0 & 0 & 0 & 1 & 0 \\ 0 & 0 & 1 & 0 & 0 & 0 & 0 & 0 & 0 \\ 0 & 0 & 0 & 0 & 0 & 1 & 0 & 0 & 0 \\ 0 & 0 & 0 & 0 & 0 & 0 & 0 & 0 & 1 \end{bmatrix}.$$

Actually, for any $1 \leq i \leq n-1$ and $1 \leq j \leq n-1$, the transformation associated with $\mathcal{K}_n$ maps the point $(\eta_i^{(n)}, \eta_j^{(n)})$ into the point $(\eta_j^{(n)}, \eta_i^{(n)})$. Therefore, the matrix-vector multiplication $\bar{p}_n \to \mathcal{A}_n \bar{p}_n$ is equivalent to performing $n-1$ second-order partial derivatives with respect to the variable x, and the multiplication $\bar{p}_n \to \mathcal{K}_n \mathcal{A}_n \mathcal{K}_n \bar{p}_n$ corresponds to $n-1$ second-order partial derivatives with respect to the variable y. The sum of the two contributions gives a discretization of the Laplace operator.
We observe that, in the construction of the matrix $D_n^\star$, we implicitly assumed that the unknown polynomial p_n vanishes on $\partial\Omega$. This condition is obtained by eliminating all the entries corresponding to the boundary nodes. The same trick was applied in (7.4.11) for $n = 3$ in the one-dimensional case.

A theoretical analysis of the convergence of p_n to U as $n \to +\infty$ can be developed. We outline the proof in the Legendre case ($\alpha = \beta = 0$). We start by recasting (13.1.1) in a weak form (see section 9.3), which is obtained by multiplying the differential equation by a test function ϕ such that $\phi \equiv 0$ on $\partial\Omega$, and integrating on Ω. With the help of *Green's formula* one gets

$$(13.2.6) \qquad \int_\Omega \left[\frac{\partial U}{\partial x}\frac{\partial \phi}{\partial x} + \frac{\partial U}{\partial y}\frac{\partial \phi}{\partial y} \right] dxdy = \int_\Omega f\phi \, dxdy =: F(\phi).$$

Similarly, problem (13.2.2) can be formulated in a variational setting. In fact, by formula (3.5.1) with $w \equiv 1$, integration by parts yields

(13.2.7) $$\sum_{\substack{i=0\\j=0}}^{n} \left[\frac{\partial p_n}{\partial x}\frac{\partial \phi}{\partial x} + \frac{\partial p_n}{\partial y}\frac{\partial \phi}{\partial y}\right](\eta_i^{(n)}, \eta_j^{(n)})\ \tilde{w}_i^{(n)}\tilde{w}_j^{(n)}$$

$$= \sum_{j=0}^{n}\left(\int_{-1}^{1}\left[\frac{\partial p_n}{\partial x}\frac{\partial \phi}{\partial x}\right](x, \eta_j^{(n)})\ dx\right)\tilde{w}_j^{(n)} + \sum_{i=0}^{n}\left(\int_{-1}^{1}\left[\frac{\partial p_n}{\partial y}\frac{\partial \phi}{\partial y}\right](\eta_i^{(n)}, y)\ dy\right)\tilde{w}_i^{(n)}$$

$$= -\sum_{j=0}^{n}\left(\int_{-1}^{1}\left[\frac{\partial^2 p_n}{\partial x^2}\phi\right](x, \eta_j^{(n)})\ dx\right)\tilde{w}_j^{(n)} - \sum_{i=0}^{n}\left(\int_{-1}^{1}\left[\frac{\partial^2 p_n}{\partial y^2}\phi\right](\eta_i^{(n)}, y)\ dy\right)\tilde{w}_i^{(n)}$$

$$= -\sum_{\substack{i=0\\j=0}}^{n} [\Delta p_n\ \phi](\eta_i^{(n)}, \eta_j^{(n)})\ \tilde{w}_i^{(n)}\tilde{w}_j^{(n)}$$

$$= \sum_{\substack{i=0\\j=0}}^{n} [f\phi](\eta_i^{(n)}, \eta_j^{(n)})\ \tilde{w}_i^{(n)}\tilde{w}_j^{(n)} \quad =: \ F_n(\phi), \qquad \forall \phi \in \mathbf{P}_n^{\star,0}.$$

Basically, the expression above follows from (13.2.6) by replacing the integrals with the help of a quadrature formula based on the points belonging to $\aleph_n$. Finally, we can estimate a certain norm of the error $|U - p_n|$ with a suitable generalization of theorem 9.4.1. The rate of convergence is obtained by examining the error due to the use of the quadrature in place of the exact integrals. For example, the error $|F(\phi) - F_n(\phi)|$, $\phi \in \mathbf{P}_n^{\star,0}$, converges to zero with a rate depending on the smoothness of the function f (see CANUTO, HUSSAINI, QUARTERONI and ZANG(1988), section 9.7). As usual, the analysis of the other Jacobi cases is more complicated. The Chebyshev case ($\alpha = \beta = -\frac{1}{2}$) is treated for instance in BRESSAN and QUARTERONI(1986b). The strategy in this class of proofs is to deal with the two variables x and y separately, in order to take advantage of the theory developed for the one-dimensional case.

Of course, alternate spectral discretizations of Poisson's equation, such as the tau method, may be considered. Further generalizations result from considering approximating polynomials of different degrees with respect to the variables x and y. The same techniques apply to other partial differential equations defined in Ω. For example, spectral methods have been successfully applied to approximate the Navier-Stokes equations, which are briefly introduced in section 13.4. However, the theoretical study of convergence can be quite involved even for simple problems.

Setting up an approximation scheme is not an easy task when Ω is not a square. In this situation, one can try to map Ω into a square and, after change of variables, discretize the new differential problem. Another approach, sometimes combined with the previous one, is to use a domain-decomposition method. The reader interested in this kind of applications can find a good list of references in the books of CANUTO, HUSSAINI, QUARTERONI and ZANG(1988), BOYD(1989).

13.3 Hints for the implementation

Besides the standard techniques used to numerically solve (13.2.3), other algorithms, related to the specific structure of the matrix $D_n^\star$, $n \geq 2$, can be devised and implemented. First, let us introduce some notations. The *tensor product* of two $n \times n$ matrices $\mathcal{W} \equiv \{w_{ij}\}_{\substack{1\leq i\leq n\\ 1\leq j\leq n}}$ and $\mathcal{Z} \equiv \{z_{ij}\}_{\substack{1\leq i\leq n\\ 1\leq j\leq n}}$ is defined to be the $n^2 \times n^2$ block matrix

(13.3.1)
$$\mathcal{W} \otimes \mathcal{Z} := \begin{bmatrix} w_{11}\mathcal{Z} & w_{12}\mathcal{Z} & \cdots & w_{1n}\mathcal{Z} \\ w_{21}\mathcal{Z} & w_{22}\mathcal{Z} & \cdots & w_{2n}\mathcal{Z} \\ \vdots & \vdots & & \vdots \\ w_{n1}\mathcal{Z} & w_{n2}\mathcal{Z} & \cdots & w_{nn}\mathcal{Z} \end{bmatrix}.$$

If $\mathcal{W}$ and $\mathcal{Z}$ are invertible, then we have the relation

(13.3.2)
$$(\mathcal{W} \otimes \mathcal{Z})^{-1} = \mathcal{W}^{-1} \otimes \mathcal{Z}^{-1}.$$

Let $\hat{I}_n$ denote the $(n-1) \times (n-1)$ identity matrix. Thus, we obtain the following expression:

(13.3.3)
$$D_n^\star = \hat{I}_n \otimes \hat{A}_n + \hat{A}_n \otimes \hat{I}_n,$$

where $\hat{\mathcal{A}}_n$ is the $(n-1)\times(n-1)$ matrix $\{\tilde{d}_{ij}^{(2)}\}_{\substack{1\le i\le n-1\\ 1\le j\le n-1}}$ (see (13.2.4) for $n=4$). In section 8.2 we claimed that for $-1<\alpha<1$ and $-1<\beta<1$, the eigenvalues of $-\hat{\mathcal{A}}_n$ are real positive and distinct, although this cannot always be proven.
Under this assumption, we can find an invertible $(n-1)\times(n-1)$ matrix $\mathcal{B}_n$ such that $\hat{\mathcal{A}}_n = \mathcal{B}_n^{-1}\Lambda_n\mathcal{B}_n$, where Λ_n is diagonal. At this point, one easily checks that $(\mathcal{B}_n\otimes\mathcal{B}_n)^{-1}D_n^\star(\mathcal{B}_n\otimes\mathcal{B}_n)$ is equal to the diagonal matrix $\hat{I}_n\otimes\Lambda_n+\Lambda_n\otimes\hat{I}_n$. This shows that the eigenvalues of $-D_n^\star$ can be recovered from those of problem (8.2.1). In particular, they are $\lambda_{n,m}+\lambda_{n,k}$, $1\le m\le n-1$, $1\le k\le n-1$. From (13.3.2), we finally obtain the following relation (see LYNCH, RICE and THOMAS(1964)):

$$(D_n^\star)^{-1} = (\mathcal{B}_n\otimes\mathcal{B}_n)(\hat{I}_n\otimes\Lambda_n + \Lambda_n\otimes\hat{I}_n)^{-1}(\mathcal{B}_n^{-1}\otimes\mathcal{B}_n^{-1}). \tag{13.3.4}$$

This suggests the possibility of performing the multiplication $\bar{f}_n \to (D_n^\star)^{-1}\bar{f}_n$ very efficiently, provided we know how to decompose the block $\hat{\mathcal{A}}_n$.

System (13.2.3) can be also solved by a preconditioned iterative method (see section 8.3). Following section 8.4, we can construct a $(n-1)\times(n-1)$ preconditioning matrix $\hat{\mathcal{R}}_n$ for $-\hat{\mathcal{A}}_n$, based on a finite-difference scheme at the collocation nodes. Then, one finds that the $(n-1)^2\times(n-1)^2$ matrix $\mathcal{R}_n := \hat{I}_n\otimes\hat{\mathcal{R}}_n+\hat{\mathcal{R}}_n\otimes\hat{I}_n$ is a good preconditioner for $-D_n^\star$, since the eigenvalues of $\mathcal{R}_n^{-1}D_n^\star$ and $\hat{\mathcal{R}}_n^{-1}\hat{\mathcal{A}}_n$ have the same qualitative behavior. In addition, since $\mathcal{R}_n$ has the same structure of $D_n^\star$, we can invert it by the procedure described above.

Another iterative technique is the *alternating-direction* method, which is detailed in PEACEMAN and RACHFORD(1955), DOUGLAS and RACHFORD(1956), VARGA(1962). Let us fix the the real parameter $\omega>0$ and denote by I_n the $(n-1)^2\times(n-1)^2$ identity matrix. The algorithm consists of two successive steps, i.e.,

$$(-\mathcal{A}_n + \omega I_n)\,\bar{q}_n^{(k)} := \bar{f}_n + (\mathcal{K}_n\mathcal{A}_n\mathcal{K}_n + \omega I_n)\,\bar{p}_n^{(k)}, \qquad k\in\mathbf{N}, \tag{13.3.5}$$

$$(-\mathcal{K}_n\mathcal{A}_n\mathcal{K}_n + \omega I_n)\,\bar{p}_n^{(k+1)} := \bar{f}_n + (\mathcal{A}_n + \omega I_n)\,\bar{q}_n^{(k)}, \qquad k\in\mathbf{N}, \tag{13.3.6}$$

where the initial guess $\bar{p}_n^{(0)}$ is assigned and $\{\bar{q}_n^{(k)}\}_{k\in\mathbf{N}}$ is an auxiliary sequence of vectors. Then we have $\lim_{k\to+\infty}\bar{p}_n^{(k)} = \bar{p}_n$, where $\bar{p}_n$ is the solution of (13.2.3).

The scheme (13.3.5)-(13.3.6) is implicit. However, we can invert the block diagonal matrix $-\mathcal{A}_n + \omega I_n$ in (13.3.5) by computing once and for all the inverse of the block $-\hat{\mathcal{A}}_n + \omega \hat{I}_n$. Similarly, by writing $(-\mathcal{K}_n \mathcal{A}_n \mathcal{K}_n + \omega I_n)^{-1} = \mathcal{K}_n(-\mathcal{A}_n + \omega I_n)^{-1}\mathcal{K}_n$, we can deal with (13.3.6). The theory shows that the fastest convergence of the method is obtained when $\omega = \sqrt{\lambda_{n,1}\ \lambda_{n,n-1}}$, where $\lambda_{n,1}$ and $\lambda_{n,n-1}$ are the smallest and the largest eigenvalues of $-\hat{\mathcal{A}}_n$ respectively.

To conclude this section, we present the results of a numerical test. In (13.1.1) we take $f(x,y) := 2\pi$, $\forall(x,y) \in \Omega$, i.e., a constant charge density in Ω. The corresponding potential U is plotted in figure 13.3.1. Besides, in table 13.3.1, we report for different n the value of the approximating polynomial p_n at the center of Ω. Chebyshev collocation nodes have been used. As n increases, these computed quantities converge to a limit, which is expected to be the potential at the point $(0,0)$.

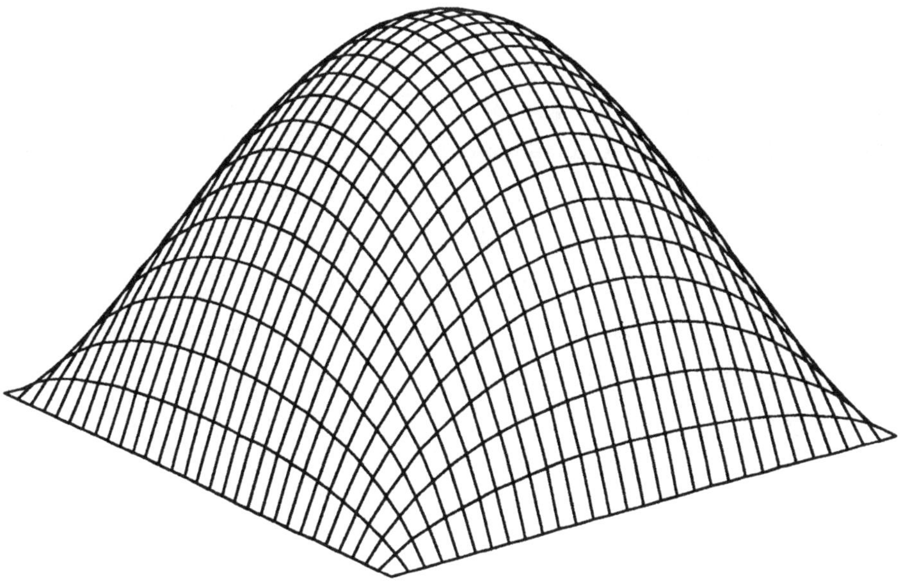

Figure 13.3.1 - Solution of problem (13.1.1) for $f \equiv 2\pi$.

n	$p_n(0,0)$
2	1.570796326
4	1.856395658
6	1.851713702
8	1.851563407
10	1.851562860
12	1.851563065

Table 13.3.1 - Approximations of $U(0,0)$.

13.4 The incompressible Navier-Stokes equations

The Navier-Stokes equations describe, with a good degree of accuracy, the motion of a viscous *incompressible* fluid in a region Ω (see for instance BATCHELOR(1967)). They find applications in aeronautics, metereology, plasma physics, and in other sciences. In two dimensions Ω is an open set of $\mathbf{R}^2$. The unknowns, depending on the space variables $(x,y) \in \bar{\Omega}$ and the time variable $t \in [0,T]$, $T > 0$, are the two components of the *velocity vector field* $\vec{V} \equiv (V_1, V_2)$, $V_i : \Omega \times [0,T] \to \mathbf{R}$, $1 \leq i \leq 2$, and the *pressure* $\mathcal{P} : \Omega \times]0,T] \to \mathbf{R}$. Assuming that the *density* of the fluid is constant, after dimensional scaling the equations become

$$(13.4.1) \qquad \frac{\partial V_1}{\partial t} + V_1 \frac{\partial V_1}{\partial x} + V_2 \frac{\partial V_1}{\partial y} - \nu \, \Delta V_1 = -\frac{\partial \mathcal{P}}{\partial x} \qquad \text{in } \Omega \times]0,T],$$

$$(13.4.2) \qquad \frac{\partial V_2}{\partial t} + V_1 \frac{\partial V_2}{\partial x} + V_2 \frac{\partial V_2}{\partial y} - \nu \, \Delta V_2 = -\frac{\partial \mathcal{P}}{\partial y} \qquad \text{in } \Omega \times]0,T],$$

$$(13.4.3) \qquad \frac{\partial V_1}{\partial x} + \frac{\partial V_2}{\partial y} = 0 \qquad \text{in } \Omega \times]0,T],$$

where $\nu > 0$ is a given parameter associated with the *kinematic viscosity*, and Δ is the Laplace operator defined in (13.1.1). The nonlinear relations (13.4.1) and (13.4.2) are known as *momentum equations*, while (13.4.3) is the *continuity* (or *incompressibility*) equation. They express the conservation laws of the momentum and the mass respectively. We note the analogy between the momentum equations and the Burgers equation, presented in section 10.4 for the case of one space variable. The functions V_i, $1 \leq i \leq 2$, require initial and boundary conditions. For example, by imposing $V_i \equiv 0$ on $\partial\Omega\times]0,T]$, $1 \leq i \leq 2$, we are specifying a *no-slip* condition at the boundary. Instead, no initial or boundary conditions are needed for the pressure, which is determined up to an additive constant.

For a theoretical study of the equations (13.4.1)-(13.4.3), the reader is invited to consult the books of LADYZHENSKAYA(1969), TEMAM(1985), KREISS and LORENZ(1989) and the numerous references contained therein. On the subjects of numerical approximation by finite-differences or finite element methods the bibliography is very rich, and a comprehensive list of references cannot be included here. We would like to remark that the numerical investigation of the equations relevant to the physics of fluids has also been a primary source of algorithms and solution techniques in the field of spectral methods. Due to the abundance of results, we are unable to mention all the contributors. A distinguished referring point is the book of CANUTO, HUSSAINI, QUARTERONI and ZANG(1988), which is specifically addressed to computations in fluid dynamics. The reader will find there material for further extensions.

* * * * * * * * * * * * * * *

REFERENCES

ADAMS R.(1975), *Sobolev Spaces*, Academic Press, New York.

AGOSHKOV V.I.(1988), Poincaré-Steklov's operators and domain decomposition methods in finite dimensional spaces, Proceedings of the I International Symposium on Domain Decomposition Methods for P.D.E., SIAM, Philadelphia.

ASKEY R.(1975), *Orthogonal Polynomials and Special Functions*, SIAM, Philadelphia.

AUBIN J.P.(1972), *Approximation of Elliptic Boundary-Value Problems*, Wiley-Interscience, New York.

BABUŠKA I. & SURI M.(1987), The p-version of the finite element method with quasiuniform meshes, M^2AN, **21**, pp.199-238.

BABUŠKA I., SZABO B.A. & KATZ I.N.(1981), The p-version of the finite element method, SIAM J. Numer. Anal., **18**, n.3, pp.515-545.

BALDOCK G.R.& BRIDGEMAN T.(1981), *The mathematical Theory of Wave Motion*, Ellis Horwood Limited, Chichester (England).

BATCHELOR G.K.(1967), *An introduction to Fluid Dynamics*, Cambridge University Press, Cambridge.

BELLMAN R.(1944), A note on an inequality of E. Schmidt, Bulletin of the Amer. Math. Soc., **50**, n.10, pp.734-736.

BERGH J. & LÖFSTRÖM(1976), *Interpolation Spaces*, Spinger-Verlag, Berlin.

BERNARDI C., DEBIT N. & MADAY Y.(1990), Coupling finite element and spectral methods: first results, Math. Comp., **54**, n.189, pp.21-39.

BERNARDI C. & MADAY Y.(1988), A collocation method over staggered grids for the Stokes problem, Internat. J. for Numer. Methods in Fluids, **8**, pp.537-557.

BERNARDI C. & MADAY Y.(1989), Properties of some weighted Sobolev spaces and application to spectral approximations, SIAM J. Numer. Anal., **26**, n.4, pp.769-829.

BERNARDI C. & MADAY Y.(1990), Some spectral approximations of monodimensional fourth-order problems, Manuscript, to appear in J. Approx. Theory.

BERNARDI C., MADAY Y. & SACCHI-LANDRIANI G.(1989), Nonconforming matching conditions for coupling spectral and finite element methods, Applied Numer. Math., **6**, pp.65-84.

BERNSTEIN S.(1912a), Sur l'ordre de la meilleure approximation des fonctions continues par des polynômes de degré donné, Mém. Acad. Roy. Belgique, 2me série, **4**, pp.1-104.

BERNSTEIN S.(1912b), Démonstration du théorème de Weierstrass fondée sur le calcul de probabilités, CKMS, **13**, pp.1-2.

BERTOLUZZA S. & FUNARO D.(1991), Remarks about the matrices of the pseudospectral approximation of Neumann problems, in the Proceedings of the IMACS Int. Sym. on Iterative Methods in Linear Algebra, Brussels, to appear.

BINI D., CAPOVANI M., LOTTI G. & ROMANI F.(1981), *Complessità Numerica*, Boringhieri, Torino.

BJØRSTAD P.E. & WIDLUND O.B.(1986), Iterative methods for the solution of elliptic problems on regions partitioned into substructures, SIAM J. Numer. Anal., **23**, n.6, pp.1097-1120.

BOYD J.P.(1980), The rate of convergence of Hermite functions series, Math. Comp., **35**, n.152, pp.1309-1316.

BOYD J.P.(1981), The rate of convergence of Chebyshev polynomials for functions which have asymptotic power series about one endpoint, Math. Comp., **37**, n.155, pp.189-195.

BOYD J.P.(1982), The optimization of convergence for Chebyshev polynomial methods in an unbounded domain, J. Comput. Phys., **45**, pp.43-79.

BOYD J.P.(1984), Asymptotic coefficients of Hermite functions series, J. Comput. Phys., **54**, n.3, pp.382-410.

BOYD J.P.(1989), *Chebyshev & Fourier Spectral Methods*, Lecture Notes in Engineering, Springer, New York.

BRACHET M.E., MEIRON D.I., ORSZAG S.A., NICKEL B.G., MORF R.H. & FRISCH U.(1983), Small scale structures of the Taylor-Green vortex, J. Fluid Mech., **130**, pp.411-452.

BRANDT A., FULTON S.R. & TAYLOR G.D.(1985), Improved spectral multigrid methods for periodic elliptic problems, J. Comput. Phys., **58**, pp.96-112.

BRAUER F. & NOHEL J.A.(1967), *Ordinary Differential Equations*, W.A. Benjamin, New York.

BRESSAN N.& PAVONI D.(1990), Truncation versus mapping in the spectral approximation to the Korteweg-de Vries equation, in *Spectral and High Order Methods for P.D.E.* (C.Canuto & A.Quarteroni Eds.), North Holland, Amsterdam, pp.443-450.

BRESSAN N.& QUARTERONI A.(1986a), An implicit/explicit spectral method for Burgers equation, Calcolo, **23**, pp.265-284.

BRESSAN N.& QUARTERONI A.(1986b), Analysis of Chebyshev collocation methods for parabolic equations, SIAM J. Numer. Anal., **23**, n.6, pp.1138-1154.

BREZIS H.(1983), *Analyse Fonctionnelle*, Masson, Paris.

BREZZI F. & GILARDI G.(1987), in *Finite Element Handbook* (H.Kardenstuncer Ed.), McGraw-Hill, New York, p.1-121.

BREZZI F., RAPPAZ J. & RAVIART P.A.(1980), Finite dimensional approximation of nonlinear problems. Part I: branches of nonsingular solutions, Numer. Math., **36**, pp.1-25.

BRIGHAM E.O.(1974), *The Fast Fourier Transform*, Prentice-Hall, Englewood Cliffs NJ.

CAI W., GOTTLIEB D. & HARTEN A.(1990), Cell averaging Chebyshev methods for hyperbolic problems, ICASE (Hampton VA) report n.90-27.

CALOGERO F.(1983), Computation of Sturm-Liouville eigenvalues via Lagrangian interpolation, Lett. Nuovo Cimento, **37**, pp.9-16.

CALOGERO F. & FRANCO E.(1985), Numerical tests of a novel technique to compute the eigenvalues of differential operators, Nuovo Cimento, **89b**, p.161.

CANUTO C.(1986), Boundary conditions in Chebyshev and Legendre methods, SIAM J. Numer. Anal., **23**, n.4, pp.815-831.

CANUTO C.(1988), Spectral methods and a maximum principle, Math. Comp., **51**, n.184, pp.615-629.

CANUTO C. & FUNARO D.(1988), The Schwarz algorithm for spectral methods, SIAM J. Numer. Anal., **25**, n.1, pp.24-40.

CANUTO C., HUSSAINI M.Y., QUARTERONI A.& ZANG T.A.(1988), *Spectral Methods in Fluid Dynamics*, Springer Series in Computational Physics, Springer-Verlag, New York.

CANUTO C. & QUARTERONI A.(1980), Propriétés d'approximation dans les espaces de Sobolev de systémes de polinômes orthogonaux, C.R. Acad. Sci. Paris, **290**, pp.A925-A928.

CANUTO C. & QUARTERONI A.(1981), Spectral and pseudo-spectral methods for parabolic problems with nonperiodic boundary conditions, Calcolo, **18**, pp.197-218.

CANUTO C. & QUARTERONI A.(1982a), Approximation results for orthogonal polynomials in Sobolev spaces, Math. Comp., **38**, pp.67-86.

CANUTO C. & QUARTERONI A.(1982b), Error estimates for spectral approximations of hyperbolic equations, SIAM J. Numer. Anal., **19**, pp.629-642.

CANUTO C. & QUARTERONI A.(1984), Variational methods in the theoretical analysis of spectral approximations, in *Spectral Methods for Partial Differential Equations* (R.G. Voigt, D. Gottlieb & M.Y. Hussaini Eds.), SIAM-CBMS, Philadelphia, pp.55-78.

CANUTO C. & QUARTERONI A.(1985), Preconditioned minimal residual methods for Chebyshev spectral calculations, J. Comput. Phys., **60**, pp.315-337.

CANUTO C. & QUARTERONI A.(1987), On the boundary treatment in spectral methods for hyperbolic systems, J. Comput. Phys., **71**, pp.100-110.

CHAN T.F., GLOWINSKI R., PERIAUX J. & WIDLUND O.B. (Editors) (1989), *Domain Decomposition Methods*, SIAM, Philadelphia.

CHAN T.F., GLOWINSKI R., PERIAUX J. & WIDLUND O.B. (Editors) (1990), *Domain Decomposition Methods for Partial Differential Equations*, SIAM, Philadelphia.

CHAN T.F. & GOOVAERTS D.(1989), Schur complement domain decomposition algorithms for spectral methods, Applied Numer. Math., **6**, pp.53-64.

CHANG K.W. & HOWES F.A.(1984), *Nonlinear Singular Perturbation Phenomena: Theory and Applications*, Springer, New York.

CHATELIN F.(1983), *Spectral Approximation of Linear Operators*, Academic Press, New York.

CHENEY E.W.(1966), *Introduction to Approximation Theory*, Mc Graw-Hill, New York.

CIARLET P.G.(1978), *The Finite Element Method for Elliptic Problems*, North Holland, Amsterdam.

CLENSHAW C.W.(1957), The numerical solution of linear differential equations in Chebyshev series, Proc. Cambridge Philos. Soc., **53**, pp.134-149.

COOLEY J.W., LEWIS P.A. & WELCH P.D.(1970), The fast Fourier transform algorithm: programming considerations in the calculations of sine, cosine and Laplace transform, J. Sound Vibration, **12**, pp.315-317.

COOLEY J.W. & TUKEY J.W.(1965), An algorithm for the machine calculation of complex Fourier series, Numer. Math., **2**, pp.197-205.

COMINCIOLI V.(1990), *Analisi Numerica, Metodi Modelli Applicazioni*, McGraw-Hill Libri Italia, Milan.

COURANT R. & HILBERT D.(1953), *Methods of Mathematical Physics*, Wiley-Interscience, New York.

DAHLQUIST G.& BJÖRCK H.B.(1974), *Analysis of Numerical Methods*, John Wiley & Sons, New York.

DAVIS P.J. & RABINOWITZ P.(1984), *Methods of Numerical Integration*, Second edition, Academic Press, New York.

DELFOUR M., FORTIN M. & PAYRE G.(1981), Finite-difference solutions of a non-linear Schrödinger equation, J. Comput. Phys., **44**, pp.277-288.

DELVES L.M.(1968), The numerical evaluation of principal value integrals, Comput. J., **10**, pp.389-391.

DETTMAN J.W.(1969), *Mathematical Methods in Physics and Engineering*, McGraw-Hill, New York.

DE VERONICO M.C.(1991), Approssimazione dell'equazione di Schrödinger mediante polinomi algebrici, Thesis, University of Pavia, Italy.

DE VERONICO M.C., FUNARO D. & REALI G.C.(1991), A novel numerical technique to investigate nonlinear guided waves: approximation of nonlinear Schrödinger equation by nonperidic pseudospectral methods, pre-print n.800, Istituto di Analisi Numerica del C.N.R., Pavia, Italy.

DEVILLE M. & MUND E.(1985), Chebyshev pseudospectral solution of second-order elliptic equations with finite element pre-conditioning, J. Comput. Phys., **60**, pp.517-533.

DITZIAN Z. & TOTIK V.(1987), *Moduli of Smoothness*, Springer-Verlag, New York.

DOUGLAS J. Jr. & RACHFORD H.H. Jr.(1956), On the numerical solution of heat conduction problems in two or three space variables, Trans. Amer. Math. Soc., **82**, pp.421-439.

DRYJA M. & WIDLUND O.B.(1990), Towards a unified theory of domain decomposition algorithms for elliptic problems, in *Domain Decomposition Methods for P.D.E.* (T.F. Chan, R. Glowinski, J. Periaux & O.B. Widlund Eds.), SIAM, Philadelphia.

DUBINER M.(1987), Asymptotic analysis of spectral methods, J. Sci. Comput., **2**, pp.3-32.

DUFF I.S., ERISMAN A.M. & REID J.K.(1986), *Direct Methods for Sparse Matrices*, Clarendon Press, Oxford.

DUFFIN R.J. & SCHAEFFER A.C.(1941), A refinement of an inequality of the brothers Markoff, Trans. Amer. Math. Soc., **50**, pp.517-528.

DURAND L.(1985), Lagrangian differentiation, Gauss-Jacobi integration and Sturm-Liouville eigenvalue problems, Int. Symp. on Orthogonal Polynomials and their Applications, Bar-le-Duc (France), october 1984, Lecture Notes in Mathematics, Springer-Verlag.

ECKHAUS W.(1979), *Asymptotic Analysis of Singular Perturbations*, North-Holland, Amsterdam.

ELLIOTT D.(1981), A convergence theorem for singular integral equations, J. Austral. Math. Soc., **B22**, pp.539-552.

ELLIOTT D.(1982), The classical collocation method for solving singular integral equations, SIAM J. Numer. Anal., **19**, pp.816-832.

ELLIOTT C.M. & OCKENDON J.R.(1982), *Weak and Variational Methods for Moving Boundary Problems*, Pitman, London.

ENGELS H.(1980), *Numerical Quadrature and Cubature*, Academic Press, London.

EPSTEIN B.(1962), *Partial Differential Equations, An Introduction*, McGraw-Hill, New York.

ERDOGAN F., GUPTA G.D. & COOK T.S.(1973), Numerical solution of singular integral equations, in *Mechanics of Fracture*, Vol.1 (G.S. Sih Ed.), Noordhoff, Leyden, pp.368-425.

ERDÖS P.(1961), Problems and results on the theory of interpolation. II, Acta Math. Acad. Sci. Hungar, **12**, pp.235-244.

ESCOBEDO M. & KAVIAN O.(1987), Variational problems related to self-similar solutions of the heat equation, Nonlinear Anal., **11**, n.10, pp.1103-1133.

FASANO A. & PRIMICERIO M.(1972), Il problema di Stefan con condizioni al contorno non lineari, Ann. Scuola Norm. Sup. Pisa, **26**, pp.711-737.

FEINERMAN R.P. & NEWMAN D.J.(1974), *Polynomial Approximation*, Waverly Press, Baltimore MD.

FLETCHER C.A.J.(1984), *Computational Galerkin Methods*, Springer-Verlag, New York.

FOX L.(1964), *An introduction to Numerical Linear Algebra*, Clarendon Press, Oxford.

FRIEDMAN A.(1959), Free boundary problems for parabolic equations. I, melting of solids, J. Math. Mech., **8**, pp.499-517.

FUNARO D.(1986), A multidomain spectral approximation of elliptic equations, Numer. Methods Partial Differential Equations, **2**, pp.187-205.

FUNARO D.(1987), A preconditioning matrix for the Chebyshev differencing operator, SIAM J. Numer. Anal., **24**, pp.1024-1031.

FUNARO D.(1988), Domain decomposition methods for pseudospectral approximations. Part I: second order equations in one dimension, Numer. Math., **52**, pp.329-344.

FUNARO D.(1990a), Computational aspects of pseudospectral Laguerre approximations, Applied Numer. Math., **6**, pp.447-457.

FUNARO D.(1990b), Convergence analysis for pseudospectral multidomain approximations of linear advection equations, IMA J. of Numer. Anal., **10**, pp.63-74.

FUNARO D.(1991), Error estimates for approximations by Laguerre polynomials in Sobolev spaces, in the Proceedings of the III Int. Sym. on Orthogonal Polynomials and their Applications (C.Brezinski, L.Gori & A.Ronveaux Eds.), J.C.Baltzer, Basel, pp.263-266.

FUNARO D. & GOTTLIEB D.(1988), A new method of imposing boundary conditions in pseudospectral approximations of hyperbolic equations, Math. Comp., **51**, n.184, pp.599-613.

FUNARO D. & GOTTLIEB D.(1989), Convergence results for pseudospectral approximations of hyperbolic systems, ICASE (Hampton VA) report n.89-59.

FUNARO D. & HEINRICHS W.(1990), Some results about the pseudospectral approximation of one-dimensional fourth-order problems, Numer. Math., **58**, pp.399-418.

FUNARO D. & KAVIAN O.(1988), Approximation of some diffusion evolution equations in unbounded domains by Hermite functions, Preprint of the Institut Élie Cartan, n.15, Nancy (France).

FUNARO D., QUARTERONI A. & ZANOLLI P.(1988), An iterative procedure with interface relaxation for domain decomposition methods, SIAM J. Numer. Anal., **25**, pp.1213-1236.

FUNARO D. & ROTHMAN E.(1989), Preconditioning matrices for the pseudo spectral approximation of first-order operators, in *Finite Element Analysis in Fluids* (T.J.Chung & G.R.Karr Eds.), UAH Press, Huntsville AL, pp.1458-1463.

FURZELAND R.M.(1980), A comparative study of numerical methods for moving boundary problems, J. Inst. Math. Appl., **26**, pp.411-429.

GARNIR H.G., DE WILDE M. & SCHMETS J.(1973), *Analyse Fonctionelle*, Birkhäuser Verlag, Basel.

GASTALDI F. & QUARTERONI A.(1989), On the coupling of hyperbolic and parabolic systems: analytical and numerical approach, Applied Numer. Math., **6**, pp.3-31.

GATTESCHI L.(1964), Proprietà asintotiche di una funzione associata ai polinomi di Laguerre e loro utilizzazione al calcolo numerico degli zeri dei polinomi stessi, Atti Accad. Sci. Torino, **98**.

GAUTSCHI W. & LI S.(1991), The remainder term for analytic functions of Gauss-Radau and Gauss-Lobatto quadrature rules with multiple end points, J. Comput. Appl. Math., to appear.

GEAR C.W.(1971), *Numerical Initial Value Problems in Ordinary Differential Equations*, Prentice-Hall, Englewood Cliffs NJ.

GHIZZETTI A. & OSSICINI A.(1970), *Quadrature Formulae*, Academic Press, New York.

GOFFMAN C. & PEDRICK G.(1965), *First Course in Functional Analysis*, Prentice-Hall, Englewood Cliffs NJ.

GOLOMB M. & SHANKS M.(1965), *Elements in Ordinary Differential Equations*, McGraw-Hill, New York.

GOTTLIEB D.(1981), The stability of pseudospectral Chebyshev methods, Math. Comp., **36**, pp.107-118.

GOTTLIEB D., GUNZBURGER M. & TURKEL E.(1982), On numerical boundary treatment for hyperbolic systems, SIAM J. Numer. Anal., **19**, pp.671-697.

GOTTLIEB D., HUSSAINI M.Y. & ORSZAG S.A.(1984), Theory and application of spectral methods, in *Spectral Methods for Partial Differential Equations* (R.G.Voigt, D.Gottlieb & M.Y.Hussaini Eds.), SIAM-CBMS, Philadelphia, pp.1-54.

GOTTLIEB D.& LUSTMAN.L.(1983), The spectrum of the Chebyshev collocation operator for the heat equation, SIAM J. Numer. Anal., **20**, n.5, pp.909-921.

GOTTLIEB D., LUSTMAN.L.& TADMOR E.(1987a), Stability analysis of spectral methods for hyperbolic initial-boundary value problems, SIAM J. Numer. Anal., **24**, pp.241-256.

GOTTLIEB D., LUSTMAN.L.& TADMOR E.(1987b), Convergence of spectral methods for hyperbolic initial-boundary value problems, SIAM J. Numer. Anal., **24**, pp.532-537.

GOTTLIEB D. & ORSZAG S.A.(1977), *Numerical Analysis of Spectral Methods, Theory and Applications*, CBMS-NSF Regional Conference Series in Applied Mathematics, SIAM, Philadelphia.

GOTTLIEB D. & TADMOR E.(1990), The CFL condition for spectral approximations to hyperbolic initial-boundary value problems, ICASE (Hampton VA) report n. 90-42.

GOTTLIEB D. & TURKEL E.(1985), Topics in spectral methods for time dependent problems, in *Numerical Methods in Fluid Dynamics* (F. Brezzi Ed.), Springer-Verlag, pp.115-155.

GREENSPAN D.(1961), *Introduction to Partial Differential Equations*, McGraw-Hill, New York.

GROSCH C.E. & ORSZAG S.A.(1977), Numerical solution of problems in unbounded regions: coordinate transformations, J. Comput. Phys., **25**, pp.273-296.

GUELFAND J.M., GRAEV M.I. & VILENKIN N.J.(1970), *Les Distributions*, 5 Volumes, Dunod, Paris.

HACKBUSCH W.(1985), *Multi-Grid Methods and Applications*, Spriger Series in Computational Mathematics, Springer-Verlag, Heidelberg.

HAGEMAN L.A. & YOUNG D.M.(1981), *Applied Iterative Methods*, Academic Press, Orlando.

HALDENWANG P., LABROSSE G., ABBOUDI S. & DEVILLE M.(1984), Chebyshev 3-D spectral and 2-D pseudospectral solvers for the Helmholtz equation, J. Comput. Phys., **55**, pp.115-128.

HALMOS P.R.(1950), *Measure Theory*, D.Van Nostrand Company Inc., New York.

HARTMAN S. & MIKUSIŃSKI J.(1961), *The Theory of Lebesgue Measure and Integration*, Pergamon Press, Oxford.

HEINRICHS W.(1988), Line relaxation for spectral multigrid methods, J. Comput. Phys., **77**, n.1, pp.166-182.

HEINRICHS W.(1989), Improved condition number for spectral methods, Math. Comp., **53**, n.187, pp.103-119.

HILDEBRANDT T.H.(1963), *Theory of Integration*, Academic Press, London.

HILL J.M. & DEWYNNE J.N.(1987), *Heat Conduction*, Blackwell Scientific Publications, Oxford.

HOCHSTADT H.(1971), *The functions of Mathematical Physics*, Wiley-Interscience, New York.

HOCHSTADT H.(1973), *Integral Equations*, John Wiley & Sons, New York.

HOFFMAN K. & KUNZE R.(1971), *Linear Algebra*, Second edition, Prentice-Hall, Englewood Cliffs NJ.

IOAKIMIDIS N.I.(1981), On the weighted Galerkin method of numerical solution of Cauchy type singular integral equations, SIAM J. Numer. Anal., **18**, n.6, pp.1120-1127.

ISAACSON E. & KELLER H.B.(1966), *Analysis of Numerical Methods*, John Wiley & Sons, New York.

JACKSON D.(1911), Über Genauigkeit der Annäherung stetiger Funktionen durch ganze rationale Funktionen gegebenen Grades und trigonometrische Summen gegebener Ordnung, Dissertation, Göttingen.

JACKSON D.(1930), *The Theory of Approximation*, American Mathematical Society, New York.

JAIN M.K.(1984), *Numerical Solution of Differential Equations*, Second edition, John Wiley & Sons, New York.

KANTOROVICH L.V. & KRYLOV V.I.(1964), *Approximate methods of Higher Analysis*, Noordhoff, Groningen.

KARAGEORGHIS A. & PHILLIPS T.N.(1989), On the coefficients of differential expansions of ultraspherical polynomials, ICASE (Hampton VA) report n. 89-65.

KAVIAN O.(1990), Manuscript.

KOLMOGOROV A.N. & FOMIN S.V.(1961), *Elements of the Theory of Functions and Functional Analysis*, Graylock Press, Albany NY.

KOPRIVA D.A.(1986), A spectral multidomain method for the solution of hyperbolic systems, Applied Numer. Math., **2**, pp.221-241.

KOSLOFF D. & TAL-EZER H.(1989), Modified Chebyshev pseudospectral method with $O(N^{-1})$ time step restriction, ICASE (Hampton VA) report n. 89-71.

KREISS H.O. & LORENZ J.(1989), *Initial-Boundary Value Problems and the Navier-Stokes Equation*, Academic Press, London.

KRENK S.(1975), On quadrature formulas for singular integral equations of the first and the second kind, Quart. Appl. Math., **33**, pp.225-232.

KRENK S.(1978), Quadrature formulae of closed type for solution of singular integral equations, J. Inst. Math. Appl., **22**, pp.99-107.

KUFNER A.(1980), *Weighted Sobolev Spaces*, Teubner Verlagsgesellschaft, Leipzig.

LADYZHENSKAYA O.A.(1969), *The Mathematical Theory of Viscous Incompressible Flow* (translated from russian), Second edition, Gordon and Breach Science Publishers, New York.

LAGERSTROM P.A.(1988), *Matched Asymptotic Expansions*, Spinger-Verlag, New York.

LANCZOS C.(1956), *Applied Analysis*, Prentice-Hall, Englewood Cliffs NJ.

LAPIDUS L. & SEINFELD J.H.(1971), *Numerical Solutions of Ordinary Differential Equations*, Academic Press, New York.

LAX P.D. & MILGRAM N.(1954), Parabolic equations, Contribution to the theory of partial differential equations, Ann. of Math. Stud., **33**, pp.167-190.

LIONS J.L. & MAGENES E.(1972), *Non-Homogeneous Boundary Value Problems and Applications*, Springer-Verlag, New York.

LOWENTHAL F.(1975), *Linear Algebra with Linear Differential Equations*, John Wiley & Sons, New York.

LUKE Y.L.(1969), *The Special Functions and their Approximations*, Academic Press, New York.

LUSTMAN L.(1986), The time evolution of spectral discretization of hyperbolic equations, SIAM J. Numer. Anal., **23**, pp.1193-1198.

LYNCH R.E., RICE J.R. & THOMAS D.H.(1964), Direct solution of partial differential equations by tensor product methods, Numer. Math., **6**, pp.185-199.

MACARAEG M. & STREETT C.L.(1986), Improvements in spectral collocation through a multiple domain technique, Applied Numer. Math., **2**, pp.95-108.

MADAY Y.(1990), Analysis of spectral projectors in one-dimensional domains, Math. Comp., **55**, n.192, pp.537-562.

MADAY Y.(1991), Résultats d'approximation optimaux pur les opérateurs d'interpolation polynomiale, C.R. Acad. Sci. Paris, t.312, Série I, pp.705-710.

MADAY Y., PERNAUD-THOMAS B. & VANDEVEN H.(1985), Une réhabilitation des méthodes spectrales de type Laguerre, Rech. Aérospat., **6**, pp.353-375.

MADAY Y. & QUARTERONI A.(1981), Legendre and Chebyshev spectral approximations of Burger's equation, Numer. Math., **37**, pp.321-332.

MADAY Y. & QUARTERONI A.(1982), Approximation of Burger's equation by pseudospectral methods, R.A.I.R.O. Anal. Numer., **16**, pp.375-404.

MALIK M.R., ZANG T.A. & HUSSAINI M.Y.(1985), A spectral collocation method for the Navier-Stokes equations, J. Comput. Phys., **61**, pp.64-88.

MARKOFF A.(1889), Sur une question posée par Mendeleieff, Bulletin of the Academy of Sciences of St. Petersburg, **62**, pp.1-24.

MARKOFF W.(1892), Über Polynome die in einen gegebenen Intervalle möglichst wenig von Null obweichen, Mathematische Annalen, Academy of Sciences of St. Petersburg, **77**(1916), pp.213-258. From the original version in russian.

MAVRIPLIS C.(1989), Laguerre polynomials for infinite-domain spectral elements, J. Comput. Phys. ,**80**, pp.480-488.

MERCIER B.(1982), Stabilité et convergence des méthodes spectrales polynômiales. Application a l'équation d'advection, R.A.I.R.O., **16**, n.1, pp.67-100.

MERCIER B.(1989), *An Introduction to the Numerical Analysis of Spectral Methods*, Springer-Verlag, Berlin.

MONEGATO G.(1982), The numerical evaluation of one-dimensional Cauchy principal value integrals, Computing, **29**, pp.337-354.

MORCHOISNE Y.(1979), Resolution of Navier-Stokes equations by a space-time pseudospectral method, Rech. Aérospat. 1979-5, pp.293-306.

MUÑOZ R.(1990), Theoretical analysis of some spectral multigrid method, in *Spectral and High Order Methods for P.D.E.* (C.Canuto & A.Quarteroni Eds.), North Holland, Amsterdam, pp.287-294.

NATANSON I.P.(1965), *Constructive Function Theory*, Ungar, New York.

NĚCAS J.(1962), Sur une méthode pour resoudre les équations aux dérivées partielles du type elliptique voisine de la variationnelle, Ann. Scuola Norm. Sup. Pisa, **16**, pp.305-326.

NEVAI G.P.(1976), Lagrange interpolation at zeros of orthogonal polynomials, in *Approximation Theory II* (G.C.Lorenz, C.K.Chui & L.L.Schumaker Eds.), Academic Press, New York, pp.163-201.

NOCHETTO R.H.(1990), Numerical methods for free boundary problems, in *Free Boundary Problems: Theory and Applications* (K.H.Hoffman & J.Sprekels Eds.), Pitman Research Notes in Mathematics.

NUSSBAUMER H.J.(1981), *Fast Fourier Transform and Convolution Algorithms*, Springer Verlag, New York.

OKIKIOLU G.O.(1971), *Aspects of the Theory of Bounded Integral Operators in L^p-Spaces*, Academic Press, London.

ORSZAG S.A.(1972), Comparison of pseudospectral and spectral approximations, Stud. Appl. Math., **51**, pp.253-259.

ORSZAG S.A.(1980), Spectral methods for problems in complex geometries, J. Comput. Phys., **37**, pp.70-92.

ORSZAG S.A.(1986), Fast eigenfunction transforms, Science and Computers Advances in Mathematics Supplementary Studies, **10**, pp.23-30.

OSBORN J.E.(1975), Spectral approximation for compact operators, Math. Comp., **29**, n.131, pp.712-725.

PATERA A.T.(1984), A spectral element method for fluid dynamics: laminar flow in a channel expansion, J. Comput. Phys., **54**, pp.468-488.

PAVONI D.(1988), Single and multidomain Chebyshev collocation methods for the Korteweg-De Vries equation, Calcolo, **25**, n.4, pp.311-346.

PEACEMAN D.W. & RACHFORD H.H.Jr.(1955), The numerical solution of parabolic and elliptic differential equations, J. Soc. Indust. Appl. Math., **3**, pp.28-41.

PELZ R.B.(1990), Pseudospectral methods on massively parallel computers, in *Spectral and High Order Methods for P.D.E.* (C.Canuto & A.Quarteroni Eds.), North Holland, Amsterdam, pp.493-503.

POGORZELSKI W.(1966), *Integral Equations and their Applications*, Vol.1, Pergamon Press, Oxford.

PRANDTL L.(1905), Über flüssigkeiten bei sehr kleimer reibung, Proceedings of the Third International Mathematics Congress, Heidelberg, pp.484-491.

QUARTERONI A.(1990), Domain decomposition methods for systems of conservation laws: spectral collocation approximations, SIAM J. Sci. Stat. Comput., **11**, pp.1029-1052.

QUARTERONI A. & SACCHI-LANDRIANI G.(1988), Domain decomposition preconditioners for the spectral collocation method, J. Sci. Comput., **3**, n.1, pp.45-76.

RALSTON A.(1965), *A First Course in Numerical Analysis*, Mc Graw-Hill, New York.

RAO M.M.(1987), *Measure Theory and Integration*, Wiley-Interscience, New York.

RHEE H.K., ARIS R. & AMUNDSON N.R.(1986), *First-Order Partial Differential Equations. Volume I: Theory and Application of Single Equations*, Prentice-Hall, Englewood Cliffs NJ.

RICHARDSON L.F.(1910), The approximate arithmetical solution by finite differences of physical problems involving differential equations with an application to the stresses in a masonry dam, Phylos. Trans. Roy. Soc. London, A **210**, pp.307-357.

RIVLIN T.J.(1969), *An Introduction to the Approximation of Functions*, Blaisdell Publishing Company, Waltham MA.

RIVLIN T.J.(1974), *The Chebyshev polynomials*, Wiley-Interscience, New York.

RØNQUIST E.M. & PATERA A.T.(1987), A Legendre spectral element method for the Stefan problem, Internat. J. Numer. Methods Engrg., **24**, pp.2273-2299.

RUBINSTEIN L.I.(1971), *The Stefan Problem*, Translation of Mathematical Monographs, **27**, AMS, Providence RI.

RUDIN W.(1966), *Real and Complex Analysis*, McGraw Hill, New York.

SCHWARZ H.A.(1890), *Gesammelte Mathematische Abhandlungen*, Springer, Berlin.

SCHWARZ L.(1966), *Théorie des Distributions*, Hermann, Paris.

SIMMONS G.F.(1972), *Differential Equations with Applications and Historical Notes*, McGraw-Hill, New York.

SINGLETON R.C.(1969), An algorithm for computing the mixed radix fast Fourier transform, IEEE Trans. Audio Electroacoustic, **AU15**, pp.91-97.

SMOLLER J.(1983), *Shock Waves and Reaction-Diffusion Equations*, Springer-Verlag, New York.

SNEDDON I.N.(1957), *Elements of Partial Differential Equations*, McGraw-Hill, New York.

SOBOLEV S.L.(1964), *Partial Differential Equations of Mathematical Physics*, Pergamon Press, Oxford.

SOD G.A.(1985), *Numerical Methods in Fluid Dynamics, Initial and Initial Boundary-Value Problems*, Cambridge University Press, Cambridge.

SOLOMONOFF A. & TURKEL E.(1989), Global properties of pseudospectral methods, J. Comput. Phys., **81**, pp.239-276.

SRIVASTAVA H.M. & MANOCHA H.L.(1984), *A Treatise on Generating Functions*, Ellis Harwood, Chichester (England).

STEKLOFF W.(1916), On approximate evaluation of definite integrals by means of formulas of mechanical quadratures. I: convergence of formulas of mechanical quadratures (in Russian), Bulletin de l'Académie Impériale des Sciences, Petrograd, **10**, pp.169-186.

STRANG G.(1972), Variational crimes in the finite element method, in *Mathematical Fundation of the Finite Element Method with Applications to Partial Differential Equations* (A.K. Aziz Ed.), Academic Press, New York, pp.689-710.

STRANG G. & FIX G.J.(1973), *An Analysis of the Finite Element Method*, Prentice-Hall, Englewood Cliffs NJ.

STREETT C.L., ZANG T.A. & HUSSAINI M.Y.(1985), Spectral multigrid methods with applications to transonic potential flows, J. Comput. Phys., **57**, pp.43-76.

STROUD A.H. & SECREST D.H.(1966), *Gaussian Quadrature Formulas* Prentice-Hall, Englewood Cliffs NJ.

SWARZTRAUBER P.N.(1986), Symmetric FFTs, Math. Comp., **47**, pp.323-346.

SZEGÖ G.(1939), *Orthogonal polynomials*, American Mathematical Society, New York.

TADMOR E.(1986), The exponential accuracy of Fourier and Chebyshev differencing methods, SIAM J. Numer. Anal., **23**, n.1, pp.1-10.

TAL-EZER H.(1986a), Spectral methods in time for hyperbolic equations, SIAM J. Numer. Anal., **23**, pp.11-26.

TAL-EZER H.(1986b), A pseudospectral Legendre method for hyperbolic equations with an improved stability condition, J. Comput. Phys., **67**, pp.145-172.

TAL-EZER H.(1989), Spectral methods in time for parabolic equations, SIAM J. Numer. Anal., **26**, n.1, pp.1-11.

TEMAM R.(1985), *Navier-Stokes equations*, Second edition, North-Holland, Amsterdam.

TEMPERTON C.(1983), Self-sorting mixed-radix fast Fourier transforms, J. Comput. Phys., **52**, pp.1-23.

TEMPLE G.(1971), *The Structure of Lebesgue Integration Theory*, Oxford University Press, London.

TIMAN A.F.(1963), *Theory of Approximation of Functions of a Real Variable*, Pergamon Press, Oxford.

TODD J.(1963), *Introduction to the Constructive Theory of Functions*, Birkhäuser Verlag, Basel.

TREFETHEN L.N.(1988), Lax-stability vs. eigenvalue stability of spectral methods, in *Numerical Methods for Fluid Dynamics III* (K.W.Morton & M.J.Baines Eds.), Clarendon Press, Oxford, pp.237-253.

TREFETHEN L.N. & TRUMMER M.R.(1987), An instability phenomenon in spectral methods, SIAM J. Numer. Anal., **24**, n.5, pp.1008-1023.

TREVES F.(1967), *Topological Vector Spaces, Distributions and Kernels*, Academic Press, New York.

TRICOMI F.G.(1954), *Funzioni Ipergeometriche Confluenti*, Monografie Matematiche del C.N.R., Edizioni Cremonese, Rome.

TRICOMI F.G.(1957), *Integral Equations*, Interscience Publisher, New York.

TRIEBEL H.(1978), *Interpolation Theory, Function Spaces, Differential Operators*, North Holland, Amsterdam.

TRULLINGER S.E., ZAKHAROV V.E. & POKROVSKY V.L. (Editors) (1986), *Solitons*, Modern Problems in Condensed Matter Sciences, North Holland, Amsterdam.

TURKEL E.(1980), Numerical methods for large-scale time-dependent partial differential equations, in *Computational Fluid Dynamics*, Vol.2, Hemisphere, Washington, pp.127-262.

USPENSKY J.V.(1928), On the convergence of quadrature formulas related to an infinite interval, Trans. Amer. Math. Soc., **30**, pp.542-559.

VAINIKKO G.M.(1964), Asymptotic evaluations of the error of projection methods for the eigenvalue problem, USSR Computational Math. and Math. Phys., **4**, pp.9-36.

VAINIKKO G.M.(1967), On the speed of convergence of approximate methods in the eigenvalue problem, USSR Computational Math. and Math. Phys., **7**, pp.18-32.

VANDEVEN H.(1990), On the eigenvalues of second-order spectral differentiation operator, in *Spectral and High Order Methods for P.D.E.* (C.Canuto & A.Quarteroni Eds.) North Holland, Amsterdam, pp.313-318.

VARGA R.S.(1962), *Matrix Iterative Analysis*, Prentice-Hall, Englewood Cliffs, NJ.

VULIKH B.Z.(1963), *Introduction to Functional Analysis for Scientists and Technologists*, Pergamon Press, Oxford.

WATSON G.(1966), *A treatise on the Theory of Bessel Functions*, Cambridge University Press, Cambridge.

WEIDEMAN J.A.C. & TREFETHEN L.N.(1988), The eigenvalues of second-order spectral differentiation matrices, SIAM J. Numer. Anal., **25**, n.6, pp.1279-1298.

WEINBERGER H.F.(1965), *A first Course in Partial Differential Equations*, John Wiley & Sons, New York.

WIDDER D.V.(1975), *The Heat Equation*, Academic Press, New York.

WOUK A.(1979), *A Course of Applied Functional Analysis*, John Wiley & Sons, New York.

YOSIDA K.(1960), *Lectures on Differential and Integral Equations*, Wiley-Interscience, New York.

ZAMPIERI E.(1989), Iterative methods for domain decomposition using spectral tau and collocation approximation, Calcolo, **26**, n.1, pp.61-91.

ZANG T.A. & HUSSAINI M.Y.(1986), On spectral multigrid methods for time-dependent Navier-Stokes equations, Appl. Math. Comput., **19**, pp.359-372.

ZANG T.A., WONG Y.S. & HUSSAINI M.Y.(1982), Spectral multigrid methods for elliptic equations, J. Comput. Phys., **48**, pp.485-501.

ZANOLLI P.(1987), Domain decomposition algorithms for spectral methods, Calcolo, **24**, pp.201-240.

ZYGMUND A.(1988), *Trigonometric Series*, Second edition, Volumes I & II combined, Cambridge University Press, Cambridge.

Index

Lecture Notes in Physics

For information about Vols. 1–369
please contact your bookseller or Springer-Verlag

Vol. 370: H.-D. Doebner, J.-D. Hennig (Eds.), Quantum Groups. Proceedings, 1989. X, 434 pages. 1990.

Vol. 371: K.W. Morton (Ed.), Twelfth International Conference on Numerical Methods in Fluid Dynamics. Proceedings, 1990. XIV, 562 pages. 1990.

Vol. 372: F. Dobran, Theory of Structured Multiphase Mixtures. IX, 223 pages. 1991.

Vol. 373: C. B. de Loore (Ed.), Late Stages of Stellar Evolution. Computational Methods in Astrophysical Hydrodynamics. Proceedings, 1989. VIII, 390 pages. 1991.

Vol. 374: L. Ting, R. Klein, Viscous Vortical Flows. V, 222 pages. 1991.

Vol. 375: C. Bartocci, U. Bruzzo, R. Cianci (Eds.), Differential Geometric Methods in Theoretical Physics. Proceedings, 1990. XIX, 401 pages. 1991.

Vol. 376: D. Berényi, G. Hock (Eds.), High-Energy Ion-Atom Collisions. Proceedings, 1990. IX, 364 pages. 1991.

Vol. 377: W. J. Duschl, S.J. Wagner, M. Camenzind (Eds.), Variability of Active Galaxies. Proceedings, 1990. XII, 312 pages. 1991.

Vol. 378: C. Bendjaballah, 0. Hirota, S. Reynaud (Eds.), Quantum Aspects of Optical Communications. Proceedings 1990. VII, 389 pages. 1991.

Vol. 379: J. D. Hennig, W. Lücke, J. Tolar (Eds.), Differential Geometry, Group Representations, and Quantization. XI, 280 pages. 1991.

Vol. 380: I. Tuominen, D. Moss, G. Rüdiger (Eds.), The Sun and Cool Stars: activity, magnetism, dynamos. Proceedings, 1990. X, 530 pages. 1991.

Vol. 381: J. Casas-Vazquez, D. Jou (Eds.), Rheological Modelling: Thermodynamical and Statistical Approaches. Proceedings, 1990. VII, 378 pages. 1991.

Vol. 382: V.V. Dodonov, V. I. Man'ko (Eds.), Group Theoretical Methods in Physics. Proceedings, 1990. XVII, 601 pages. 1991.

Vol. 384: M. D. Smooke (Ed.), Reduced Kinetic Mechanisms and Asymptotic Approximations for Methane-Air Flames. V, 245 pages. 1991.

Vol. 385: A. Treves, G. C. Perola, L. Stella (Eds.), Iron Line Diagnostics in X-Ray Sources. Proceedings, Como, Italy 1990. IX, 312 pages. 1991.

Vol. 386: G. Pétré, A. Sanfeld (Eds.), Capillarity Today. Proceedings, Belgium 1990. XI, 384 pages. 1991.

Vol. 387: Y. Uchida, R. C. Canfield, T. Watanabe, E. Hiei (Eds.), Flare Physics in Solar Activity Maximum 22. Proceedings, 1990. X, 360 pages. 1991.

Vol. 388: D. Gough, J. Toomre (Eds.), Challenges to Theories of the Structure of Moderate-Mass Stars. Proceedings, 1990. VII, 414 pages. 1991.

Vol. 389: J. C. Miller, R. F. Haglund (Eds.), Laser Ablation-Mechanisms and Applications. Proceedings. IX, 362 pages, 1991.

Vol. 390: J. Heidmann, M. J. Klein (Eds.), Bioastronomy - The Search for Extraterrestrial Life. Proceedings, 1990. XVII, 413 pages. 1991.

Vol. 391: A. Zdziarski, M. Sikora (Eds.), Ralativistic Hadrons in Cosmic Compact Objects. Proceedings, 1990. XII, 182 pages. 1991.

Vol. 392: J.-D. Fournier, P.-L. Sulem (Eds.), Large-Scale Structures in Nonlinear Physics. Proceedings. VIII, 353 pages. 1991.

Vol. 393: M. Remoissenet, M.Peyrard (Eds.), Nonlinear Coherent Structures in Physics and Biology. Proceedings. XII, 398 pages. 1991.

Vol. 394: M. R. J. Hoch, R. H. Lemmer (Eds.), Low Temperature Physics. Proceedings. XXX,XXX pages. 1991.

Vol. 395: H. E. Trease, M. J. Fritts, W. P. Crowley (Eds.), Advances in the Free-Lagrange Method. Proceedings, 1990. XI, 327 pages. 1991.

Vol. 396: H. Mitter, H. Gausterer (Eds.), Recent Aspects of Quantum Fields. Proceedings. XIII, 332 pages. 1991.

Vol. 398: T. M. M. Verheggen (Ed.), Numerical Methods for the Simulation of Multi-Phase and Complex Flow. Proceedings, 1990. VI, 153 pages. 1992.

Vol. 399: Z. Švestka, B. V. Jackson, M. E. Machedo (Eds.), Eruptive Solar Flares. Proceedings, 1991. XIV, 409 pages. 1992.

Vol. 400: M. Dienes, M. Month, S. Turner (Eds.), Frontiers of Particle Beams: Intensity Limitations. Proceedings, 1990. IX, 610 pages. 1992.

Vol. 401: U. Heber, C. S. Jeffery (Eds.), The Atmospheres of Early-Type Stars. Proceedings, 1991. XIX, 450 pages. 1992.

New Series m: Monographs

Vol. m 1: H. Hora, Plasmas at High Temperature and Density. VIII, 442 pages. 1991.

Vol. m 2: P. Busch, P. J. Lahti, P. Mittelstaedt, The Quantum Theory of Measurement. XIII, 165 pages. 1991.

Vol. m 3: A. Heck, J. M. Perdang (Eds.), Applying Fractals in Astronomy. IX, 210 pages. 1991.

Vol. m 4: R. K. Zeytounian, Mécanique des fluides fondamentale. XV, 615 pages, 1991.

Vol. m 5: R. K. Zeytounian, Meteorological Fluid Dynamics. XI, 346 pages. 1991.

Vol. m 6: N. M. J. Woodhouse, Special Relativity. VIII, 86 pages. 1992.

Vol. m 7: G. Morandi, The Role of Topology in Classical and Quantum Physics. XIII, 239 pages. 1992.

Vol. m 8: D. Funaro, Polynomial Approximation of Differential Equations. X, 305 pages. 1992.

GPSR Compliance

The European Union's (EU) General Product Safety Regulation (GPSR) is a set of rules that requires consumer products to be safe and our obligations to ensure this.

If you have any concerns about our products, you can contact us on ProductSafety@springernature.com

In case Publisher is established outside the EU, the EU authorized representative is:

Springer Nature Customer Service Center GmbH
Europaplatz 3
69115 Heidelberg, Germany

Batch number: 10640288

Printed by Printforce, the Netherlands